AF556241

Conversion to Organic Agriculture

Dr. A.K. Singh, Ph. D.
Associate Professor (Agriculture)
North Eastern Regional Institute of Water and Land Management
Tezpur, Assam (India)
Member, International Competence Centre for Organic Agriculture (ICCOA)
Associate Trustee, Indian Organic Certification Agency (INDOCERT)

International Book Distributing Co.
(Publishing Division)

Published by

INTERNATIONAL BOOK DISTRIBUTING CO.
(Publishing Division)
Khushnuma Complex Basement
7, Meerabai Marg (Behind Jawahar Bhawan)
Lucknow 226 001 U.P. (INDIA)
Tel. : 91-522-2209542, 2209543, 2209544, 2209545
Fax : 0522-4045308
E-Mail : ibdco@airtelbroadband.in

First Edition 2007

Price: Rs. 1495/-

ISBN 81-8189-171-6

Composed & Designed at :
Panacea Computers
3rd Floor, Agrawal Sabha Bhawan
Subhash Mohal, Sadar Cantt. Lucknow-226 002
Phone : 0522-2483312, 9335927082
E-mail : prasgupt@rediffmail.com

Printed at:
Salasar Imaging Systems
C-7/5, Lawrence Road Industrial Area
Delhi - 110 035
Tel. : 011-27185653, 9810064311

Foreword

After introduction of high-yielding varieties of seeds of various crops, complemented with intensive cultivation practices using fertilizers, pesticides and other inputs, India has undoubtly taken a leap in agricultural production, but there is a flip side to this story too. Our annual agricultural production which stood at 0.03% from 1900 to 1950, increased to 3.3% during 1960-70 and thereafter started declining and stood at less than 2% during 1990-98 and 2002-03, compared with our population growth of 1.8%, proving that conventional agriculture is not sustainable in spite of availability of adequate HYV seeds, chemical fertilizers, pesticides and irrigation. The productivity of the soil declined because of loss of organic matter. Agriculture became a business venture instead of a way of life and urged by short-term profitability. The stress was on increasing output of specific crops depending on high external non-renewable inputs and borrowed money. Pest and disease brought despair to farmers forcing them to use calendar based synthetic pesticides affecting wild life, biodiversity, human and animal health. Therefore, there is an urgent need for organic farming for reducing the problems created by green revolution technologies. Organic farming centres around a living system where the soil, plant and animal including man are bound to the 'wheel of life' where the process of growth and decay balance one another.

One of the common questions faced with in recent years, when the Central and State governments announced implementation of organic agriculture in the country, is, how to go about converting from conventional to organic agriculture. Somewhere we have to make a start to compile all the sustainable organic methods for further refining and improvement. I am pleased to say that the author has taken a timely step in this direction, of putting the latest information on organic farming conversion techniques into a book. It provides us with an easy and concise source of information for conversion of conventional agriculture to organic agriculture. An excellent guide-cum-reference book for organic farmers, and researchers students, planners, extension officers, NGOs, officers of the agriculture department and allied departments. I hope, that all organic agriculture enthusiusts will read this informative book and derive much satisfaction, as I have from the valuable publication.

Defence Research Laboratory Tezpur, Assam
DRDO, Ministry of Defence, Govt. of India

DR. S.N. DUBE
Director

Preface

Organic agriculture offers trade opportunities for farmers in the developing and developed countries. According to the ITC and UNCTAD/GAT, more than 130 countries produce certified organic foods. This market of organic products is expected to grow globally in the coming years with high growth rates of over 15 per cent. This organic market expansion makes it possible for farmers to reap the benefits of a trade with relatively high price premiums.

Organic farm production is an environment-friendly approach that optimises the health and productivity of interdependent communities of animals and people. To the extent feasible, the system relies on crop rotations, crop residues, animal manures, legumes, green manures, off-farm organic wastes, bio-fertilizers and using natural biological and environmental-friendly spray to maintain soil productivity and to control insects, pests, weeds and plant diseases. Organic farms should have high standards of biodiversity and animal welfare.

Organic farming system largely excludes the use of synthetically compound fertilizers, pesticides, herbicides, hormones, growth promoters, food irradiation, use of genetically-modified crops and livestock feed additives. Organic farming help to reduce the ground and surface water contamination, thus contributing to food safety in a larger sense and sustainable agriculture. An organic farm must spend some two or three years keeping strict rules (called 'Standards') to clear break from chemicals and gives the farmer the chance to rebuild soil fertility through natural or biological means.

Planting an organic garden requires planning. A properly planned and planted garden will naturally resist disease, deter insect pests, and be healthy and hardy. It is important to understand the magnitude of the organic project before it begins. The biggest barrier most farmers face when switching to organic production is the change in thinking that must occur to make it successful. The organic farmer must understand how everything is inter-related and how one set of circumstances will influence other factors in farming. A healthy soil is much more important in an organic system than a conventional system.

This technical guidebook presents some aspects of effective planning for conversion of conventional farming to organic farming through improved soil fertility management, crop management, pest and disease management, animal management, etc. The guide attempts to provide some insight into the problems which may arise during organic conversion on farmers fields. The guide is intended for researchers, students, planners, agriculture extension officers, progressive farmers and others actively involved in on-farm organic conversion.

Dated : 13th April 2006 **A.K. Singh**

Tezpur

Contents

Chapter 1

Organic Farming: Opportunity and Challenges

If we ponder on the natural phenomenon of productivity, plants exhibit primary productivity. Every plant has a gross primary production from which a considerable quantity of produced energy has to be utilised by the plant itself for its very own maintenance. The extra leaves the plant produce to protect the grain as it matures was considered as surplus and useless by the human mind. The question then rose to have plants shed away such extra leaves so that the energy utilised by the plant for its own maintenance be reduced, thereby increase the yield of the crop in terms of net primary productivity. The result was that grains got directly exposed and attracted the insects that were only happy to consume them. "If the only tool we have is the hammer, we tend to see every problem as a nail"; we had enough chemicals as a tool and we started using them as pesticides. Today with overwhelming resistance by pests, the quantity and diversity of pesticides in the form of insecticides, fungicides and weedicides have increased. The way these are sprayed without proper regulations is a matter of concern. Moreover, this has also been responsible to eliminate or check the population of useful insects, spiders and birds, who are the natural biological control agents. The soil organisms ranging from microorganisms to the earthworms, which are an important component of the soil ecosystem, have also reduced. The soil is hungry for the soil organic matter.

Biotechnology has provided ample scope to not only utilise knowledge in an appropriate manner to avail benefits from existing organisms, but also to interfere with the genetic constituents of an organism and creating modifications through gene manipulation or genetic engineering. Today genetically modified (GM) technology is projected as a solution for food security in developing nations. Theory of the Multinational Corporations is "if you are against hunger, you must be for GMO" for which there is absolutely no justification. Genes once released into the environment may not be possible to be recalled. GM technologies place biodiversity in the region at risk; and India is one of those rich bio-reserves. There is plenty of natural, normal good food in the world to nourish human population. The European Union and countries such as U.K., Australia and New Zealand do not accept several of these GM (genetically modified) crops.

Damage done by chemical agriculture can be overcome by bioremediation over a period of time. Organic farming is one of the proven bioremediation method for getting quality and healthy food. Organic agriculture is an environment-friendly ecological production system that promotes and enhances biodiversity, biological cycles and biological activities. It is based on minimal use of off-farm inputs, avoiding the use of synthetic fertilizers and pesticides and management practices that restore, maintain and enhance ecological balance. Producing Organic is a commitment to a system which ensures production of healthy and nutritious food year after year without environmental degradation. The primary goal of organic agriculture is to optimise the health and productivity of interdependent communities of soil life, plants, animals and people.

Organic production systems are based on specific and precise standards (IFOAM Basic Standards) of production. The product follows the defined standards throughout production, handling, processing and marketing stages, and certified by a duly constituted certification body or authority. Organic standards are legally defined as set out under EU Council Regulation (1804/1999, 2092/91), National Programme on Organic Production (NPOP), Codex Alimentarius Commission (23rd & 24th Session 1999 & 2001) etc. The "organic" label given by certifying agency is not a health claim, it is a process claim.

To become organic, a farmer or operator must become registered with one of the certification bodies and the land/animal has to be converted. It is the responsibility of the certification body to inspect the operator or food producer on a regular basis to ensure that the producer is complying with organic standards.

1. THE SCOPE FOR ORGANIC FOODS

In recent years, organic farming experienced a tremendous growth in many countries. Organic agriculture is currently practiced in more than 120 countries. It is estimated that world wide about 17 million hectares are managed organically. The share of land area under organic management (percent) per country is highest in some European countries, where it reaches up to some considerable share of the total agricultural land. The success of organic agriculture in these countries is mainly due to the increased consumer awareness of health and environmental issues. Notably, organic agriculture is also gaining importance in a number of developing countries including China, Egypt, India, Philippines, Sri Lanka and Uganda.

The worldwide markets for organic foods are expanding, with three major markets-Europe, United States and Japan (Asia) and recording annual growth rates of 15 per cent to 30 per cent for the past five years. As per the survey of IFOAM, FIBL and SOEL (2005), more than 26 million hectares of cultivated lands are certified organic. According to International Trade Centre (ITC) projections, the organic market size in the year 2010 would be around US $ 46 billion in the European

Union, US $ 45 billion in the United States and US $ 11 billion in Japan. A worldwide organic farm as per the SOEL survey (February 2005) is presented in Table 1.1.

Table 1.1: Organic farms worldwide

Country	Number of organic farms
Mexico	120,000
Indonesia	45,000
Italy	44,043
Uganda	33,900
Kenya	30,000
Tanzania	30,000
Peru	20,000
Austria	19,056
Spain	17,028
Germany	16,476
Brazil	14,003
India	5,147

Source: SOEL Survey, February 2005

India's exports of organic food item handled by Agricultural and Processed Food Products Export Development Authority (APEDA) have increased from an estimated Rs. 26 million in 1998-99 to an estimated level of Rs. 71.23 million during 2003-04. The Indian organic foods industry is expected to witness a boom period. There is huge export opportunity for the Indian organic farming industry both in niche markets as well as mainstream market the world over.

National Programme for Organic Production (NPOP)

Present Status in India (2003-04)	
Total production	119656 Tonnes + 1657000 nos. of seedlings & cuttings + 264000 litres effective micro organisms
Total quantity exported	6792 Tonnes
Total quantity exported	Rs.7123 Lakh
Total area under certified organic cultivation	2508826 ha (This includes wild herbs collection from forest area of MP & UP of 2432500 ha)
Number of items exported	31

The major organic products sold in global market include off-season fruits and vegetables, dried fruits and nuts, cocoa, spices, herbs, oil crops and derived products, sweeteners, dried leguminous crops, meat, dairy products, alcoholic beverages, processed food and pot plants, etc. The market for organic grains and cereals is also growing. Fresh produce, packaged grocery items (like cereals, sauces) and bulk/packaged items (pasta, grains, beans) are among the top categories in organic product stores in USA. Japanese organic consumers buy mostly frozen vegetables, dried fruits, vegetable juice, soybeans and fresh produce. Cereals and baked foods, fresh produce especially vegetables, and milk and dairy products hold the largest organic market shares by product category in Europe. Australia and Egypt are already an important producer and exporter of organic products, as are Madagascar, South Africa, Israel and Turkey. Exotic fruits, herbs and spices, nuts, dried and fresh fruits, essential oils, oilseeds, vegetables and cotton are some of most important items produced organically by these countries. Among the other significant producing countries are China, India, the Republic of Korea and Sri Lanka. These countries have potential to produce cocoa, coffee, tea, essential oils, herbs, spices, peanuts, rice, pulses and seeds, vanilla, fruits, vegetables, etc.

Countries that have a significant presence in the food processing industry, such as Germany, Italy, Sweden and France, face greater demand for organic ingredients. European Union regulations require that 70 per cent to 95 per cent of certified organic processed items be composed of organic gradients. For many countries, this will mean greater reliance on imports to meet demand. This presents a good opportunity for developing countries especially for those which can supply tropical produce, and off-season produce like spices, herbs, dried and powdered fruits, sugar, cocoa, etc.

2. ORGANIC FARMING SCENARIO IN INDIA

Organic Agriculture is not a new concept to India. At the beginning of the 19th century, Sir Albert Howard, one of the most important pioneers of organic farming, worked in India for many years, studying soil-plant interactions and developing composting methods. In doing so, he capitalized substantially on India's highly sophisticated traditional agricultural systems, which had long applied many of the principles of organic farming (e.g. crop rotations with legumes, mixed cropping, botanical pesticides etc.). Though the introduction of Green Revolution agricultural technology in the 1960s reached the main production areas of the country, there were still certain areas (especially mountain areas) and communities (especially certain tribes) that did not adopt the use of agro-chemicals. Therefore, some areas can be classified as *'organic zone'*. An increasing number of farmers have consciously abandoned agro-chemicals and now started to produce crops organically, as a viable alternative to Green Revolution agriculture.

India is classified into 21 agro-ecological zones based on temperature, soil conditions, and rainfall. Hence, each zone has a comparative advantage for the production of different products, e.g. tea, spices and fruits in the north-eastern region; spices and coffee in the southern region; rice, wheat and fruits in the northern region; cotton and herbs in central and western region. Products with potential in the domestic market are fruits, vegetables, rice and wheat. Products with potential in the export market are tea, rice, fruits, vegetables, cotton, herbs and spices. Further, India has time-tested indigenous farming systems and use of indigenous technical knowledge in agriculture and allied sector.

India has also the following advantages:

- India is strong in high quality production of tea, spices, rice specialties, ayurvedic herbs etc.
- India has a rich heritage of agricultural traditions which are suitable for designing organic production systems.
- In several regions of India agriculture is not very intensive as regards the use of agro-chemicals (mountain areas).
- The labour is relatively cheap compared to agro-chemicals.
- The NGO sector in India is very strong and has established close linkages to large numbers of marginal farmers.

Certified organic farming in India is still in a nascent stage. There are numerous initiatives of farmer groups, NGOs and corporate bodies producing organically and selling the products on domestic and international markets. In addition to improved food quality and environment, the potential for organic agriculture to reduce production costs, to stabilize yields and to increase farmers' income, especially in marginal production regions, is considered as high under the Indian scenario. In 2000, the Indian Government started to promote organic agriculture on various levels under the National Programme for organic Production (NPOP).

According to official statistics, until February 2001 there were only 304 organic farms in India and the figure has increased to 5147 Farms during February 2005 (Table 1.1), accounting for barely 0.05% (76,326 hectares) of total agricultural land (SOEL survey, February 2005). However, the database is still very poor and it can be assumed that the real figures are much higher. Presently, organic production for the whole of India is around 14,000 tonnes. Organic products produced in India (Table 1.2) are tea, spices, rice, coffee, cashew nuts, oil seeds, wheat, pulses, cotton, herbal extracts, fruits and vegetables. Considering the increase in organic farming activity in India, the major organic products for which a growing demand is anticipated would be tea, spices and bananas.

Table 1.2: Major products produced in India by organic farming

Type	Products
Cereals	Rice, wheat
Spices	Cardamom, black pepper, white pepper, ginger, turmeric, vanilla, mustard, tamarind, clove, cinnamon, nutmeg, chilly
Pulses	Red gram, black gram
Fruits	Mango, banana, pineapple, passion fruit, sugarcane, orange, cashew nut, walnut
Vegetables	Okra, brinjal, garlic, onion, tomato, potato
Oil seeds	Sesame, castor, sunflower
Beverages	Tea, coffee
Others	Cotton, herbal extracts

Source: Org-Marg, 2002 (Field survey and the publication-Organic and Biodynamic farming, Government of India, Planning Commission)

2.1 Indian Organic Products in the Markets

The domestic market for organic products is not yet as developed as the export market. The products available in the domestic market in organic quality are rice, wheat, tea, coffee, pulses, fruits and vegetables. As most organic production originates from small farmers, wholesalers and traders account for a 60% share in the distribution of organic products. Large organized producers distribute their products through supermarkets as well as through self-owned stalls.

Considering the profile of existing consumers of organic products, supermarkets and restaurants are the major marketing channels for organic products. Major markets for organic products lie in metropolitan cities - Mumbai, Delhi, Kolkata, Chennai, Bangalore and Hyderabad to name a few.

Recently, KVIC (Khadi and Village Industries Commission) with its 7000 outlets in India has taken the initiative to market organic products in the domestic market. In 2002, KVIC successfully launched its organic product line 'Desi Ahaar' in some of its outlets - and was sold out in short time. Besides KVIC there is a number of formal and informal initiatives to sell organic products in the domestic market. In the long term, the domestic market is expected to be much more important for Indian organic farmers than export markets.

Products with potential in the domestic market are fruits, vegetables, rice and wheat. Products with potential in the export market are tea, rice, fruits and vegetables, cotton, wheat and spices. Products for which production in India has a comparative advantage are given in Table 1.3.

Table 1.3: Major locations for organic productions in India

Product	Season	States	Major Locations
Tea	Throughout the year	Assam, West Bengal, Uttranchal	Darjeeling, Guwahati, Dehradun
Spices	Throughout the year	Kerala, Tamil Nadu, Karnataka	Cochin, Coimbatore, Idduki, Coorg
Coffee	Throughout the year	Kerala, Tamilnadu, Karnataka	Coimbatore, Coorg, Wayanadu, Peeremade
Rice	Kharif* and Rabi*	Punjab, Haryana, Assam, Maharashtra, Tamil Nadu	Amritsar, Jalandhar, Siphajhar, Ratnagiri, Kanchipuram, Thiruvallur
Wheat	Kharif and Rabi	Punjab, Haryana, Uttar Pradesh	Ambala, Patiala, Bhatinda, Faridkot
Vegetables	Throughout the year	All India Various locations	
Fruits	Throughout the year	All India Various location	
Cotton	Kharif	Maharashtra, Gujarat Madhya Pradesh,	Akola, Amravati, Amreli, Kheda, Indore

Source: Org-Marg, 2002.

**Kharif is essentially from May to September and Rabi is from November to March*

2.1.1 *Future demand and growth forecast*

Indian organic producers and exporters are well aware of the demand for organic products in developed countries. The export sales are likely to reach 21,525 tonnes during 2006-07.

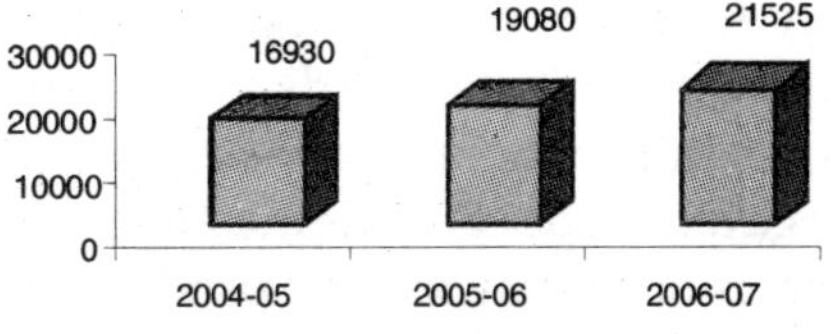

Future demand – Export
(Source: Org-Marg 2002)

Products available for the export market are rice, wheat, tea, spices, coffee, pulses, fruits and vegetables, cashew nuts, cotton, oil seeds and medicinal herbs. The channels adopted for the export of organic products, except for tea, are mainly through export companies. Organic tea is produced by major well organised tea estates which are exporting tea directly. In the case of other organic products, predominantly small farmers are involved in producing organic products. Hence, these products are exported through exporters.

Demands of organic products from India:

- Basmati rice (if there is enough volume)
- Spices (from small quantities to 50 tonnes)
- Tea (only if high quality, up to 40 tonnes)
- Coffee (only if high quality)
- Mango (fresh, dry and puree, quantities up to 50 tonnes)
- Pineapple (juice, dry, small quantities up to 50 tonnes)
- Bananas (fresh and dry, small quantities up to 70 tonnes)
- Organic vanilla (only if good price, high quality)
- Cashew nuts (only if good price and high quality)
- Protein grains (if there is not a problem with the quality).

Leading organic food consumers are from countries (Av US $ 50 pc) like Swiss, Denmark, Sweden and Austria. In Asia, Japan is the major consumer. Malaysia, Taiwan, Singapore, Korea do import some organic produce. Organic products are mainly exported to the following countries:

Europe	Netherlands, United Kingdom, Germany, Belgium, Sweden, Switzerland, France, Italy, Spain
Americas	USA, Canada
Middle East	Saudi Arabia, UAE
Asia	Japan, Singapore
Africa	South Africa
Australia	Australia

Trade with organic products are showing growth rates of 15–20%, which are rarely found in food markets. Trade growth has been hindered by yet evolving national organic standards, certification and accreditation programs, and inefficient market infrastructures for organic products in many countries. Growth forecast of organic products in domestic market is presented in Table 1.4. Frame conditions for sales of Indian organic products and export and import in the selected markets is given in Tables 1.5 and 1.6.

Table 1.4: Growth forecast of organic products (domestic market)

Product	Growth in 5 years (%)
Spices (all)	14
Pepper	5
Turmeric	4.5
Tea	13
Rice	10
Fruits (all)	8
Banana	15
Mango	5
Orange	5
Pineapple	5
Herbal extracts	7
Cotton	7
Coffee	5
Oil seeds	5
Honey	5
Groundnut	5
Baby food	5
Coconut	5

Source: Org-Marg, 2002

Table 1.5: Frame conditions for sales of organic products

Product	Sales	Potential	Availability
Tea	Good-moderate	Good-moderate	Good-moderate
Rice	Good	Good-moderate	Moderate-poor
Protein grains	Good-moderate	Good	Poor
Spices	Good-moderate	Good-moderate	Moderate
Vanilla	Good-moderate	Good-moderate	Moderate-poor
Mango	Moderate-good	Moderate-good	Moderate-poor
Pineapple	Good-moderate	Good-moderate	Moderate-poor
Bananas	Good-moderate	Moderate	Moderate-poor
Nuts	Good	Good	Poor

Source: FiBL, 2002

Table 1.6: Imported Indian organic products in the selected markets

Product	Volumes (tonnes)	Season	% of turnover
Tea (green and black)	65	Throughout the year	10-15
Spices (pepper, ginger and hibiscus)	132	Throughout the year	10-15
Fruits (mango and pineapple)	160	May-July, Nov-May	6 -10
Nuts (walnut and cashew nut)	28	October and July	10-15
Sesame	600	Throughout the year	10-15

Source: FiBL, 2002

Organic farmers' association

In some countries, several organic farmer organisations have joined together to form a strong network able to formulate their interests, coordinate the organic market and develop a credible label. In India there were several local initiatives to form organic farmers organisations (e.g. the Bio-dynamic Association), but so far none of them managed to cover a majority of the organic sector. In June 2003, an All India Organic Farmers' Association was founded which might achieve this objective.

National government support

Organic farming has been identified as a major thrust area of the 10th five-year plan of the central government. The Indian Central Government set up a National Institute of Organic Farming in October 2003 in Ghaziabad, Uttar Pradesh. The purpose of this institute is to formulate rules, regulations and certification of organic farm products in conformity with international standards. One billion rupees have been allocated to the National Institute of Organic Farming (NIOF) alone for the 10th five-year plan. The Planning Commission has set up a working group, and the Department of Commerce has established National Organic Standards. A large amount of financial resources are being allocated for organic agriculture in the 11th plan also by the Government of India. International Competence Centre for Organic Agriculture (ICCOA), Association for Promotion of Organic Farming (APOF) and other concerned organizations are organizing meetings with organic stakeholders to prepare a note to facilitate the government on the allocation of the resources for 11th five year plan.)

Prospects for organic farming in India

Organic agriculture in many ways would be an eminently preferable model for the development of agriculture in India, especially where yields under conventional farming are low. Organic agriculture offers multiple benefits. These include economical aspects (e.g. price premiums, high demand), natural resource conservation (e.g. improved soil fertility and water quality, prevention of soil erosion, preservation of natural and agro-biodiversity) and social benefits (e.g. generation of rural employment and corresponding lower urban migration, improved household nutrition and local food security, reduced dependence on external inputs). Organic systems can achieve similar yield levels as conventional ones, and can even contribute to increase productivity in the long term, as experienced by many organic farmers.

Chapter 2

Organic Food Quality

Food quality and safety are of concern to every individual. Consumers expect their food to be enjoyable, nutritious and safe. Food-borne illness and food borne injury are at best unpleasant; at worst, they can be fatal. Effective hygiene control, therefore, is vital to avoid the adverse human health and economic consequences of food-borne illness, food-borne injury, and food spoilage. Everyone, including farmers and growers, manufacturers and processors, food handlers and consumers, has a responsibility to assure that food is safe and suitable for consumption.

There are many types of safety hazards associated with foods that can arise during the production of foods or their subsequent handling, processing and packaging. Microbiological hazards include bacteria, protozoa, parasites, viruses, and fungi or their toxins. Naturally occurring toxicants in the environment such as zinc, arsenic and cyanide or in the food itself such as solanin and histamine, may also constitute food safety hazards. Toxic industrial chemicals or radioactive wastes are other potential sources of food contamination. Examples of such industrial contaminants are: arsenic, cadmium, copper, lead, mercury and polychlorinated biphenyls (PCBs). Contaminants may enter the food chain due to excess or improper fertilizer use; examples are cadmium, nitrates and nitrites. Misuse of authorised pesticides or veterinary drugs may also create chemical hazards in food. Contaminated animal feed or improper animal feeding practices could also lead to unsafe food.

On a global basis, a rating of health risks arising from foods showed that risks due to food additives and pesticide residues are relatively minor (both acute and chronic effects) as compared with microbiological and other naturally occurring toxins. Epidemiological investigation of food-borne illness in Europe showed that only 0.5% of outbreaks was due to chemical substances. The problem of food contamination by chemicals was perceived as an important public health concern.

Besides safety, quality attributes include: nutritional value; organoleptic properties such as appearance, colour, texture, taste; and functional properties. Consumers, the food industry and government regulators are also concerned with these quality criteria. Quality can be considered as a complex characteristic of food that determines its value or acceptability to a consumer. From a regulatory or consumer protection point of view, "quality" refers to the basic objective requirements which must be met under existing laws and regulations to assure that foods are safe, not contaminated, adulterated or fraudulently presented. Furthermore, recent international agreements emphasise the need for food safety measures to be based on risk analysis following principles and procedures elaborated by relevant international organizations. International food safety standards and food hygiene requirements are equally valid for conventionally and organically produced food.

1. CHEMICAL FOOD CONTAMINANTS

With respect to chemicals, organic agriculture differs from conventional agriculture as it refrains from using synthetic agricultural inputs, such as synthetic pesticides, herbicides, fertilizers, fungicides, veterinary drugs (e.g. antibiotics, growth hormones), synthetic preservatives and additives, and irradiation. Thus, potential hazards posed by synthetic input residues are prevented, to the extent possible. This underlies consumer expectations that organic foods are healthier.

1.1 Pesticide Residues

Studies carried out to investigate the relative presence of pesticide residues on organically as opposed to conventionally-grown products confirm the reduced presence of pesticide residues in organic food although organic food may not be defined as pesticide-free. Organic certification schemes specify that land must be free from chemical inputs for 2 or 3 years prior to organic production. However, the possible presence of pesticide residues from previous land use means that low levels of pesticides or other contaminants can occasionally be found in certified organic food. The presence of pesticides by such means does not necessarily preclude the food being described as organic, providing all other certification requirements have been fulfilled. The presence of pesticide residues at low levels on organic produce may also be explained in terms of chemical sprays drifting from conventionally-managed farms. While pesticide residues are demonstrably lower in organic products, it is important to emphasise that consumers of conventionally grown products are protected against health risks associated with these chemical residues through the adoption by governments of appropriate legislation and regulations that require farmers to adhere to good agricultural practices and establish maximum residue limits (MRLs), based on rigorous risk analysis, which do not pose significant health risks. Implementation of national food control programmes ensures adherence to the MRLs. European data on pesticide residues in total diet studies show that calculated intakes are very low, often below 1% of the Acceptable daily intake (ADI) as determined by toxicological studies. Some long-term effects of certain persistent pollutants (POP) pesticides are given below:

Table 2.1: Long-term effect of certain persistent pollutants pesticides

Trade Name	Long-term effects
Endosulfan	Nervous system damage
DDT	Cancer, damage to liver, nerve, brain, toxic to wildlife
Aldrin/Dieldrin/Endrin	Cancer (suspect), baby birth defects, toxic to wildlife
BHC/Lindane	Cancer (proven), miscarriage, leukaemia, toxic to fish
Chlordimeform	Cancer (suspect), bladder damage, toxic to wildlife
Heptachlor/Chlordane	Leukaemia, toxic to fish

Table 2.2: Common pesticides used on certain crops across the World

Food product	Contaminant Pesticides
Apples	Diphenylamine, Captan, Endosulfan, Phosmet
Oranges	Chlorpyrifos, Parathion
Grapes	Captan, Dimethoate, Dicloran, Carbaryl
Cherries	Parathion, Malathion
Bananas	Diazion, Thiabendazone, Carbaryl
Cabbage	Methamidophos, BHC
Cauliflower	Methamidophos, Endosulfan, Diazion
Tomatoes	Chlorpyrifos
Green beans	Methamidophos, Endosulfan
Carrots	DDT, Parathion, Dieldrin
Cucumbers	Methamidophos, Endosulfan, Dieldrin, Chlorpyrifos
Spinach	Endosulfan, DDT
Onions	DDT, Malathion
Potatoes	DDT, Dieldrin, Aldicarb, Chlordane
Sweet Potatoes	DDT, Dieldrin, Dicloran, BHC
Watermelon	Captan, Carbaryl
Corn	Carbaryl, Dieldrin, Lindane, Chlorpyrifos

International guidelines for organically-produced foods include lists of substances that can be used for plant pest and disease control if the need for such is recognised by the certification body. In organic management, biological control is the preferred method of pest management.

1.2 Nitrates from Fertilizer

Nitrate content of organically-grown crops, particularly nitrophilic leaf, root and tuber crops, is reported to be significantly lower than in conventionally-grown products. The occurrence of nitrates in certain crops is favoured by the use of highly soluble mineral fertilizers. The nitrate that is leached causes the following hazardous effects:

- Excess nitrates applied move below the root zone or into groundwater. They remain there for extended periods of time and draining of such water causes disease called "Methemoglobinemia", a disease that interferes with oxygen carrying capacity of blood.
- Excess use of urea in rice fields promotes the growth and spread of vectors causing of human disease called Japanese encephalitis (JE). Children between the age group of 4-14 years mainly affected.
- Nitrosomine illness is caused by the presence of secondary amines, which causes cancer in human beings. Feroxyl nitrates, alkyl nitrates, vapours of HNO_3 and nitrate aerosoles cause respiratory illness.
- *Eutrophication,* process of enrichment of surface water bodies such as lakes, reservoirs and streams with nutrients, particularly phosphorus and nitrogen results in intense proliferation of algae and higher aquatic plants in excessive quantitative deteriorating water quality.

Some surveys, however, indicate that factors other than farming system are of importance in determining nitrate levels. It has been reported, for example, that governments of Germany and France have encouraged conversion to organic farming in certain areas in a bid to improve water quality, particularly in relation to its nitrate content (FAO, 1997). The negative food safety implication of high nitrate content in foods is that under certain conditions they may be converted to nitrosamines which are carcinogens. Nitrates can also impair the ability of the blood to carry oxygen, and may pose a risk of methemoglobinemia. These risks are subject to risk management and reduction in the conventional food supply.

1.3 Environmental Contaminants

Besides pesticide residues, there are several other chemical hazards associated with foods. Those chemical contaminants coming from general environmental pollution are found equally in organically and conventionally-grown products. This should be expected since persistent pollutants in the soil such as chlorinated hydrocarbons and certain heavy metals cannot be avoided or eliminated through the use of organic farm management techniques. High contamination occurs mainly due to industrial activity - whether from mining and smelting activities, the energy sector, agricultural practices or disposal of hazardous and municipal wastes. A more widespread use of organic agriculture may contribute to a reduction of environmental degradation, which may ultimately result in reduced levels of certain contaminants in food.

The use of bio-solids from wastewater treatment facilities (often referred to as sludge) on fields to produce food crops creates concerns about contamination of food by heavy metals, toxic organic compounds (such as dioxin, PCBs) and persistent microbial pathogens. The Codex and EU organic standards prohibit the use of sewage sludge. Some European countries (e.g. Denmark, Sweden) prohibit sludge application of any kind to lands for grazing. Other countries (e.g. Germany, Netherlands) regulate its use even for conventional agriculture. The regulations of the National Organic Programme also ban the use of sewage sludge in organic agriculture.

1.4 Veterinary Drugs and Contaminants in Animal Feeds

According to EC regulations (EC No 1804/1999), animal health management in organic systems precludes the preventative use of chemically synthesised allopathic medicinal products. Animal health management in organic production systems should mainly be based on prevention by measures such as appropriate selection of breeds and strains, a balanced high-quality diet and a favourable environment. Public health concerns about microbial resistance to certain antibiotics as well as the high level of media coverage given to the issue of Bovine Somatotropin (BST) use seem to have propelled consumer demand for organic dairy products which is growing at a rate of 35% annually in the United States. In several countries of the EU there is high market demand for organic milk and further rapid growth in demand is projected. In Denmark, organic milk currently accounts for 20% of total milk production. The Government's aim is to raise the share of organic milk in the national milk market to 50 % in 5 years and 100 % in 10 years. All milk used in Denmark's school feeding programmes is organically produced.

Contaminants in animal feeds, such as pesticide residues, agricultural and industrial chemicals, heavy metals and radioactive nuclides, can give rise to safety hazards in foods of animal origin. As EC regulations (EC No 1804/1999) require that livestock, claimed to be produced organically, is fed on organically-produced feed stuffs, the potential for contamination with pesticide residues and other agricultural chemicals is greatly reduced compared to conventional farming methods.

2. MICROBIOLOGICAL FOOD CONTAMINANTS

2.1 Contamination from Natural Fertilizers

Animal manure and other organic waste are the main fertilizers used in organic farming. These natural fertilizers are also widely used in conventional agriculture along with chemically synthesised fertilizers. Microbiological contamination arising from the use of natural fertilizers and measures needed to address it, must focus on organic agriculture.

Untreated or improperly treated manure or bio-solids used as fertilizers or soil nutrient agents, whether in organic or non-organic agriculture, can lead to contamination of products and/or water sources. Animal and human faecal matter are known to contain a range of human pathogens. Properly treated manure or bio-solids are effective and safe fertilizers. Growers need to follow good agricultural practices for handling these natural fertilizers to minimise microbial hazards. Recent research indicates that pathogenic organisms can survive up to 60 days under compost conditions.

The Codex General Principles of Food Hygiene provide the basic rules for ensuring food safety for all foods. Organic production, as with all other types of food production, must follow the provisions outlined in this international code of practice. These include a requirement that producers should implement measures to control contamination from fertilizers used in primary production and to protect food sources from faecal and other contamination.

Much of the research on the composting of manure and application of manure to field crops has focused on the effects of different practices on soil fertility and crop quality. Further research is required on pathogen survival in untreated manure, treatments to reduce pathogen levels in manure, and assessing the risk of cross-contamination of food crops from manure under varying conditions. Recent research suggests for example that some pathogens, such as the hepatitis A virus, have a higher thermal threshold than others. In addition, the time and temperature required eliminating or reducing microbial hazards in manure or other organic materials may vary depending on regional climate and the specific management practices of an individual operation.

Good hygienic and good agricultural practices are also required to protect contamination of food by untreated manure. Additional research is also needed to determine how pathogens in manure may spread in the field. However, for some operations, drift and runoff from adjacent fields may result in microbial hazards. Growers may consider scheduling application of manure on adjacent fields to maximise the time between manure application to those fields and harvest of fresh market products.

2.2 *E. Coli* Contamination

The US Centre for Disease Control (CDC) identifies the main source for human infection with *E. coli* as meat contaminated during slaughter. Virulent strains of *E. coli,* such as *E. coli* 0157:H7, develop in the digestive tract of cattle, which is mainly fed with starchy grain as research at Cornell University has demonstrated. Cows mainly fed with hay generate less than 1 % of the *E. coli* found in the faeces of grain-fed animals. It is one of the most important goals of organic farming to keep the nutrient cycles closed. Therefore, ruminants like cattle and sheep are fed with diets with a high proportion of grass, silage and hay. It can be concluded that organic farming potentially reduces the risk of *E. coli* infection.

2.3 Mycotoxins

Mycotoxins are toxic by-products of certain moulds that can grow on certain food products under suitable conditions. Aflatoxins are the most toxic of these compounds and can induce liver cancer at very low doses if ingested over a prolonged period of time. Since fungicides are not allowed in organic production and given that mycotoxins constitute a major health hazard, their relative presence in foods produced organically or conventionally has been the subject of many studies. From these studies it cannot be concluded that organic farming leads to an increased risk of mycotoxin contamination. It is important to emphasise that good agricultural, handling and storage practices are required in organic as in conventional agriculture to minimise the risk of mould growth and mycotoxin contamination. Good practices in animal feeding for all producers require that ingredients used for animal feeding should be checked to ensure that adequate quality standards are maintained and that mycotoxins and other possible contaminants are not present at higher than acceptable levels. Good animal feeding practices also require that feed is stored in such a way as to avoid contamination. As organically raised livestock are fed greater proportions of hay, grass and silage, there is reduced opportunity for mycotoxin-contaminated feed to lead to mycotoxin-contaminated milk.

2.4 Food Irradiation

Irradiation of food assists in the control of insects, parasites, pathogenic bacteria and various deteriorative changes that occur in some foods. When food irradiation is carried out according to internationally accepted guidelines, there is no associated food safety risk. Despite the accumulated body of scientific evidence demonstrating the safety of this processing technology (International Consultative Group on Food Irradiation–ICGFI), it is still not fully accepted by all consumer segments. Consumers, who are reluctant to accept this technology, welcome the assurance that organic foods cannot be subjected to ionising radiation. However, in many countries labelling regulations require that conventionally-produced food that has been subjected to irradiation, be labelled as such. Consumers of conventional foods are thus assured of their right to choose whether they want to purchase foods treated by irradiation.

2.5 Genetically Engineered Organism (GMO)

The relatively recent introduction of genetically engineered seeds is another issue of interest to the organic community. International guidelines for organic foods ban the use of genetically modified organisms, as genetic engineering is considered incompatible with the principles of organic production. Public concern has been widely expressed over the possible negative impact of this technology on the environment as well as on human and animal health, even though there is no clear scientific evidence of this. Choosing "organic" is a way for consumers to ensure avoidance of GMOs. Labelling of conventionally produced foods indicating the use of GMOs is gaining increasing acceptance worldwide. In this case, consumers of conventionally produced foods will also be assured their right to exercise choice over the use of GMOs.

3. CODEX GENERAL STANDARD (Contaminants and Toxins)

Codex Alimentaris (Latin, meaning food law or code) is an international food standards-setting body established by two United Nations organizations, the Food and Agriculture Organization (FAO) and the World Health Organization (WHO) in 1962. It has 170 member countries, all of which are members of FAO or WHO or both. The main work on standard setting for cleanliness, pesticides residues, etc. for agricultural crops and animal products is carried out in the more than 20 Codex Committees and Task Forces. The Codex Commission (CAC) adopts the standards proposed by these Committees and Task Forces.

This Standard contains the main principles and procedures which are used and recommended by the Codex Alimentarius in dealing with contaminants and toxins in foods and feeds, and lists the maximum levels of contaminants and natural toxicants in foods and feeds which are recommended by the CAC to be applied to commodities moving in international trade.

Codex protects the health and economic interests of consumers and encourages fair international trade in food. Through adoption of food standards, codes of practice, and other guidelines developed by its committees, and by promoting their adoption and implementation by governments, Codex seeks to ensure that the world's food supply is sound, wholesome, free from adulteration, and correctly labelled.

3.1 Contaminant

Any substance not intentionally added to food, which is present in such food as a result of the production (including operations carried out in crop husbandry, animal husbandry and veterinary medicine), manufacture, processing, preparation, treatment, packaging, transport or holding of such food or as a result of environmental contamination. The term does not include insect fragments, rodent hairs and other extraneous matter. This standard applies to any substance that meets the terms of the Codex definition for a contaminant, including contaminants in feed for food-producing animals. Some exceptions are-

- Contaminants having only food quality significance, but no public health significance, in the food(s).
- Pesticide residues, as defined by the Codex definition that are within the terms of reference. Pesticide residues arising from pesticide uses not associated with food production may be considered for inclusion in the General Standard for Contaminants.
- Residues of veterinary drugs, as defined by the Codex definition, that is within the terms of reference.
- Microbial toxins, such as botulinum toxin and staphylococcus enterotoxin, and microorganisms that are within the terms of reference.
- Processing aids (that by definition are intentionally added to foods).

3.2 Natural Toxins

The Codex definition of a contaminant implicitly includes naturally occurring toxicants such as are produced as toxic metabolites of certain microfungi that are not intentionally added to food (mycotoxins). Microbial toxins that are produced by algae and that may be accumulated in edible aquatic organisms such as shellfish (phycotoxins) are also included in this standard. Mycotoxins and phycotoxins are both subclasses of contaminants. Inherent natural toxicants that are implicit constituents of foods resulting from a genus, species or strain ordinarily producing hazardous levels of a toxic metabolite(s), i.e. phytotoxins are not generally considered within the scope of this standard.

Foods and feeds can become contaminated by various causes and processes. Contamination generally has a negative impact on the quality of the food or feed and may imply a risk to human or animal health. Contaminant levels in foods should be as low as reasonably achievable. The following actions may serve to prevent or to reduce contamination of foods and feeds:

- Preventing food contamination at the source, e.g. by reducing environmental pollution.
- Applying appropriate technology in food production, handling, storage, processing and packaging.
- Applying measures aimed at decontamination of contaminated food or feed and measures to prevent contaminated food or feed to be marketed for consumption.

To ensure that adequate action is taken to reduce contamination of food and feed a Code of Practice should be elaborated comprising source-related measures and good manufacturing practice as well as good agricultural practice in relation to the specific contamination problem.

HACCP-based approach is recommended to enhance food safety as described in *Hazard Analysis and Critical Control Point (HACCP) System and Guidelines for its Application.* The controls described in this "General Principles" document are internationally recognized as essential to ensure the safety and suitability of food for consumption. The general principles are commended to governments, industry (including individual primary producers, manufacturers, processors, food service operators and retailers) and consumers alike.

4. ORGANIC FOOD

In view of the reduced use of chemically synthesised inputs in organic farming, many studies have been carried out to investigate safety and quality implications of the production system. It has been demonstrated that organically-produced foods have lower levels of pesticide and veterinary drug residues and, in many cases, lower nitrate contents. Animal feeding practices followed in organic livestock production, also lead to a reduction in contamination of food products of animal origin.

4.1 Nutritional Quality

Many sensory analysis studies have been carried out to investigate differences in selected organoleptic parameters between organically and conventionally-grown products. Certain studies have shown significant differences for selected products, as in the case of the sensory differences between organically and conventionally grown apples of the Golden Delicious variety. The organically-grown apples were found to be firmer and received higher taste scores than conventionally-grown apples. The content of flavanoids in the organic apples was higher. Another study showed that organic tomatoes were sweeter and conventional carrots had more "carrot taste".

Comparative studies between organically-grown and conventionally-grown wheat have shown that the former has lower protein content. This may have undesirable consequences on the baking quality of the wheat flour as the wheat proteins play an important role in the rheological behaviour of dough. Another recent study has shown that there was better loaf browning in bread made from organically grown wheat due to higher alpha-amylase and sugar contents which favour the Maillard reaction.

Some studies have focussed on the differences in the storage life of organically-grown products in comparison with conventionally grown products. Results indicate that better storage capacity of some organic products may be a quality difference resulting from production effects, e.g. less storage losses due to fungi in organic carrots and less deterioration of cooking properties in organic potatoes.

4.2 Organic Food Processing and Quality

At the level of food processing both the CAC (Codex Alimentaris Commission) and the EU (European Union) organic food guidelines restrict the use of food ingredients of non-agricultural origin used in the processing of organic foods. Allowed ingredients and additives are included in annexes to the two guidelines. Both international organic guidelines preclude the use of irradiation on food or any ingredient thereof. Besides these considerations, there are no differences between the processing of conventional and organic foods. Food safety concerns that have been investigated in relation to the contamination of processed foods with chemical compounds, such as bisphenol A, phthalates and heavy metals, contained in certain packaging materials or processing contact surfaces are equally relevant to organic and conventionally produced food. It may also be useful to note that current consumer trends towards 'natural' and 'minimally processed' food may overlap with the demand for organically-grown foods. A major function of processing operations is to render a food microbiologically stable for a defined period. With minimally processed foods unacceptable levels of microbiological contamination will occur if adequate care is not exercised in their processing and handling.

Other aspects of organic food quality

The understanding of food quality has been expanded beyond mere definition by chemical content, technical characteristics for processing and storage, appearance and taste. Particularly in organic agriculture, but not exclusively so, other considerations like ethical values and production principles (environmental impact such as energy efficiency, non-pollution, animal welfare, aim for sustainability and social impact) are gaining weights as integral product values. In this context, organic agriculture's contribution to cleaner drinking water, e.g. UK's environmentally sensitive areas and Germany's water protection areas, and to higher weed, insect and bird diversity or general environmental quality are positive values that are appreciated by consumers.

It is through inspection and certification, therefore, that the consumer is assured that the essential elements constituting 'organic' production are met and that foods labelled as 'organic' are really what they claim to be. Growing consumer demand and expanding economic interests in organic production have led to increasing distances between producer and consumer. This highlight even furthers the importance of external control and certification procedures to ensure consumer protection.

Chapter 3

Quality Control Standards and Certification

In order to assure the consumer that a product is produced organically, a kind of quality control is needed. The organic quality is based on standards, inspection, certification and accreditation. All organic food is produced and handled according to strict rules called 'Organic standards'. These standards cover all aspects of food production from animal welfare and wildlife conservation, to not allowing artificial food additives. All organic farms are visited at least once a year by a Certifying Inspector to check that standards are being met. Organic standards do not define a quality status which can be measured in the final products (e.g. quantity of pesticides residues, heavy metals, etc). They define the way of production (e.g. that no chemical pesticides and fertiliser are used). There are organic standards on the national as well as international level. For certification, the standards of the target market or importing country are relevant. Certain private labels such as Naturland, Demeter or BIO SUISSE have additional requirements on top of their national and international standards.

1. AIMS OF ORGANIC PRODUCTION AND PROCESSING

Organic Production and Processing is based on a number of principles and ideas. All are important and this list does not seek to establish any priority of importance. The principles as per IFOAM include:

- To produce sufficient quantities of high quality food, fibre and other products.
- To work compatibly with natural cycles and living systems through the soil, plants and animals in the entire production system.
- To recognize the wider social and ecological impact of and within the organic production and processing system.
- To maintain and increase long-term fertility and biological activity of soils using locally adapted cultural, biological and mechanical methods as opposed to reliance on inputs.
- To maintain and encourage agricultural and natural biodiversity on the farm and surroundings through the use of sustainable production systems and the protection of plant and wildlife habitats.

- To maintain and conserve genetic diversity through attention to on-farm management of genetic resources.
- To promote the responsible use and conservation of water and all life therein.
- To use, as far as possible, renewable resources in production and processing systems and avoid pollution and waste.
- To foster local and regional production and distribution.
- To create a harmonious balance between crop production and animal husbandry.
- To provide living conditions that allow animals to express the basic aspects of their innate behaviour.
- To utilise biodegradable, recyclable and recycled packaging materials.
- To provide everyone involved in organic farming and processing with a quality of life that satisfies their basic needs, within a safe, secure and healthy working environment.
- To support the establishment of an entire production, processing and distribution chain which is both socially and ecologically responsible.
- To recognize the importance of, and protect and learn from, indigenous knowledge and traditional farming systems.

2. IMPORTANT ORGANIC STANDARDS

2.1 IFOAM Basic Standards

The most important organic standards are the IFOAM Basic Standards which also describes the principle of organic farming and provide recommendations on how to achieve the minimum requirements. Being the 'mother of organic standards', *IFOAM Basic standards* are not standards for certification but *standards for standard setting on the national or international level.* It provides a framework for certification bodies and standard setting organizations worldwide to develop their own certification standards. They are regularly reviewed and updated in a democratic process by the IFOAM members from all over the world.

2.2 Codex Alimentarius Guidelines

The *Guidelines* include general sections describing the organic production concept and the scope of the text; description and definitions; labelling and claims (including products in transition/conversion); rules of production and preparation, including criteria for the substances allowed in organic production; inspection and certification systems; and import control. The Codex Alimentarius Commission at its 23rd Session in 1999 adopted the *Guidelines for the Production, Processing, Labelling and Marketing of Organically Produced Foods*, with the exception of the provisions for livestock and livestock products. The Codex Alimentarius Commission at its 24th Session in 2001 adopted the sections concerning livestock and livestock products and beekeeping and bee products for inclusion in the *Guidelines for the Production, Processing, Labelling and Marketing of Organically-Produced Foods.*

2.3 EU Regulation (EEC) No 2092/91

The European Union was one of the first to set up a policy on organic farming by adopting EU Council Regulation (EEC) No 2092/91, amended by Council Regulation (EC) No 1804/1999. With this regulation, the Council created a community framework defining in detail the requirements for agricultural products or foodstuffs bearing a reference to organic production methods. The regulation is set up primarily as a labelling regulation, meant to regulate the internal market for organic products but it also describes the organic production standards and the inspection and supervision requirements. As it deals with virtually all agricultural products and with all aspects of primary food production and food processing, the remit of the regulation is very broad. At its creation in 1991, the regulation took into account, to a large extent, the existing private production rules. At that time, only a few Member States had developed national legislation. While the original regulation was relatively short and covered only plant production, in 1999 it was extended substantially to cover animal production in a relatively detailed way. Being a Council regulation, this legislation is directly applicable in all Member States.

2.4 Indian National Standards for Organic Production (NSOP)

In India, standards for organic agriculture were announced in May 2001, and the National Programme on Organic Production (NPOP) is administered by Agricultural and Processed Food products Export Development Authority (APEDA) using the IFOAM Basic Standards (Edition 2000) under the Ministry of Commerce. Definite principles, basic standards of production, documentation, inspection and certification guidelines are approved by the National Standards Committee constituted by the members of IFOAM in India. Certain responsibilities to define details are delegated to the accredited certification bodies.

The government has set the frame conditions in which the organic sector of a country operates. Most important are the content and legal status of organic standards, the regulations concerning the use of organic claims and labels, the legislation on consumer protection and the accreditation system. As per the national accreditation policy, all the certifying agencies operating in India are to obtain accreditation from anyone of the accrediting agency appointed by the Government of India, viz., Spices Board, Coffee Board, Tea Board or APEDA. *Until now the Indian standards are only compulsory for products to be exported. It is planned to apply the same standards also for the domestic market.*

The Certification according to NPOP is still awaiting international recognition and efforts are on to achieve this at the earliest. This would certainly be an added advantage for the export segment of Indian farmers. The latest copy of Indian organic standards can be procured from Agricultural and Processed Food products Export Development Authority (APEDA), Ministry of Commerce, Govt. of India, New Delhi.

3. COMMON ISSUES IN THE VARIOUS STANDARDS

3.1 Maintenance of Organic Management

The standard's requirements should be met during the conversion period. All the standard's requirements should be applied on the relevant aspects from the beginning of the conversion period onward. If the whole farm is not converted, the certification programme should ensure that the organic and conventional parts of the farm are separate and inspectable. Converted land and animals should not get switched back and forth between organic and conventional management.

3.2 Soil Healths and Water Quality

Necessary measures have to be taken to maintain and improve landscape and enhance biodiversity quality. Land preparation by burning vegetation is restricted to the minimum. Crop production, processing and handling systems should return nutrients, organic matter and other resources removed from the soil through harvesting by the recycling, regeneration and addition of organic materials and nutrients. Grazing management should not degrade land or pollute water resources. Relevant measures should be taken to prevent or remedy soil and water salinization. Operators should not deplete nor excessively exploit water resources, and should seek to preserve water quality. They should where possible recycle rainwater and monitor water extraction. Clearing of primary forest is not permitted. The farmer should take defined and appropriate measures to prevent soil and water erosion.

3.3 Biodiversity

Diversity in plant production and activity should be assured by minimum crop rotation requirements and/or variety of plantings. Minimum rotation practices for annual crops should be established unless the operator demonstrates diversity in plant production by other means. Operators are required to manage pressure from insects, weeds, diseases and other pests, while maintaining or increasing soil organic matter, fertility and microbial activity. For perennial crops, the certifying body should set minimum standards for orchard/plantation floor cover and/or diversity or refuge plantings in the orchard.

3.4 Soil Productivity

The productivity of the farmland should be preserved and promoted only by applying the compost derived from the remainders of the agricultural products produced in the said fields, etc. and methods effectively utilising biological functions of the organism inhabiting and growing in the fields. Material of microbial, plant or animal origin should form the basis of the fertility program.

Nutrients and fertility products should be applied in a way that protects soil, water, and biodiversity. The fertility and the biological activity of the soil must be increased, in the first instance, by cultivation of legumes, green manures or deep rooting plants in appropriate rotation programme.

Mineral fertilizer (ground rock, lime, dolomite, etc), microbial fertilizers, biodynamic preparations and botanical preparations only to be used as supplement. No synthetic/chemical fertilizers are permitted. In cases where the productivity of the farmland cannot be preserved and promoted only by the methods utilising the biological functions of the organism inhabiting and growing in the said fields or in the circumference, utilise only the fertilizers and the soil improvement materials noted in the organic standards. The maximum amount of manures may not exceed 170 kg nitrogen/year/hectare of agricultural area used (EEC2092/91).

3.5 Pest, Disease and Weed Management

Control of noxious animal, insects, pests, disease, weeds and plant in fields, etc. should be executed only by the cultivation method (suitable crop lists and variety, the adjustment of the cropping time, and other cultivation management of the agricultural products), physical method (using light, heat, sound, etc., or manual or mechanical methods), biological method (by introducing microorganisms suppressing the proliferation of microorganisms being the cause of diseases, predators, plants repelling noxious animal and plant, or by improving the environment suited for growing them), or an appropriate combination of these methods. In cases of being critical, the agricultural chemicals noted in the organic standards may be used. Preventive methods should be adopted to maintain the plant health. Ionising radiation should not be executed for the disease and pest control, the preservation of the foods, removal of pathogens or sanitation. The products which can be used under the supervision and inspection by certification body for pest and disease controls are for example: bio-pesticides (Bt, *trichoderma,* NPV, pseudomonas etc.), Bordeaux mixture, sulphur, soft soap, and most plant-based products (neem, rotenone, pyrethrum, etc.). Maximum use of copper under NPOP is 8.0 kg/ha/yr. EU Regulation restricted the use of copper up to 6 kg/ha/yr on a 5-year average.

3.6 Contamination Control

Buffer zones or border crops must be established to avoid potential contamination and limit contaminants in organic products. In the transportation, selection, processing, cleaning, storage, packaging, and other processes, control in such a manner as not being mixed with other agricultural products than the organic agricultural products. For synthetic structure coverings, mulches, fleeces, insect netting and silage wrapping, only products based on polyethylene and polypropylene or other polycarbonates are permitted. These should be removed from the soil after use and should not be burned on the farmland. All equipment from conventional farming systems should be thoroughly cleaned of potentially contaminating materials before being used on organically-managed areas.

3.7 Collection of Wild Harvested Product

Wild harvested products should only be certified organic if derived from a stable and sustainable growing environment. Harvesting or gathering the product should not exceed the sustainable yield of the ecosystem, or threaten the existence of plant or animal species. Products can only be certified organic if derived from a clearly defined collecting area, which is not exposed to prohibited substances, and which is subject to inspection. The collection area should be at an appropriate distance from conventional farming, pollution and contamination. The operator managing the harvesting or gathering of the products should be clearly identified and be familiar with the collecting area in question. The collection does not affect the stability of the natural habitat or the maintenance of the species in the collection area. The harvesting of wild crop should be in a manner that such harvesting will not destroy to the environment and will sustain the growth and production of the wild crop.

3.8 Animal Management

The establishment of organic animal husbandry requires an interim period, the conversion period. Management of the animal environment takes into account the behavioural needs of the animals and provides for sufficient free movement, sufficient fresh air and natural daylight; protection against excessive sunlight, temperatures, rain and wind; enough lying and/or resting area; natural materials for all animals requiring bedding; ample access to fresh water and organic feed; adequate facilities for expressing behaviour in accordance with the biological and ethological needs of the species.

Uses of preventive antibiotics or growth promoters are not permitted. No chemicals (including urea) or artificial vitamins should be added to the feed. At least 50 % of the feed should come from the own farm itself or produced in cooperation with other organic farms in the region. If certified organic feed is not available, a maximum of 20 % of non-organic feed can be fed (max. 15 % for ruminants), provided it does not contain prohibited chemical additives or genetically modified products. The percentage should be reduced within 5 years to 15 % (10 % for ruminants).

The use of conventional veterinary medicines (e.g. antibiotics) is allowed when alternative treatment is not sufficient. Synthetic growth promoters and hormones are not allowed. Embryo transfer techniques, hormonal heat treatment and genetically engineered animals are not allowed. Breeding systems should be based on breeds that can reproduce successfully under natural conditions without human involvement. Artificial insemination is allowed. Landless animal husbandry systems are not allowed. Herd animals should not be kept individually. The certification programme may allow exceptions e.g. male animals, smallholdings, sick animals and those about to give birth. Poultry and rabbits should not be kept in cages.

Breeding stock brought-in from conventional farms with a yearly maximum of 10 % of adult animals of the same species on the farm.

3.9 Social Justice

Social rights and justice are integral part of organic agriculture. The laws relating to labour welfare and rights of children should be honoured. All employees and their families should have access to potable water, food, housing, education, transportation and health services. All employees should have equal wages when doing the same job. They must have equal opportunities irrespective of colour, creed and gender. Social security needs (include maternity, sickness and retirement benefits) should be met. Labour conditions regarding noise, dust, light and exposure to chemicals should be within acceptable limits; and they should have adequate protection. The rights of indigenous people should be respected.

4. SOME KEY ISSUES IN THE VARIOUS STANDARDS

In organic farming system, certain minimum requirements are to be met to fulfil its objectives. The requirements of all the Regulations are based on the same principles, there is slight variation from each other.

4.1 Conversion

When a farmer switches over to the system of organic farming from the conventional system of farming, it is known as conversion. The time between the start of organic management and certification is called conversion period. The conversion period is decided based on the past use of the land and ecological situation. In NPOP, Regulation (EEC) 2092/91 and most other standards, the conversion period for annual crops is 24 months and for perennial crops is 36 months from the start of organic management. However, the inspection activities must take place from the very beginning. The produce of the first 12 months has to be marketed only *as conventional.* The products from 12 to 24 months for annual and 24 to 36 months for perennials can be marketed as *'organic-in-conversion'* if a buyer wants such a product. Only after 24 months (for annuals) and after 36 months (for perennials) the products can be sold as organic with the relevant certificates. However, for USDA (United States) NOP, the conversion period is three years irrespective of whether the crop is annual or perennial and there is no in conversion certificate during the transition period.

Animal products to be sold as organic only after the farm or relevant part of it has been under conversion for at least 12 months. For dairy and egg production the animal production standard should have been met for not less than 30 days.

4.2 Planting Material

Species and varieties cultivated should be adapted to soil and climatic condition and resistant to pests and diseases. According to all the regulations, planting materials must be from certified organic sources. Use of conventional seedlings/ seeds is not allowed in Regulation (EEC) 2092/91 from January 2005. In case the organic planting materials are not available, conventional untreated seeds/planting materials can be used one time with prior approval from the certification body proving the unavailability of the product in the market (only accepted by NSOP, India). Use of genetically engineered seeds or planting materials such as tissue culture, pollen culture and transgenic plants are not allowed.

4.3 Mixed Farming and Cropping Pattern

Animal husbandry, poultry, fisheries, etc. should be practised in addition to agricultural farming. Shifting cultivation is not allowed. Crop rotation should be followed if annual crops are grown. Intercropping should be practised when perennial crops are grown. Crop rotation should cover green manure as well as fodder crops. In case of perennial crops, cover crops should be grown to protect the soil. Mono-cropping should be avoided. Per acre of farm land, at least 3 trees must be grown. Farmers should avoid overgrazing and over manuring of land by keeping appropriate number of animals per farm land. Maximum number of animals per acre of farm land is: 2 milk cow or buffalo/ha, or 5 calves, or 13 sheep or goat, or 14 pigs, or 580 chickens.

4.4 Soil and Water Conservation

Measures like stone pitching/contour wall construction are to be taken up to prevent soil erosion. In case of saline soils, saline resistant varieties may be grown. Judicious irrigation is to be practised. Mulching is required. Pollution of surface and ground water should be prevented. Clearing of primary forest is prohibited. Cleaning of land through straw burning should be restricted to minimum.

4.5 Documentation

Documentation of farm activities is must for acquiring certification when organic crops are raised. The following documents/records are to be maintained.

a) Field map
b) Field history sheet
c) Activity register
d) Input record
e) Output record
f) Harvest record
g) Storage record
h) Sales record
i) Pest control records
j) Movement record
k) Equipments cleaning records
l) Labelling records.

4.6 Parallel Production

Parallel production is growing the same variety or indistinguishable variety of crops by the same operator in different qualities. This is strictly not permitted for annual crops. In case of perennial crops the entire unit will have to be converted in five years. During this period, strict separation must be maintained and clearly documented. This is not a major concern in case of USDA-NOP as long as sufficient separation is maintained.

4.7 Processing, Grading, Labelling and Packaging

Processing technologies like solar drying, freeze-drying, hot air chambers are permitted. Irradiation of agricultural produce is not permitted. No synthetic additives are to be added during processing.

Japan (JAS) standard requires grading procedure. Grading is an internal verification that a specific lot of product meets the JAS standards and that internal operational procedures were followed in production before the JAS seal is affixed.

The label should convey clear accurate information on the organic status of the product. (i.e. conversion in progress or organic). The labels for organic and conversion in progress products should be distinguishable by different coloured labels. The details like name of the product, quantity of the product, name and address of the producer, name of certification agency, certification, lot number etc. are to be given in the label.

For Example,

Information required on the label

Crop	OG (Organic Ginger)
Country	I (India)
Field No.	05
Date of harvest	32 (1st Feb)
Year	2005
Lot number	OGI 0532 2005.

Lot number is helpful in tracing back the product particularly the field number in which it is grown in case of contamination. Lot number should include the crop, country; field number, date of harvest (in Julian calendar) and production year.

For packing, recycling and reusable materials like clean jute bags, should be used. Use of biodegradable materials can also be used. Unnecessary packaging material should be avoided. Organic and non-organic products should not be stored and transported together except when labelled.

4.8 Export

Each individual consignment/invoice requires a transaction certificate. This achieves two objectives. Firstly, the certification agency is able to keep track of the material produced and marketed from each certified production unit. Secondly, the organic quality of the product is proved at the point of entry into the country of import. In EU, import permission has to be obtained for each product for every specific producer, exporter and importer.

The list of permitted and restricted products under certified organic farming (NSOP India) is given in Appendices (No 1 to 8)

5. CERTIFICATION PROCESS

The difference between organic farming and other sustainable system like eco-farming, permaculture etc. is that the former is characterised by certification, which is a procedure bound by rules and regulations. All products sold as 'organic' must be certified. The farmer, processor/manufacturer, distributor and retailer have to go through a process of certification. Certification for organic farms is an essential pre-requisite for marketing their produce, especially in the international markets. This is done mainly to ensure genuineness of organically farm products reaching the market and consumer. The certificate issued by an accredited agency is the attestation of the organic method of production of the products.

Any person or organisation intending to produce or process organic products must be subject to an inspection and certification procedure by an approved inspection body. Certified organic refers to agricultural products that have been grown and processed according to uniform standards, verified by independent state or private organisations accredited by the standard setting organisation. Anyone contravening this Regulation could be subject to prosecution by the Trading Standards Officers. The labelling and marketing of organic products are controlled by the National Regulation.

The certification process actually starts in the conversion process. The steps to be followed for certified organic production are as follows:

- Fulfil the requirement of conversion period.
- Develop a comprehensive operations plan for the farm including an Ecosystem Management Plan.
- Convert the farm completely or comply with strict regulations in the cases of split and parallel operations (organic and conventional).
- Farm follow-up inspections by certifier.
- Continuously update the chosen certifier on major changes.
- Keep track of inspection certificates issued by certifier.

5.1 Procedure of Certification

The producer or farmers makes contact with certifying agency. Certification agency provides information on standards, fees, application, inspection, certification and appeal procedures. The producer then submits application along with field history, farm map, record keeping system etc. The contract indicating scope, obligation, inspection and certification, sanction and appeals, duration, fee structure is executed. The costs of certification depends on size of farm, type of production system, group of farmers, extent of animal husbandry, location of unit, travel time to reach the inspection site, and costs for travel, food and accommodation during inspection.

5.2 Inspection

If an organic farmer wants his products to be certified, he has to undergo an inspection at least once a year. Inspectors verify that organic practices such as long-term soil management, buffering between organic farms and neighbouring conventional farms, and record-keeping are being followed. Processing inspections include review of the facility's cleaning and pest control methods, ingredient transportation and storage, and record-keeping and audit control. The inspector evaluates the performance of the farm activities with the help of the farmer's statements and records and by viewing the fields, animals and farm buildings. He can take samples for laboratory testing and may conduct unannounced inspections. The inspector transmits his/her findings to the certification body as a written report.

5.3 Certification

A defined procedure in which a certification body assesses a farm or company and assures in writing that it meets the requirements of the organic standards. Certification includes annual submission of an organic system plan and inspection of farm fields and processing facilities. The certification body compares the results of the inspection with the requirements of the organic standards. A certification committee decides whether certification may be granted or not and then the agency issues approval or denial of certificate. **Certificate is given for current year's harvest only and hence annual certification is required.** In case of denial of the certificate, if the operator has valid reasons not to accept the certification decision, he/she can request for reconsideration of the decision in writing.

5.4 Accreditation

In order to make sure that the certification programme is competent to carry out inspection and certification, a third level of quality control is needed. Authorised bodies regularly evaluate certification programmes and check their proper functioning according to certain criteria. In case the certification body complies with the criteria, they accredit the certification programme.

Accreditation Agencies in India: In India, there are, at present, six accreditation agencies approved by the central government's Ministry of Commerce (MoC). They are the Agricultural and Processed Food Products Export Development Authority (APEDA), Coffee Board, Spices Board, Tea Board, Coconut Development Board and Cocoa & Cashewnut Board. Several other Indian initiatives have also applied for accreditation. Increasing competition can be expected in the Indian certification market in the near future.

5.5 Type of Certification

Foreign Certification: Presently, a lot of the export-oriented organic projects in developing countries are inspected and certified by certification bodies based in the importing countries. These international certification bodies usually have long experience in organic agriculture and its certification. Importers often prefer their services as they are well known and provide their services worldwide. The main disadvantage is that costs of certification are high due to frequent plane trips and western salaries have to be paid.

Co-certification: During the last few years, most Western certification programmes started to build up local branch officers for conducting the inspections and to work with local inspection staffs. Still, the inspection work is supervised by the head office, but the number of required plane trips is less. Local Inspectors find it easier to inspect farms, as they usually speak the same language and are familiar with the local conditions.

Indigenous Certification: Indigenous certification bodies can usually offer cheaper inspection fees as less travelling is required and only local salaries have to be covered. Indigenous certification may especially support the development of domestic market for organic products. However, for export purposes local certification bodies have to achieve international recognition which means to meet the different requirements of different import countries.

In 2002, a group of farmer organisations, NGOs and companies established an Indian Organic Certification Agency named INDOCERT as a charitable trust in order to provide affordable certification services of high quality. Under the guidance of FiBL and bio.inspecta AG and with support from SECO (State Secretariat for Economic Affairs), the office started offering inspection and certification services in August 2002. Since then, even the rates of the international certification bodies have come down significantly. Through re-certification arrangements with bio.inspecta, INDOCERT certification from the beginning enable access also to export markets. Recently, INDOCERT is accredited by the DAP, Germany for certification for EU Regulation 2092/91. This certification agency is the first fully Indian Certification Body listed by FOODPLUS to offer EUROGAP Certification in India. INDOCERT expects to gain US and JAS (Japan) accreditation by 2006. There are some other indigenous certification initiatives, but so far none of them gained momentum.

It must be understood that, for export to the EU market, a certificate from a certification body, which is accepted by EU competent authorities, is essential. To regulate the export of certified organic products, the Director General of Foreign Trade, Government of India has issued a public notice according to which no certified organic products may be exported unless they are certified by an inspection and certifying agency duly accredited by one of the accreditation agencies designated by the Government of India. For the domestic market also it is expected that certification will be made compulsory.

Certification Bodies in India

- ECOCERT International (based in France and Germany, branch office in Aurangabad, Maharashtra).
- IMO Control Pvt. Ltd.-Institute for Marketecology (based in Switzerland, office in Bangalore, Karnataka).
- LACON GmbH (based in Germany, office in Aluva, Kerala)
- SGS India Pvt. Ltd. (based in Switzerland, offices in Delhi and other cities).
- BIOINSPECTA (based in Switzerland, branch office in Cochin, Kerala)
- SGS India Pvt. Ltd (based in India, office in Bangalore)
- APOF Organic Certification Agency (AOCA) (based in India, office in Gurgaon, Haryana)
- SKAL International (based in the Netherlands, branch office in Mumbai).
- INDOCERT (based in India, office in Aluva, Kerala)
- India Society for Certification (ISCOP) (based in India, office in Coimbtore).

The above certification bodies are fully accredited under the Indian National Program for Organic Production (NPOP).

5.6 Smallholder Group Certification

Where large numbers of smallholders are to be inspected by a foreign certification body, the involved costs can be very high. In such cases, smallholder group certification can be done. It is done for defined groups (may be up to 900-1000 farmers) of collaborating organic producers, processors, and exporters with similar farming and production system located in geographical proximity. Only one application is required to certify the entire farms of such smallholder group and the certification and inspection fee is shared by each individual farmer/operator. It should be based on internal regulations and defined sanctions in case of non-compliance. The group certification necessarily insists on internal control system (ICS).

The **Internal Controls System** operates like a small internal control body: internal standards, a written commitment of the participating farmers, internal inspectors inspecting the farms at least once a year and an internal system of sanctions against defaulting farmers. This is a documented quality assurance system that allows the certification body to delegate the inspection of individual group members to a body identified from within the operator group. The external certification programme evaluates the proper functioning of the ICS and re-inspects at random a certain percentage of the farms. Contracted party is the farmer group, project or corporate which also is the owner of the certificate. The external auditor can invariably insists on physical inspection of all individual holdings extending up to and above four hectares. It may also be noted that the number of units, above four hectares should not exceed 50 per cent of the total area under the group certification.

The following criteria may used for identification of farmers for group certification:

Marginal Farmers	: Below 1 hectare
Small Farmers	: 1.00 - 1.99 hectare
Semi medium Farmers	: 2.00 - 3.99 hectare
Medium Farmers	: 4.00 - 9.99 hectare
Large Farmers	: Above 10.00 hectare

5.7 Cost of Inspection and Certification

Most certifiers are charging inspection and certification fees based on the number of person-days involved, plus fees for the issue of certificates. Sometimes, different fees are applied for small farmers, large farmers, and processors or traders. All the accredited certifiers publishes tariff for inspection and certification. The rate varies between the types of growers. An example of the fee structure of a certification body operating in India is given below:

Table 3.1: Cost of inspection (indicative) and certification in India

Category	Details	Fees (Rs)
Small farmers and co-operatives	Travel and inspection*	12000/day
	Report preparation	5000 flat fee
	Certification	5000/certificate
Estate manufacturers and exporters	Travel and inspection	19200/day
	Report preparation	5000 flat fee
	Certification	5000/certificate
Large and medium-sized processors	Travel and inspection	16800/day
	Report preparation	5000 flat fee
	Certification	5000/certificate

Source: Org-Marg, 2002 **Travel and inspection fees depends on the location*

Over the past few years, many international certifiers opened branch offices in India, operated by Indian staff. Thus, the costs for certification came down considerably. At the same time, local certification bodies started to emerge, partially with the aim of further reducing certification costs. It can be assumed that certification costs will soon reach the lowest possible level while still ensuring quality requirements for inspection and certification work.

Organically-certified export items requires 'transaction certificate' for each transaction. This needs to be obtained by the exporter. Similarly processing/ packing unit also needs to be certified before it is operationalized.

The quality system is administered through established documented procedures which are periodically audited to ensure that production, processing, handling, management, certification, accreditation, and other systems too meet the specified requirements and outcomes by standardised protocols.

Who can apply for Organic Certification?

Any growers who have the inclination for organic farming can apply for organic certification. Organic farming is an alive production system and not just simple replacement of fertilizers and pesticides by manure and predators. It is an ongoing dynamic process for making soil healthy which is a vital living matter on earth. Hence, before going for organic cultivation the grower should go for training on different components of organic farming and acquire sufficient knowledge on this system.

After getting training, a plan of farmer's land should be prepared which includes land preparation, crop selection, nutrient requirement, mixed farming, crop diversification, etc. Thereafter farmer has to approach an accredited certification agency (e.g. INDOCERT) whose names are given in the APEDA website (www.apeda.com). All the accreditation agencies have their own application forms and agreement forms. The application form invariably contain the details of personal information of the farmer, the details of farming unit like extent of area, name of the owner, products to be certified, whether internal control system are working in the area of group farming, details of processing units, travel time to visit each of the farming unit etc. Apart from the above general information, specific information on farming units like name and location, system on quality and management, persons responsible for- quality control management, grading, management control of production processes, levels of grading etc. also need to be submitted. Farm maps showing the plot wise details like type of crop, date of planting etc. should be provided along with application. The above details with respect to the processing and management control of production process units also need to be provided.

5.8 Views on National Program on Organic Certification

At present, India is in the nascent stage of organic farming. Hence, the Government of India has planned a National Program for Organic Production and Certification. The objective of the program is to provide an institutional mechanism for implementing national standards for organic products through a National Accreditation Policy and Program. Some of the favourable responses of the national government are-

- Comprehensive policy, but yet to recognised by international certification agencies
- Attracts more certification agencies in India; this will reduce the cost for certification, which is a major constraint for small farmers
- Different agencies have different parameters, therefore standardisation would help
- Certification bodies should tie up with other certification agencies in other countries
- National Certification would centralise the certification procedure.

Cost and quality emerge as the major constraints for certification. Implementation of national certification agencies that should provide comprehensive policy prices (reduction in cost of certifications), international standardisation and simplified certification procedure with less documentation are some of the important strategies of the national government to improve/promote the organic farming in India.

Chapter **4**

Organic Conversion Plan: Crop Husbandry

Organic food production is as much a state of mind as it is a production system. The biggest barrier most people face when switching to organic production is the change in thinking that must occur to make it successful. The organic gardener must understand how everything is interrelated and how one set of circumstances will influence other factors in how the plant grows. A strong, healthy plant is much more important in an organic system then a conventional system. A two to three year leaning curve must be accepted by anyone intending to switch from conventional to organic gardening. What works one year may fail the next.

An organic production system is designed to:

- enhance biological diversity within the whole system
- increase soil biological activity
- maintain long-term soil fertility
- recycle wastes of plant and animal origin in order to return nutrients to the land

ORGANIC CROP PRODUCTION
Fundamental Principles and Practices

Biodiversity/ Sustainability	***Natural Plant Nutrition***	***Natural Pest Management***	***Integrity***
Rotation	Rotation	Rotation	Buffer
Green Manure	Green Manure	Green Manure	Records
Cover Crops	Animal Manure	Cover Crops	
Animal Manure	Composting	Composting	
Composting	Natural Fertilizers	Intercropping	
Intercropping	Foliar Fertilizers	Biocontrol	
Biocontrol		Farmscaping	
Farmscaping		Sanitation	
Buffers		Tillage	
		Fire	
		Natural Pesticides	

- rely on renewable resources in locally organized agricultural systems, thus minimizing the use of non-renewable resources
- promote the healthy use of soil, water and air as well as minimize all forms of pollution thereto that may result from agricultural practices
- handle agricultural products with emphasis on careful processing methods in order to maintain the organic integrity and vital qualities of the product at all stages.

Organic farming is neither merely replacing the chemicals with organics nor it is going back to the traditional agriculture. It is the conversion of soil from non-living to living and to sustain the same. Normally it takes 3-4 years before the soil is converted to living. Once the soil is converted to organic anything grown on such land the produce becomes organic provided the principles of organic standards are followed in the process of production, processing, transport and storage.

In India, organic farming looks different in high production areas compared to marginal regions. While organic farms in the fertile midlands use latest technology to achieve optimum performance, the small dairy farmers in the mountain areas are almost 'organic by default', using similar technology as their ancestors. Though agro-chemicals have reached most of the remote regions of India, application of fertilizers and pesticides there is usually much lower because they are not remunerative. Conversion to organic farming thus seems easier, and the potential to increase poor farmers' income is bigger than in the fertile plains. Therefore, the Indian government has put its focus for organic agriculture on marginal regions.

In the Indian context, organic farming can be significant in two distinct ways:

1) Organic farming can help to reduce production costs (especially where labour is cheap compared to input costs) and to increase or stabilize yields on marginal soils. This is especially relevant for smallholders in marginal areas where Green Revolution agriculture has lead to a depletion of soil fertility and to high debts because of increase in input costs.
2) In areas where farmers have access to established organic markets within the country or abroad, products can achieve a higher price compared to the conventional market. Especially in the trend of decreasing prices for agricultural products, this can be an important way to stabilize or even increase incomes.

Crop production and animal husbandry can be converted as per basic organic standards. The conversion to organic farming will not only improve the farm ecosystem, but also assure the economic survival of the farm. The whole farm family should get ready for the conversion in many aspects, too. The adjustments, which are required on the farms for conversion and the related chances and risk, have to be analysed carefully.

1. WHAT DO THE ORGANIC STANDARDS SAY ON CONVERSION?

- Standards requirements must be applied from the beginning of the conversion.
- Total crop production and animal husbandry should be converted to organic management.
- Duration of the conversion period is 24 months before the start of production cycle for annual crops, 36 months before harvest for perennial crops and 12 months for animal products. With regard to dairy & egg production and meat production this period is 30 days and 12 months respectively. All the organic standard ought to be met before sowing/planting of annual/perennial crop.
- Start of conversion period is calculated from the date of application to the certification body, when farmers commit themselves in following the standards.
- During conversion period, products can be labelled as "Produce of organic agriculture in the process of conversion" provided organic standards have been met for all at least 12 months.
- Converted land and animals should not get switched back and forth between organic and conventional management.
- Another prerequisite for certification of a farm is proof of sufficient knowledge and ability in the field of organic agriculture. Apart from practical work experience, participation in workshops, training, seminar, etc. is essential to get the up-to-date technologies, marketing, policy and modification (if any) in standards, etc.

2. PREPARATION FOR THE CONVERSION

It is important that all persons to be involved in the farm should have a clear understanding on what would be organic management. For some adoptions on the farm level, new materials are needed and therefore it requires some investments. As the quantity of the production may also decreases at least in the first 2-3 years of the conversion, farmers need to find ways to overcome constraints.

2.1 Conversion Plan

Conversion to organic agriculture basically concerns the whole farm operation and the entire land surface. The conversion of the entire farm must occur under economically acceptable basic conditions. It can therefore take place gradually to cover ever greater areas of the farm land and operation cultivated in accordance with the standards.

There should be a clear plan of how to proceed with the conversion. This plan should be updated as necessary and cover all aspects relevant to the organic standards. The plan should indicate that the totality of crop production and animal production in the operation would be converted to organic management. Standards should determine how organic and non-organic production and product could be clearly separated and distinguishable in production and documentation, to prevent unintentional mixing of inputs and products. Independent sections of the operation unit should be converted in such a way that the standards are completely met on each section before it is certified as organic.

A farm may be converted by gradual introduction of organic practices over the whole farm, or by application of organic principles to only a portion of the operation at first. Where conversion is carried out gradually, it is imperative for the areas under various stages of conversion to be clearly and explicitly distinguishable and separated.

If new fields (e.g. purchase of land or lease of land) are taken under organic cultivation on a farm that is in conversion or already certified, these areas too have to comply with the usual conversion period. These new areas have to be clearly distinguishable and separated according to their stage of conversion. Any changes which might negatively influence the product quality, especially sources of possible contamination, have to be documented for to fulfil certification criteria.

All activities including crop production, animal husbandry and general environmental maintenance should be organized such that all the elements of the farm activities interact positively. Practical farming skills, based on knowledge, observation and experience are therefore important for organic growers. Conversion may be accomplished over a period of time. A conversion period enables the establishment of an organic management system and builds soil fertility.

2.2 Step-by-Step Conversion

If an immediate conversion of the whole operation would impose totally unacceptable risks, farm operations producing wine, fruits or ornamental plants may carry out a step-by-step conversion to organic production, on condition that the whole farm is converted to fulfil organic standards within five years.

The criteria for step-by-step conversion are:

a) A binding conversion plan with full written details of the conversion steps, with a timetable.

b) Evidence of the inspectability regarding production techniques, avoidance of drift and separate flows of produce.

c) The production procedures and flows of produce on the whole farm should be documented. The conversion plan should also include the farming procedures in the non-organically run areas. The rule is: as quickly as possible and as organic as possible.

d) Total separation of the different farm areas. The overlap at the borders between organic and non-organic farm areas is to be kept as small as possible.

e) Any interim regression to non-organic methods on the organically-farmed areas is out of the question.

3. CROP HUSBANDRY

3.1 Length of Conversion Period (Crop Production)

The conversion period should be long enough to improve soil fertility significantly and to re-establish the balance of the ecosystem. The length of the conversion period should be adapted to:

- the past use of the land
- the ecological context and its implications
- the experience of the operator.

The length of the conversion period should be defined to provide for a period of at least 24-36 months.

- Plant products from annual production should only be considered organic when a conversion period of at least 24 months has elapsed prior to the start of the production cycle. In the case of perennials (excluding pastures and meadows) a period of at least 36 months prior to harvest should be required.
- There should be at least a 12-month conversion period prior to pastures, meadows and products harvested therefrom, being considered organic.
- The conversion period may be reduced or extended by the standard-setting organization depending on conditions such as past use of the land, management capacity of the operator and environmental factors. Minimum conversion period must be equal or exceed 12 months. A complete document of previous use of land is required by the certification body.

3.2 Framework of the Organic Crop Production

The framework of the organic production can only be defined after initial discussions with farmers, extension agents and research personnel in the area, and also after a rapid reconnaissance survey of the area. The survey should be conducted by a multidisciplinary team of plant and livestock scientists, social scientists and economists. Information on major crops and livestock, cropping patterns, livestock characteristics, management practices, use of FYM, sources of organic materials and soil amendments on the farm is essential. The area under conversion should be characterized according to the physical and biological environment, socio-economic environment, farming systems and land use. In addition to these characteristics, there should be discussions with the farmers to determine their production objectives and constraints, as well as to obtain a general idea of farmers' perceptions and living conditions.

Once the area has been described, it may be possible to identify predominant sub-groups or strata within the population. These strata can be targeted for further improvement in organic farming. One possibility is to stratify the population. The main purpose of stratification is to subdivide the participants into homogeneous subsets which can provide more reliable information about the scope of the technologies and their applicability over a range of contrasting situations. In selecting the strata the following factors should be considered:

- What factors are likely to affect the response of the enterprise to the technology being introduced?
- What factors are likely to affect the farmers' adoption of the technology?
- Is there the information available to allow stratification of the participants according to these factors?

Some examples of variables which could be used for stratification include: gender, farm size, system of land tenure (i.e. owns, rents, leases, etc.), cropping system (e.g. intercrop vs monocrop), soil type, land cover, system of water control, and slope. This list is by no means exhaustive but gives an indication of the type of information needed for stratification.

The following sources may provide information for identifying important strata:

- censes
- baseline surveys
- rapid reconnaissance surveys
- previous studies of the area under consideration
- discussions with researchers, farmers groups and village leaders.

Stratification is best done before selecting the farm for organic conversion. If it is not possible to do so, then some attempt should be made to stratify the farmers who have agreed to participate in the conversion process.

3.3 Identification and Scale of the Problem

For more obvious problems, perhaps relating to direct inaccessibility of resources and inputs, the difficulty lies in identifying the solution. Some of these problems may be due to an unavailability or reduction in the supply of plant nutrient sources in remote regions, or the increased cost of tillage operations, or greater division of an already scarce water resource. Solutions to these problems are usually found in cultural practices that were once traditional methods of soil management.

Some problems extend across farms while others are localised and specific to individual farms. Where there is an organized body of farmers or where there are agricultural extension workers, the scale of the problem can be determined by participatory rural appraisal or other appropriate survey methods. This should be a necessary prerequisite in developing a programme of conversion of organic farming.

3.4 Site Selection

The choice of sites and locations is made after the framework of the organic conversion has been clearly defined. Important considerations in site selection include:

- The organic garden should have a southern exposure (in hilly area) or be in an open field if at all possible. There should be a minimum of six hours of direct sunlight at the chosen location.
- The site should be free of old bunds, roads, cross drainage ways or any other such conditions which are likely to influence the crop yield.
- A well-drained site even after a heavy rain is ideal. Poor drainage may be improved by digging ditches, installing a tile drain field, or adding organic matter.
- Nearby trees and shrubs may have extensive root systems that may interfere with water and nutrient uptake of plants at your site. Locate the site to minimize or avoid this problem. As a last resort, consider removal of some trees and shrubs that may interfere with production.
- Land with a slope of 1.5 percent or greater (18- inch elevation change in 100 feet) should be avoided or terraced to prevent runoff and soil erosion. Contour planting, which is setting the rows to follow the contour of the land, can also help with runoff problems.
- The site should also have a water supply nearby. Sites with serious weed problems such as nuts edge, Bermuda grass, or kudzu should be avoided unless adequate measures are taken to control them. This does not preclude using these sites, but considerable work is required to remove and control these weeds.
- Location of organic production area should be at appropriate distances from contamination sources and conventional farming area. The appropriate separation by maintaining distance between organic and conventional production system is required to provide protection from pollution and contamination.
- Fencing the site is essential if you have a significant wild animal population nearby. Domestic animals such as dogs, rabbit may also become a problem because many like to dig. Fences as high as 6 feet may be required to control animals. Finally, for convenience, a location near the house is desirable.

3.5 Plot Size and Shape

The size of your organic garden will determine, in part, many aspects of your garden plan. Large gardens where tractors is used can be worked more easily with long rows; small gardens may be worked more easily in small beds with footpaths surrounding them.

Plot size is usually limited by the land available for the farming, and by the amount of labour and inputs of other available resources. To the extent possible, plots size should be kept around 50-100 m^2. Very large plots are not required. If land is available, it is preferable to increase number of plots rather than increase the plot size. Plots need not necessary be square or rectangular. Allowance should be made for guard rows to avoid inter plot interference. However, guard rows may be unnecessary, especially if plots are not contiguous or are surrounded by the crop which has minimal edge effects.

3.6 Areas Set Aside to Enhance Biodiversity (AEB)

A Farmer has the duty to retain, augment or create ecologically diversified areas (the so called "compensatory ecological habitats") as nearly natural as possible, and to care for them properly. These habitats should constitute at least 7% of the agricultural land. They must be situated in the same parts of the farm which are used for agricultural purposes and be owned or rented by the farmer.

The following elements may be included:

- unimproved meadows and pasture land
- wet meadows, wetland
- wildflower strips between fields
- fallow land with rotation or wild flowers (for at least 16 months)
- single high stem fruit trees/isolated native trees in suitable places (100 sqm per tree) and avenues of trees
- hedges, copses and embankment copses
- ditches, ponds and pools
- waste ground, piles and stacks of stones
- drystone walls
- unmade natural paths
- other compensatory ecological habitats.

All elements defined in the National Ordinance must be handled at least in accordance with the requirements of this directive. In the distribution of the ecologically diversified habitats among various farmers, the various components are to be divided up by the responsible authorities, and the individual farmers are to retain the divided areas. Farms which have several production units, which are outside the regular farm area must identify the *compensatory ecological habitats* for each production unit in proportion to its size.

Minimum proportion of land with unimproved meadows/pastures: The proportion of less frequently cut meadows and unimproved meadows, grazed pastures, woodland clearings or wet meadows must comprise at least 5% of the permanent grassland, lays (including unimproved meadows on fallow fields) and wet meadows.

Border areas: Alongside paths, grass strips at least a 0.5 m wide area should be left. These strips can only be counted as compensatory ecological habitats when they are within the farm area, fulfil the relevant conditions for unimproved or less intensively used meadows, and are at least 3 m wide. In areas carrying perennial crops, the first 3 m of such border strips lying across the main direction of cultivation are counted as headlands of the cultivated area. They cannot be counted as unimproved or less intensively used meadows. Alongside or around surface water, wood sides, hedges and copses or trees in fields, grass strips of at least 3 m wide must be left. No fertilizers and no crop treatments may be applied. Grazing is permitted as appropriate for the location.

3.7 Crop production in Former Forest Areas

Clearing of virgin forests: Organic farming does not support any clearing of virgin forests for agricultural exploitation. Approvals of organic projects on cleared areas of virgin forests (so called "high value conservation areas") are therefore excluded.

Exceptions: If one of the following conditions is met, the application for an approval may be considered:

a) The clearing of the now agriculturally used area has been carried out before the year 1970,

b) The clearing has not been carried out under the responsibility of the present owner, or

c) The clearing has not been carried out in conflict with the laws of the respective country.

Clearing of secondary forests: If a secondary forest has been cleared or is being cleared for the purpose of agricultural use, the following information must be provided to the certifying agency:

1. A map of the area that is to be cleared, including a precise survey of the respective surfaces (max. scale: 1:50,000) and the indication of protection areas or "high value conservation areas", must be presented.
2. An inventory of the tree population and a list of endangered species must be presented, including a detailed description of the neighbouring areas (use, inventory).
3. The previous uses of the area subject to clearing must be indicated.
4. The clearing must be carried out in compliance with the national laws of the respective country, and there must be a governmental authorization.
5. The right to exploit the forest can be delegated officially by the indigenous population.
6. Before the clearing is carried out, a conception for sustainability must be submitted. It includes:

a) maximum surface that may be cleared per year
b) measures to protect the endangered species and water resources and prevent the erosion.
c) conservation and description of the retreat areas for indigenous species, wildlife strips, conservation strips along with water surfaces
d) ensuring of the provision of firewood for the local population

4. GETTING READY FOR THE CONVERSION

Farms wishing to convert to organic agriculture need to submit full details of their previous farming methods and soil analyses (nutrient reserves) report to the inspection/certification bodies.

4.1 Analyse the Situation on the Farm

The situation of the farm should be analysed very carefully considering the requirements of organic farming. Support from the field advisers or experienced organic farmers can be of great help in this analysis. Following information should be collected for farm planning:

Field history: At least 24-36 months of histories in terms of crops planted, crop yield, input applied are required for all fields. If farm is newly purchased or rented from previous farmers than get all supporting information about soil, plant or water tests, etc.

Following data on physico-chemical-biological environment should be collected:

Physico-chemical-biological environment

Climate	e.g. Agro-ecological zone, rainfall, temperature, wind, sunny days
Soil	Physical environment: e.g. Soil texture, soil pH, bulk density, colour, etc. at two depths: 10 and 15-30 cm; Chemical environment: e.g. N, P, K, organic matter and heavy metals, at 15-30 cm Biological environment Weeds, insects, diseases, other pests, crop yields, C:N ratio, etc.
Topography	e.g. slope and position on slope, (e.g. steep on top, medium in the middle, etc.), floodplain
Irrigation	e.g. Water source and quality (pH, heavy metals, D.O., TDS, EC, arsenic, nitrates, fluoride, B.O.D., C.O.D., etc.), means and frequency of delivery, on-farm practices, water table, etc.
Farm site	e.g. Field history, crop varieties used, plant arrangement, fertilizer use, pest control , etc.

Soil information: Soil and soil management is the foundation of organic production. Organic growing systems are soil based, care for the soil and surrounding ecosystems and provide support for a diversity of species, while encouraging nutrient cycling and mitigating soil and nutrient losses.

Availability of all plant nutrients in an optimum concentration is essential for the normal growth and development of plants. Soil is the store-house for the supply of plant nutrients. Potential productivity of soils varies considerably because of differences in amount of plant nutrients. However, a detailed examination of plant nutrients (chemical properties) and physical properties and their correct evaluation would enable one to identify the constraints and to suggest the ways and means by which the inherently low productive soil could be made more productive. In actual practice, the physical and chemical analyses of soils indicate the conditions such as mechanical impedance, low water and nutrient retention, excessive and low permeability, acidity, alkalinity and lack of nutrients limiting crop production.

Examine the entire soil profile to ascertain probable root distributions, clay pans, high water table, or generally poor physical condition which may greatly influence the crop productivity under organic farming. Composite soil sampling and its laboratory analysis (physical, chemical and biological properties) of the organic production area should be done to determine the proper crop management practices for organic conversion.

Each crop grown depletes soil of its nutrients by a certain amount. Hence management of soils either by replenishing nutrients by external sources or through proper crop management practices is necessary. In organic farming, only organic substances are allowed to feed the soils. These organic substances may be organic manure (e.g. FYM, compost), green manure (e.g. sunhemp, cowpea, dhaincha), concentrated organic manure (e.g. oil cakes, neem cakes), bio-fertilisers (e.g. Azotobactor, Rhizobium, Azolla, Blue Green Algae), vermin-compost, etc.

Complete information about the availability of organic fertilizer both inside and outside the organic farm unit help farmer to determine what organic source of nutrient or soil amendment is best for his specific fields for his specific crops in his specific situations. The rate of application of an organic fertilizer element or amendment considered should be optimum and economically feasible for one season. To determine the base rate, use information from soil analysis, existing experimental data, and local experience. Also take into consideration other soil conditions, crop to be grown, climatic conditions, irrigation possibility, management practices, economic considerations, etc.

Taking advantage of soil management practices such as deep tillage, compaction, addition of organic matter through organic fertilizer/compost, green manuring, inclusion of leguminous plants in crop rotation, soil productivity can be improved. In case of problem soils having excessive acidity or alkalinity, amendments like lime, gypsum, etc could be used for reclamation.

Seed and seedlings: Information on availability of organically grown seeds and seedlings is required. Seed or seedlings must be produced according to organic standards. Species and varieties cultivated in organic agriculture systems are selected for adaptability to the local soil and climatic conditions and tolerance to pests and diseases. All seeds and plant material should be organically certified.

- A wide range of crops and varieties should be grown to enhance the sustainability, self-reliance and biodiversity value of organic farms.
- Plant varieties should be selected to maintain genetic diversity.
- Organically grown varieties, and varieties known to be suited to organic cultivation should be preferred.
- Farmer should use organically bred varieties.

Organic varieties should be obtained by an organic plant breeding program. Organic plant breeding and variety development is sustainable, enhances genetic diversity and relies on natural reproductive ability. Organic plant breeding is a holistic approach that respects natural crossing barriers and is based on fertile plants that can establish a viable relationship with the living soil. Organic seed and plant materials should be propagated under organic management for at least one generation, in the case of annuals, and for perennials, two growing periods, or 12 months, whichever is the longer, before being certified as organic seed and plant material.

Non-organic seeds or perennial plants (planting stock) must be managed organically for at least one year prior to harvest of crop. List all seeds/seedlings planned for use in the current season, including seeds planted previously on proposed organic fields. If you are using conventional treated planting materials than a note from suppliers regarding its non-availability in organic status is required.

Diversity in crop production: To maintain the diversity there should be enough greenery hedge, shrubs and trees around the farmland for sheltering birds and natural enemies, which maximise the pest and microbial activity in the soil and enhance the growth. Diversity in crop production is achieved by a combination of:

- a diverse and versatile crop rotation that includes green manure, legumes and deep rooting plants
- Appropriate coverage of the soil with diverse plant species for as much of the year as possible.
- proper management from insects, weeds, diseases and other pests, while maintaining or increasing soil organic matter, fertility, microbial activity and general soil health.

Pest, disease and weed management: Organic farming systems apply biological and cultural means to prevent unacceptable losses from pests, diseases and weeds. They use crops and varieties that are well-adapted to the environment and a balanced fertility program to maintain fertile soils with high biological activity, locally adapted rotations, companion planting, green manures, and other recognized organic practices as described in the standards. Growth and development should take place in a natural manner.

Protection of natural enemies of pests through provision of favourable habitat, such as hedges, nesting sites and ecological buffer zones that maintain the original vegetation to house pest predators diversified ecosystems. These will vary between geographical locations. For example, buffer zones to counteract erosion, agro-forestry, rotating crops, intercropping etc.

Pests, diseases and weeds should be managed by the knowledgeable application of one, or a combination, of the following measures:

- choice of appropriate species and varieties
- appropriate rotation programs
- mechanical cultivation
- thermal weeding
- natural enemies including release of predators and parasites
- acceptable biodynamic preparations from stone meal, farmyard manure or plants
- mulching and mowing
- mechanical controls such as traps, barriers, light and sound

Protection from contamination of heavy metals/other pollutants: All relevant measures should be taken to ensure that organic soil and food is protected from contamination. Farmer should take reasonable measures to identify and avoid potential contamination. Accumulation of heavy metals and other pollutants should be limited and the appropriate remedial measures should be implemented where possible. Contamination that results from circumstances beyond the control of the operation does not necessarily alter the organic status of the operation.

Water management: Efficient management of irrigation water signifies the utilization of every unit of water economically to obtain maximum returns. The goal of water management practices is to give timely irrigation with adequate quantity of water. When water is to be used for crop irrigation purposes, following five factors should be considered:

- The total salt content and chemical composition of the water
- The climate (rainfall and temperature)
- The prevalent soils and drainage conditions

- The principle crops to be irrigated
- Crop cultural practices, mainly irrigation method

A source of water may be suitable or unsuitable for irrigation after it has been examined in the light of above five factors. The standards for irrigation water quality and irrigation water requirement are discussed separately in "*Water Management in Organic Farms*" chapter of this book.

5. CHALLENGES IN ORGANIC CONVERSION

In India, many organic farmers have gained practical experience with different production methods, and suppliers of organic inputs instruct them on how to use their products. Also there is a vast heritage of traditional farming practices, which are suitable for modern organic farming.

Organic farming systems in general do not simply replace agro-chemicals with organic manures and bio-pesticides. First of all the system approach puts its focus on crop rotation, mixed cropping, recycling of biomass on the farm, ecological balance, bio-control measures and preventive methods for pest and disease management.

The existing system on your farm will dictate the extent to which technical challenges are a constraint to conversion. Generally the more diverse the farming system in terms of enterprises, crops, habitat and market outlet the easier it will be to convert. The range of existing crops grown is important, as it is likely that the crops you grow well conventionally will be your best crops organically. However, the demands of rotation may mean that you have to learn how to grow unfamiliar crops, such as fertility building grass/clover leys and other crops.

Soil and fertility management are definitely challenges and are perhaps not given the importance they deserve as the cornerstone of the organic system. The organic standards encourage the building of fertility through inclusion of grass/clover leys and green manures in the rotation and cycling of nutrients within the system. Grass leys are, however, not always managed well and green manures are under-utilised leaving soils vulnerable to leaching over winter. Research has shown that green waste compost can be very effective as stable source of nutrients, if a local source can be found. Maintaining good soil structure is also a challenge and the effects of compaction can manifest themselves more clearly in organically managed soils.

Weeds are often the biggest technical concern of growers considering conversion, but the extent to which they are a problem will depend on soils and past cropping history. Investment in machinery will probably be needed but the availability of new kit such as tined and finger weeders means that weeds can be managed effectively on a field scale, especially in transplanted crops. Timing of weeding is also very crucial. The biggest problems with weeds can be in early crops where stale seedbeds are not possible and crop establishment is slower.

Rotation and diversity is the key and there are also technical solutions to many of the pest and disease problems. Many pest and disease problems on newly converted farms can be site-specific and related to pest management, so it is important to know your fields well. Although there is many permitted biocides available in the standards, the approach of substituting 'organic' inputs for conventional ones is a dubious one and should be considered only as a last resort. The risks involved with field scale production for the multiples make this approach understandable, however the organic principles of encouraging diversity and natural predators can do work. Some of the important biological control of crop pests is discussed in subsequent chapter of this book. In general, organic growers find that pest and disease problems occurs less than they expected.

The decision to convert is not one to be taken lightly. The additional management input required on farms should not be under-estimated and can be up to 50 per cent more. In the first two years following conversion farmers in the study have obtained crop yields from average 25-30 per cent below conventional yields and 15-20 per cent below typical yields from established organic farms. Higher prices often enable organic growers to obtain gross margins that are higher than conventional, but the economics of the whole rotation must be considered because fewer cash crops can be grown in an organic system. Research findings indicating that over-all farm income levels have the potential to be higher following conversion.

Considering the potential environmental benefit of organic production, its suitability for the integrative role of agriculture in rural development and its aptness to current farming input and production levels in many countries, organic agriculture should be considered as a development vehicle in the sub-region. It must be noted that highly variable farm structures, environmental conditions and infrastructure require local feasibility evaluation to determine the attractiveness of the organic model.

5.1 Commitments

Farming organically involves committing to two principles: ecological production and maintaining organic integrity. Ecological production entails using farming and ranching techniques and materials that conserve and build the soil resource, pollute little, and encourage development of a healthy diverse agroecosystem, which supports natural pest management. These techniques and materials include diverse crop rotations, green manuring, cover crops, livestock manure, composting, mineral-rich rock powders, etc. *Maintaining organic integrity* consists of actions that prevent contamination of organic production with prohibited materials, and that prevent the accidental mixing (commingling) of organic and conventional products. Farmers accomplish this, first of all, by not using prohibited synthetic fertilizers and pesticides; they also take precautions against pesticide drift from off-farm and other sources of contamination. Many kinds of equipment and storage areas employed in organic production must either be dedicated to organic use or properly cleaned between conventional and organic use. A considerable amount of paperwork and documentation is required to ensure organic integrity; it is one of the necessary "burdens" of being a certified organic farmer or rancher.

Prospective organic producers should understand in advance that prohibited substances (synthetic fertilizers and pesticides, etc.) must not have been used on the land for three full years preceding harvest of the first organic crop. Farms or specific fields that do not yet meet this requirement may be considered as in *transition,* though this term does not have legal status at this time.

Organic livestock producers must make a further commitment- to manage and raise their livestock in ways that are not cruel and that take account of the animal's natural behaviour. This includes providing pasture for ruminants and outdoor access for all livestock, and agreeing to restrictions on physical alterations.

Chapter 5

Practices of Soil Building

Switching from conventional to more sustainable organic farming systems and practices involves more than simple substitution of practices, such as replacing chemical fertilizers with organic amendments. However, it requires a fine tuning of the system with nature. Organic amendment increase or at least maintain its fertility and biological activity.

Practice of integrated soil building is an initial requirement to enhance overall soil productivity for transition to organic farming systems. This can not be done quickly. To increase soil humus (the building block of productivity) needs careful and sustained management of organic matter. Humus cannot be brought from outside. It has to be created within the soil by the action of soil microorganisms in presence of adequate soil organic matter. Therefore, it is important for a farmer to have a comprehensive understanding of the soil productivity and its relationship with various farming practices. A poor selection of the practices may cause a significant reduction of soil productivity.

1. SOIL PRODUCTIVITY

One of the major changes, which have to be achieved during the transition, is an approach to manage soil productivity. Elimination of chemical fertilizers from a conventional farming system may cause a significant reduction in crop yield and result in loss of net farm income. In order to avoid such problems, it is necessary to begin with adopting integrated soil building to improve soil productivity of a farm.

Soil productivity is influenced by various natural and artificial factors. As of the natural factors, the soil productivity includes:

Physical properties	:	Water-holding capacity, air permeability and tillability
Chemical properties	:	Nutrient supplying ability and buffering capacity
Biological properties	:	Microbial activity and disease suppression

As of the artificial factors, land, soil and cultural management significantly influence the soil productivity. Additionally both weather and cultural management directly affect crop growth. The soil productivity is highly flexible in response to changes in these natural and artificial factors, rather than being fixed and independent. The soil productivity can be enhanced only when proper land, soil and cultural management practices are carefully selected and integrated. Therefore, a comprehensive, rather than fragmented, understanding of these factors is required in order for a farmer to manage soil productivity.

To sustain the life in soil, several techniques are involved including conservation of soil and moisture, maintenance of a minimum soil organic matter, prevent loss of soil nutrient, adopting tillage practices suited to the local condition, maintain mulching, cover crop and supply of adequate quantity of FYM to improve soil texture and structure. Various soil building practices and their direct and indirect influences to the physical, chemical and biological properties of soils are explained as under:

1.1 Land Improvement

The following land improvement can be necessary for development:

- Deepening of top soil (deep ploughing, ridging, breaking up of soil crust)
- Soil layer mixing
- Subsoil break
- Improvement of texture and structure (stone or rock removal, organic matter addition)
- Drainage (excess water removal)
- Irrigation (with quality water)
- Supplementary irrigation (sprinkle, furrow irrigation)
- Soil dressing

1.2 Mineral Amendment

The following mineral amendment is required to improve the soil environment:

- Dolomite
- Lime
- Fossil shell
- Coral sand
- Rock Phosphate
- Guano-phosphate
- Zeolite

1.3 Elimination of Excessive Nutrients

- Cleaning crops
- Inundation

1.4 Organic Matter Management

Mobilisation of organic matter depends on the ingenuity of organic farmer. In organic farming, soil is fed and not the crop. The soil has to hold the organic matter and create condition for the microorganisms to work on it to release the nutrient. This nutrient should be held on safely and also release as and when plants needed. Enriching and maintenance of organic matter content may be done by adopting following practices:

- Composting
- Organic amendments
- Crop rotation
- Mixed/Inter-cropping
- Mixed planting
- Green manuring
- Cover cropping
- Organic mulching

The organic matter incorporated into soils becomes available as energy and material sources for various soil organisms. The activated and diversified soil organisms, through their food chain, decompose and reproduce various organic and inorganic substances. These substances are fixed or held by either negatively charged clay particles or intermediate organic by-products. Many of these substances are beneficial to crops as nutrients and growth regulating substances.

Some of the organic substances bond soil particles to form soil aggregates. The aggregated and structured soils provide favourable habitats for various micro-flora and fauna and also important for retention and transmission of water, air and heat. Such diversified biological community also suppresses an outbreak of pathogens to crops. Highly structured and aggregated soil also improves water holding capacity and air permeability. These comprehensive changes improve root development of the crops and in turn their yields and quality.

The humus portion of soil organic matter is, by virtue of its very high cation exchange capacity and chelating power for some of the essential trace elements. Humus adds substantially to the buffering capacity of soil, making it less amenable to pH changes by acids or bases. Organic matter also useful for reclamation of salt-affected soils. It counteracts the unfavourable effect of exchangeable sodium. The decomposition of manure and plant residues liberates carbon dioxide and organic acids which help to dissolve any insoluble calcium salts in soil solution and neutralize the alkali present. A combination of organic manure and phosphatic fertilizer (Rock Phosphate) application were found very effective in building up the soil structure.

Following figure summarizes physical, chemical and biological changes which concurrently proceed when proper organic matter management is conducted:

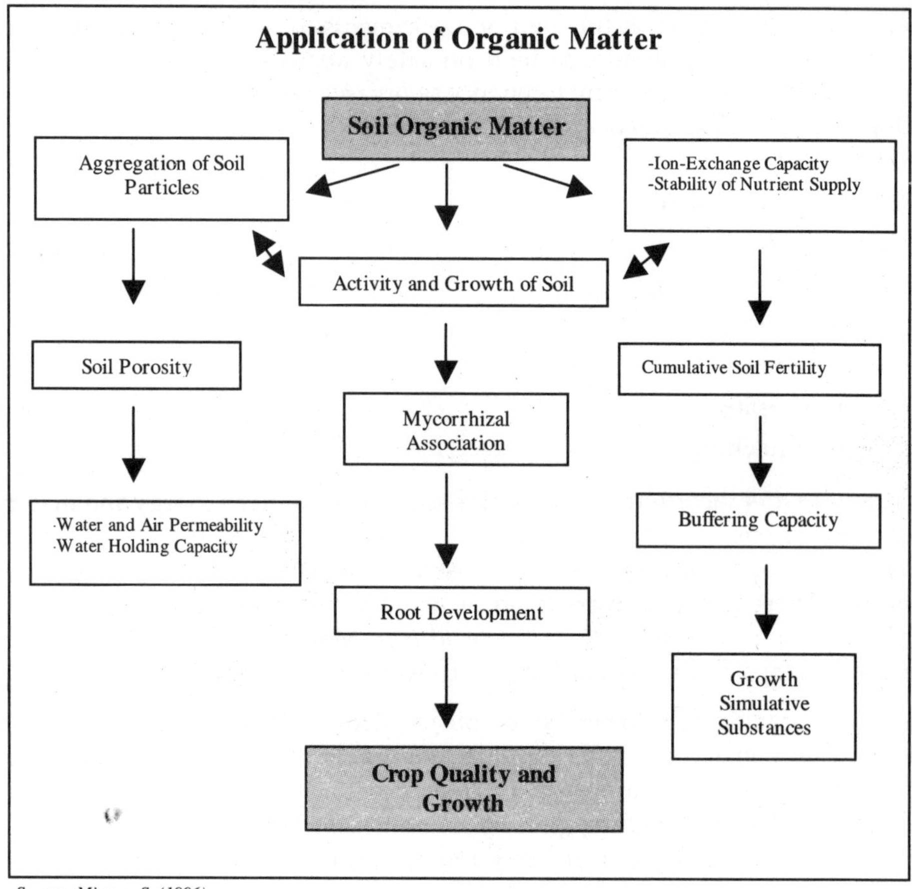

Source: Mizuno, S. (1996)

Figure 5.1: Changes in soil properties induced by organic matter management

Fresh organic matter added to the soil is easily degraded by soil biological activity. The end product of organic matter breakdown is stable humus. This process is always dependent on the matter from which it is broken down, the environmental conditions under which it is processed and the microbes involved in the decomposition. This decomposition process can be managed by composting. Compost and humus have the potential to provide a sustainable basis for primary production that will lead to an optimum production system especially on marginal soils.

Compost quality depends on the composting materials, method and conditions. The ratio of carbon to nitrogen (C/N ratio) affects the process and product. The end product can serve different purposes depending on the requirements of the crop and/or soil.

Table 5.1: Organic matter

	Decomposition (per year)	Soil Structure	Plant nutrients
Fresh organic matter (high C/N ratio)	50–80%	+	+++
Dynamic humus/feeding humus	5–50%	++	++
Stable humus (C/N ratio~ 10)	2–5%	+++	+

Source: Werff (1992)

Fresh organic matter has a breakdown rate of 50–80% after the first year of application. Stable humus breaks down much more slowly i.e. 2-5% a year. This means that fresh organic matter contributes to the nutrient availability and the moisture availability of the soil but does little to improve the structure of the soil.

1.5 Nitrogen Dynamics

In conventional agricultural farming system, nitrogen is mainly supplied by synthetic chemical fertilizers. Synthetic fertilizers are hydrolysed into available form of N for crops almost instantaneously after applying into soils, during the growing period. Much of the nitrogen from the fertilizer is lost from the plant-soil system via leaching, volatilisation and denitrification. Herbicides are also applied to eradicate weeds so as to maximize fertilizer efficiency. When such farming systems are practiced in a long run, soils are degraded and the nitrogen cycling capacity in the soil reduced.

During the transition from such a system by adopting soil building practices, nitrogen comes from a variety of sources, including composts, organic amendments, leguminous green manure plants and soil organisms. The decomposition of these materials increases the labile pool of soil organic matter and improves structure and ecological functions of soil. It may reduce loss of nitrogen by increased cation and anion exchange capacities, extensive root system and biomass. The weeds are managed to maintain as low as do not cause a significant inhibition of crop growth. The weeds can become an important part of a system as N reserves.

Large amount of compost may be required during early stage of the transition. However, mineralization of N from previous application significantly reduces the following application. It can be further reduced as proper crop rotation and habitat management for beneficial are adopted in the system.

1.6 Current Soil Condition

In order to adopt effective soil building practices for the transition, it is important for a farmer to understand the current soil conditions and such factors that may limit the soil productivity.

Field study: Information such as weather, environmental factors, land and soil characteristics, soil management practices at present and in past, yields and quantity of the harvests in recent years and average yields of the region are obtained from a farmer and a local agricultural service. This information is highly crucial for the diagnosis since these factors fundamentally determine soil productivity of the field in question.

Soil profile study: In order to examine soil physical characteristics such as texture, structure, colour, thickness, hardness and water distribution of soil layers, and the development of crop root systems, soil profile study should be conducted.

Soil profile is studied in the field from a freshly exposed pit. The profile characteristics studied in the field consists of locating the soil horizons, based on colour differences. Where vertical colour differentiation is not possible, horizons are differentiated on the variations in other soil characters. Each of these horizons is described in terms of thickness of ploughed layer, hardness of layer, texture, structure, soil type, consistence, pH, carbonate, clay film, pores, wetness, water permeability, etc. The external soil characteristics studied are form, gradient of slope, drainage condition, and groundwater level. Soil profile survey is usually conducted by the soil survey specialists.

Analysis and data interpretation: Soil samples are generally collected from two layers; one from 0 to 15 and the other from 15 to 30 cm. Selected soil properties, viz. mechanical composition, bulk density, cation exchange capacity, base saturation, calcium carbonate, electrical conductivity, organic carbon, humus, total nitrogen, C/N ratio, exchangeable Ca, Mg, K and Na, available phosphoric acid, phosphate absorption capacity, ammonium-nitrogen, nitrate-nitrogen, soil nitrogen mineralization, micronutrients (iron, manganese, copper, zinc, boron, molybdenum), heavy metals (lead, arsenic, nickel, chromium), available silicic acid, etc. should be measured.

As a parameter related to biological process, potentially mineralizable N is measured by incubation of a soil sample for period of 4 weeks under 30 degree C. This measurement is very useful to evaluate cumulative fertility of the soil.

The information obtained through the field study, soil profile observation and soil analysis are compiled and reviewed by Soil Scientist. Description of the soil condition and technical recommendations are formulated.

Periodic soil diagnosis: Since the rate of changes in soil properties are usually very moderate, it is recommended to receive soil quality diagnosis every 2-3 years to monitor the responses to the modification of soil management practices. Book keeping of the changes in crop yields and qualities are very essential as this helps to further improvement of the farming systems and practices. The standard soil test for organic farming will include an analysis of the following components:

Table 5.2: Standard soil test in organic agriculture

Soil pH	Ammonium Nitrogen
Percent Organic Matter	Nitrate Nitrogen
Cation Exchange Capacity	Phosphorus
Neutralizable Acidity	Potassium
Calcium	Heavy metals (Zn, Cu, Hg, Mn, Pb, Cd, As,
Magnesium	Ni, etc.)

Soil treatment plan: To make the farmers understand about the diagnostic results and recommendations, through discussions with farmers is indispensable. It is a good opportunity to review the results and recommendations by gaining more specific information from the farmers. Opinions and suggestions from other farmers and extension specialists are also useful to improve the diagnosis and the farming systems and practices.

Based on such site-specific information and more extensive interaction among farmers and research specialists, the most suitable soil building practices to each field, can be integrated.

Some of the example of such plan based on diagnostic report is given below:

Table 5.3: Plan based diagnostic Report of Soils

Example	Possible Reason	Recommendations
Soil having high EC and available P. Soil pH is comparatively low.	N and P fertilizers may be extensively used to produce high value crops such as vegetables or large amount of organic amendments through bone meal and fish meal which have high N and P contents. Due to the large amount of nitrogen remaining in the nitric acid state, EC value is high.	Material used for organic amendments be switched to such plant materials as oil cakes and amount of application be reduced. Application of compost made from grass and straws in combination with deep tillage may be effective to improve the soil condition. Use of composts and other organic materials that have high nitrogen and phosphorus content (bone meal, fish meal) should be avoided.
Ca^{++} is excessively accumulated and pH level has exceeded the upper limit. P contents are relatively high. Mg^{++} is high.	Application of poultry excrements or calcareous material for many years. High Ca availability may prohibit the uptake of Mg^{++} by crops although it may show fairly good Mg availability.	Matured compost made from plant materials be used instead of Ca-rich materials. Oil cake can be used as organic amendments for a nutrient management.
High humus content, High CEC, low pH level, high P-fixation, high level of C and low plant nutrients except for K.	Only compost made of straws, grasses, and dead leaves might be continuously used.	Matured compost made from fish meals, poultry excrements and bone meals should be utilised. If it is necessary, the pH level can be amended using lime or dolomite.

2. ASSESSING SOIL QUALITY

Soil quality is an assessment of how well soil performs all of its functions. It cannot be determined by measuring only crop yield, water quality, or any other single outcome. The quality of a soil is an assessment of how it performs all of its functions now and how those functions are being preserved for future use.

Soil quality cannot be measured directly, so we evaluate indicators. Indicators are measurable properties of soil or plants that provide clues about how well the soil can function. Indicators can be physical, chemical, and biological characteristics. Useful indicators:

- are easy to measure
- measure changes in soil functions
- encompass chemical, biological, and physical properties
- are accessible to many users and applicable to field conditions
- are sensitive to variations in climate and management.

Indicators can be assessed by qualitative or quantitative techniques. After measurements are collected, they can be evaluated by looking for patterns and comparing results to measurements taken at a different time or field. Here are some examples of indicators of soil quality:

Table 5.4: Indicators of soil quality

Indicator	Relationship to Soil Health
Soil organic matter (SOM)	Soil fertility; structure, stability and nutrient retention; soil erosion.
PHYSICAL: Soil structure, depth of soil, infiltration and bulk density; water holding capacity	Retention and transport of water and nutrients; habitat for microbes; estimate of crop productivity potential; compaction, plough pan, water movement; porosity; workability.
CHEMICAL: pH; electrical conductivity (EC); extractable N-P-K	Biological and chemical activity thresholds; plant and microbial activity thresholds; plant available nutrients and potential for N and P loss.
BIOLOGICAL: Microbial biomass C and N; potentially mineralizable N; soil respiration.	Microbial catalytic potential and repository for C and N; soil productivity and N supplying potential; microbial activity measure

The ultimate purpose of researching and assessing soil quality is not to achieve high aggregate stability, biological activity, or some other soil property. The purpose is to protect and improve long-term agricultural productivity, water quality, and habitats of all organisms including people.

2.1 Managing for Soil Quality

Each combination of soil type and land use calls for a different set of practices to enhance soil quality. Yet, several principles apply in most situations.

1. *Add organic matter*: Regular additions of organic matter are linked to many aspects of soil quality. Organic matter may come from crop residues at the surface, roots of cover crops, animal manure, green manure, compost, and other sources. Organic matter, and the organisms that eat it, can improve water holding capacity, nutrient availability, and can help to protect soil erosion.

2. *Avoid excessive tillage*: Tillage has positive effects, but it also triggers excessive organic matter degradation, disrupts soil structure, and can cause compaction.

3. *Avoid pollutants*: Manure and other organic matter also can become pollutants when misapplied or over-applied. On the positive side, manure can increase plant growth and the amount of organic matter returned to the soil.

4. *Increase ground cover*: Bare soil is susceptible to wind and water erosion, and to drying and crusting. Ground cover protects soil, provides habitats for larger soil organisms, such as insects and earthworms, and can improve water availability. Cover crops, perennials, and surface residue increase the amount of time that the soil surface is covered each year.

5. *Increase plant diversity*: Diversity is beneficial for several reasons. Each crop contributes a unique root structure and type of residue to the soil. A diversity of soil organisms can help control pest populations, and a diversity of cultural practices can reduce weed and disease pressures. Diversity across the landscape and over time can be increased by using buffer strips, small fields, contour strip cropping, crop rotations, and by varying tillage practices. Changing vegetation across the landscape or over time increases plant diversity, and the types of insects, microorganisms, and wildlife that live on your farm.

Careful observation of the farmland and farming practices is required to design a practical transition scheme. Some problems might take longer and require considerable investment.

Chapter 6

Soil Fertility Management

Soil fertility is maintained and enhanced by a system which optimizes soil biological activity and the physical and mineral nature of the soil as the means to provide a balanced nutrient supply for plant and animal life as well as to conserve soil resources. Production should be sustainable with the recycling of plant nutrients as an essential part of the fertilizing strategy.

The agronomic productivity of land is highly dependent on the effective soil fertility management. In addition to climatic conditions that facilitate growth, plants require both a medium for physical support and anchoring of the root system and also the provision of water and essential nutrients. However, soils vary considerably in their potentials to retain and supply the required water and nutrients. Soil fertility management is geared towards modifying the plant nutrition regime and improving the capacity of the soil in order to ensure the provision of these essential inputs. While crop yield depends on the complex interaction of all plant growth and development factors, the role of prudent soil management is particularly important to ensure a sustained level of productivity for any given cropping system.

The principles of fertility management in organic farming include:

- Biodegradable material of microbial, plant or animal origin produced from organic practices should form the basis of the fertility program.
- Nutrient losses from the farm to the natural environment should be minimized. Nutrients should be used in such a way and at appropriate times and places to optimise their effect.
- Accumulation of heavy metals and other pollutants must be prevented.
- Naturally occurring mineral fertilizers and brought-in fertilizers of biological origin permitted under the National Organic Standards should be regarded as only one component of the nutrient system, and as a supplement to, and not a replacement for, nutrient recycling.
- Manures containing human faeces and urine must not be used unless free of human pathogens. Careful attention to hygiene is required and it is recommended that they are not applied directly to vegetation for human consumption or to soil that will be used to grow annual plants within the next six months.

1. WHAT DO THE ORGANIC STANDARDS SAY?

Soil fertility should be maintained/enhanced through raising green manure crops, leguminous crops etc. The residues of plants after harvest should be incorporated into the soil as far as possible. Biodegradable materials of microbial, plant or animal origin should be applied as manures. (e.g. compost, vermicompost, farm yard manure, biofertilizer etc.) Use of synthetic/chemical fertilizers is not permitted. The mineral based materials like rock phosphate, gypsum, lime, etc. can be applied in limited quantities when there is absolute necessity.

The following products are permitted for use in manuring/soil conditioning in organic fields:

- Farm yard manure, slurry, green manures, crop residues, straw and other mulches from own farm
- Saw dust, wood shaving from untreated wood
- Calcium chloride, limestone, gypsum and chalk
- Sodium chloride
- Bacterial preparations (e.g. biofertilizer without GMO).
- Bio-dynamic preparations (free from GMO)
- Plant preparation and botanical extracts (e.g. neem cake)
- Vermicompost

The following products should be used when they are absolutely needed and taking into consideration of factors like contaminations, depletion of natural resources, nutritional imbalances, etc. *If proposing for certification, the certification agency may be consulted before using these inputs.*

- Farm yard manure, slurry, urine, straw etc. from other farms
- Blood meal, bone meal, fish meal without preservatives
- Minerals like Basic slag, Sulphate of potash etc
- Trace elements
- Wood ash from untreated wood
- Vermicompost from other farms

Note: The total amount of manure as defined in EU Regulation 2092/91 applied on the holdings should not exceed 170 kg of nitrogen/yr/ha of agricultural area used.

2. PHYSICAL SOIL MANAGEMENT PRACTICES

Physical soil management practices are directed towards:

(i) modifying the rooting environment, thereby enhancing the uptake of water and plant nutrients

(ii) reducing competition from weeds, and
(iii) minimizing surface soil loss.

Physical soil management practices includes-
(i) Tillage
(ii) Mulching
(iii) Soil amendment
(iv) Conservation practices.

Tillage loosens the surface layer of the soil and helps to improve tilth/structure. This allows easier movement of water into the soil, which in turn leads to better plant growth. Tillage operations also facilitate the application of soil amendments and fertilizers. Minimum tillage systems should be done where crops are planted in undisturbed soil may reduce production costs, preserve soil organic matter, reduce soil compaction and reduce soil loss on hillside farms. Mulching controls weeds and improves water use efficiency. Additional benefits of mulching include: structural improvement at the surface layer, arising out of a stimulation of microbial activity; reduction in post-plant operations, such as moulding; and reduction in the transmission of soil-borne pathogens to the plant. Soil conservation practices are necessary on hillsides and on gently sloping areas. The significant erosion that would otherwise occur would carry with it valuable topsoil and all the inputs applied to the soil surface, including the organic fertilizer and soil amendments (e.g. lime, gypsum). There are also side effects, such as silting of rivers and flooding of low-lying areas.

3. NUTRIENT MANAGEMENT PRACTICES

Nutrition management practices are directed towards:

- upgrade the efficiency of the plant nutrient
- maintain and improve the stock of plant nutrients in the soil/crop system
- limit plant nutrient losses
- provide the highest possible economic rate of return for the farmer.

Maintenance or adjustment of soil fertility and of plant nutrient supply is essential to sustain a desired level of crop production. It is achieved by optimizing the benefits from all possible sources of organic plant nutrients and by improving the overall management of the farm. Farmers can improve their production capacity and income using the best combination of agronomic efficiency and economic profitability. In addition, they will have a more sustainable soil environment for crop growth. Organic farmers must practice green manuring, compost making, vermicomposting, and use of bio-fertilizers to maintain the soil fertility.

Nutrient management practices in organic farming should be developed at two levels- farm and village/community. At the farm level, it addresses the alternative sources of plant nutrients available to the farmer and suggests a suitable mix of them to fit production objectives. Finally, the village level considers the wider farming community, encompassing communal natural resources, environmental issues and farmer groups.

Nutrient requirements vary with the crop, so heavy feeders and light feeders may be grouped separately to manage fertilization. Long-season crops such as eggplant (brinjal), tomato, pepper, and okra should be planted so they don't interfere with replanting short-season crops such as beans and cole crops (cabbage, cauliflower, knolkhol). Tall-growing crops such as pole beans, tomatoes, and corn should be planted so they don't shade shorter crops. It may not be possible to accommodate all of these recommendations in the garden, but attempt should be taken to accommodate as many as possible to insure a successful garden. Many options may exist and problems can be tackled from many different angles. This problem suggests the need for improved plant-nutrient management to:

(i) evaluate alternative, cheaper, locally available sources of plant nutrients; and
(ii) improve the efficiency of organic fertilizer use.

The list of possible improvements should include:

(i) varying the sources of plant nutrients, i.e. organic and biological;
(ii) varying rates of application of plant nutrients;
(iii) varying the time of application of plant nutrients;
(iv) varying the crop rotation to include legumes or a fallow period in the rotation.

Another example is that of very acid soil. Solutions to the problem may involve finding a range of *soil amendments*, such as lime or dolomite, for the specific cropping system. Alternatively, the farmers may opt to live with the acidity problem, if it is slight, but to identify a inter crop or cropping system that can produce well in such a situation. Within these two broad approaches, many possibilities exist.

Factors determining essential plant nutrients: A soil may actually require not just the three major elements, N, P and K, but any or a combination of all the others needed to guarantee the best level of nutrition for the crop. Apart from N, P and K, the parent material from which the soil is formed largely dictates the levels of other essential elements present. To a lesser extent, factors determining the development of the profile, such as climatic conditions and topography also influence the nutrient status of soils; these becoming more important as the soil mature. In particular, N and P are associated with the organic matter component in soils and are therefore affected by any activity that influences the status of organic matter. Despite their occurrence in a 15-centimetre (6 inch) layer at levels of thousands of kilograms per hectare, only a small fraction of this amount is actually available for plant uptake. Consequently, N, P and K may need to be supplied to meet crop needs. The actual application of these nutrients also depends on an understanding of their behaviour in the soil. Application issues include: the nutrient source, the timing of the application, the method of application, and the type of nutrient source.

Humus management and fertilization: Biologically active soil is the prerequisite for the balanced nutrition of crops. In order to ensure long lasting soil activity and thus crop yields, special attention has to be paid to the basis of soil fertility. For example,

- The humus balance has to be at least an equilibrium within the margin of varied crop rotation. For permanent crops, this has to be guaranteed by adequate measures such as undergrowth crops, catch crops, or permanent green.
- Biodegradable matter of microbe, vegetable or animal origin should form the basis of fertilization.
- Given the importance of a balanced lime level for topsoil stability, for the structure and thus the fertility of the soil, and because of acid absorption through precipitation, special attention has to be paid to an adequate lime supply with respect to the area.

Synthetic chemical nitrogen fertilizer as well as Chilean nitrate and urea must not be used. Mineral and trace element fertilizers that are not easily soluble can be used after consulting the Certifying Agency. Their application is based on the corresponding soil analyses, observation of plant growth, and the nutrition balance of the whole farm.

The amount of farm manure depends on the forage production of the farm and the resulting animal husbandry. The manure has to be processed to make it tolerable for soil and plants. In the case of semi-liquid manure, such treatment is obligatory: the use of stone or straw meal, dilution, ventilation or comparable measures. In the case of dung, a controlled process of decomposition is suggested.

Nutrient losses during storage and the application of liquid fertilizers and dung have to be reduced to a minimum. The contamination of the environment (by odours and pathogenic agents too) has to be avoided. Sufficient storage capacity must therefore be available so that manure is only applied as and when required by the crop and during the vegetation period.

The purchase of organic manure does not primarily serve the purpose of fertilization but is designed to increase the humus supply. Intensification beyond a tolerable extent (over-fertilization) has to be avoided. In case of on-farm husbandry, manure has to be distributed evenly according to the crop rotation over the areas cultivated. Furthermore care should be taken that the animal runs are not over-fertilized. The number of animals kept and the quantity of feed produced must correspond in such a way as to avoid over-use of the land e.g. over-grazing, with the consequence of damage to the soil (e.g. erosion). Ploughing nutrients back in the soil using green compost is recommended on the principle of the recycling of nutrients if it is certain they do not transport harmful residue. Waste and/or urban compost, faecal and sewage sludge are prohibited.

3.1 Soil Amendment

Choice of soil amendment

a. Which amendments are to be chosen?

b. Why will they be chosen?

c. Which amendments are readily available (especially those naturally occurring in the farming area)?

d. Are any of these amendments expensive or environmentally unsound? This can refer not only to harmful runoff but also to the use of an amendment that might spoil a natural resource.

How are the amendments to be tested?

a. Are the amendments to be tested at one level or several?

b. When and how often are they to be applied?

c. What is the method of application?

Further, two more complicated problems require consideration:

o Reduced supply and accessibility of organic fertilizers

o Declining yields due to soil acidity development and soil organic matter decline.

The former may affect farms with more than one soil type. A solution is to use available organic sources, such as chicken manure, green manure and sludge. If little is known about the response of the crop to these sources, then different (probably three or four) rates of application need to be tested. The latter could well be a single farm problem. Unless the results are likely to benefit a larger group of farmers, it is probably best to select an appropriate practices and avoid costly farm operations.

There are numerous possible management improvements for a specific problem. However, it may not be possible to apply all these improvements at a time. Therefore, the farmer must select the resources available to them for applying the inputs. For example, if organic conversion plans requires relatively expensive mulching treatments and the target group is resource poor farmers, then this option may not be affordable. However, if the mulching is not done, the farmers may run the risk of severe weed infestation and/or soil erosion and even greater economic losses in the long term. Consequently, as a compromise, a low cost mulching material could be used.

3.2 Nutrient Supply Vs Nutrient Demand

There are three scenarios exist with regard to synchronism of plant nutrients. The scenario are-

i) where low quality litter is added with low rate of nutrient release and plant uptake is faster asynchrony appears and nutrient deficiency is manifest,

ii) where high quality litter is added, nutrient release is faster than the plant uptake with a total loss of nutrients takes place. Hence, it is even suggested that high quality materials such as cattle-buffalo dung, sheep-poultry manure, etc. are mixed with carbonaceous material like straw, coir pith, trash etc. to avoid losses of nutrients during storage, and

iii) where high and low quality litter is mixed in right proportion, the plant uptake and nutrient release are nearly parallel ensuring high degree of synchrony.

In order to achieve high and sustainable crop production, a high degree of synchrony between nutrient release and plant demand is essential. In plantation crops like coconut, arecanut, coffee, tea, etc. high degree of synchrony is easily achieved due to longer available time scale. Similarly in long duration crops with expansive root system like cotton, sugarcane, etc. a fair degree of synchrony can be achieved provided the water supply is adequate. But it is only in short duration cereals like finger millet, wheat, jowar etc. that asynchrony is a major constraint to achieve sustainable yields following organic matter addition. Experiments have shown that organic matter addition in sandy soil improved root proliferation including rooting depth, root mass per unit soil surface which in turn improved the synchrony of nutrient demand with nutrient availability. A high degree of synchrony is achievable in multiple cropping systems, mixed cropping and high plant densities due to the root spread in different soil layers.

Many of the above factors and issues need to be considered when recommending appropriate soil nutrition management practices on the farm. Those outlined above also interact with and are influenced by other management practices, such as the *irrigation* and *drainage system* adopted for the farm. Therefore, some form of experimentation is required to determine the practices which best suit the farmers' conditions. Researchers can assess the crop's response to the proposed technology under farmers' conditions, which may be quite different from on-station conditions. By so doing, they are able to determine the limitations of the system and make adjustments to enhance its applicability.

4. SOIL PREPARATION

The success of any organic garden begins with the soil. A fertile, biologically active soil provides plants with enough nutrients for good growth. Organic gardening requires a long-term outlook with respect to soil preparation. In fact, the key to successful organic gardening is to feed the soil with organic matter, which feeds the plant, rather than to feed the plant with inorganic fertilizer as in conventional production.

An ideal soil would have equal parts of sand (>0.02 to 2.0 millimeters), silt (>0.002 to 0.02 millimeters) and clay (>0 to 0.002 millimeters), and contain about 5 percent organic matter. However, with work, most soils can be improved and made productive. Because it takes a long-term outlook to build a good soil, don't be disappointed if results are less than ideal in the first year or two. New sites should have all plant matter removed or turned under. Areas with invasive weed plants should have the plants removed to the compost pile and the soil turned under to expose roots and rhizomes to desiccation. In addition, soil solarization can help to control these hard-to-control weeds.

Some soils may have hardpans, which are impervious layers several inches under the soil. These hardpans are often found on old farmland or new home sites where equipment has compacted the soil. In either case, these hardpans must be broken up. On clay soils this can be very difficult. Soils should be turned to 10 to 12 inches deep. One method is to double dig the garden. Dig a trench 6 to 8 inches deep along one side of the garden, placing the soil on the outside edge of the garden. Then use a spade or garden fork to loosen the soil 6 inches deep at the bottom of the trench. Soil adjacent to the trench on the inside edge of the trench is moved to fill the existing trench, creating a new trench in its place. Again with a spade or garden fork, loosen soil in the bottom of this trench to a 6-inch depth. Continue in this fashion until the entire garden has been double dug. The soil from the first trench can then be moved into the last trench. This method of garden preparation will leave a deep turned soil but is very labour intensive. Alternatives include use of equipment such as tractor-mounted plows or a rototiller set to the deepest depth.

Organic matter should be added during this deep-turning process. Organic matter in soil is important for two reasons. First, as it breaks down, it releases nutrients that crops can utilize, and second, it improves the water- and nutrient-holding capacity of the soil. The amount of organic matter to add varies with the chosen material, the type of soil, and weather conditions. On sandy soils in tropical and subtropical regions, as much as 40-50 tonnes per hectare may be required to gain a benefit from the addition of organic matter. On heavier soils in regions with cooler climates and less rainfall, as little as 4-5 tonnes per hectare may be sufficient. Most organic sources have high water content, as much as 50 percent or more. In addition, many have high ash (non-organic residues) content, as high as 25 percent or more. Organic matter with 50 percent water content and 25 percent ash would require about 30 tonnes applied to one hectare to raise the organic fraction of the soil 1 percent. This may be impractical both in terms of obtaining the necessary organic matter and the fact that organic matter must be added each year to sustain the increase. The only way to tackle these problems is making use of biofertilizer.

Low rates (4-5 tonnes per hectare) of organic matter can have a noticeable improvement in soil tilth. Additions of 10-15 tonnes of organic matter per hectare per year can have a beneficial effect on soil tilth and plant growth. It is highly recommended that organic matter should be tested for its nutrient content so that application rates can be adjusted accordingly.

The major cause of poor fertility is low pH as found in an acidic soil. Adjusting the pH near 6.5 stimulates the activity of microorganisms that help decompose organic matter, and unlocks nutrients bonded to the soil. In acidic soils, apply agricultural limestone, or other materials with liming capabilities, according to the recommendations made by a soil test and organic standards. Use dolomitic limestone if the soil test indicates magnesium is low; if magnesium is high, use calcitic limestone.

5. FERTILIZING THE ORGANIC GARDEN

Organic gardening emphasizes the use of organic soil amendments to improve the nutrient content and physical characteristics of the soil. Synthetic fertilizers are not used. The presence of decomposing organic matter in soils has long been recognized as a nutrient source for plants and useful in maintaining and improving structure and tilth in clay soils and in improving the water-holding capacity of sandy soils. Moreover, organic matter contains natural organic complexes that make micronutrients, such as zinc and iron, more available to plants. A wide variety of plant and animal organic materials are useful. The gardener often is able to productively use materials that would otherwise be discarded, thus eliminating some potential environmental pollutants.

Major nutrients: Of the three major nutrients, crops respond most often to the addition of nitrogen and phosphorus; a response to the addition of potassium is rare. Although organic materials contain all of the major nutrients, they are present in widely varying ratios. Thus, few organic materials can be regarded as complete or balanced sources of plant nutrients. For example, even though manures are good nitrogen sources, they are relatively low in phosphate. Therefore, supplement manures with steamed bone meal. Continued heavy applications of manures may increase soil salts to harmful levels, a condition that can be avoided by periodically testing the soil. Because of *potential food safety problems, never use fresh manure on food gardens.*

A gardener always should have some idea of the content and availability of plant nutrients in the materials that are being added to the soil. Many materials contain nutrient elements in forms plants can't use and these may be very slowly converted to available forms. Some organic materials may require special handling, most often in the form of composting, before use in the garden. For instance, sawdust, straw or other stem plant tissue with low nitrogen content or slowly available nitrogen may actually cause a temporary deficiency of nitrogen in crops if directly incorporated into the soil. This happens because the microorganisms decomposing this material require nitrogen for their own tissues and thus compete with crops for it. Nitrogen deficiency can be avoided by composting the low nitrogen material (with added nitrogen in the form of dried blood or poultry manure) before adding it to the soil. Coarse material, such as corn stalks and other plant residues decompose slowly and also may create problems in soil preparation and cultivation if added directly to the soil. Shredding alone is not as effective as composting to handle these materials, but it is helpful. Because direct application of high nitrogen materials, such as dried blood and poultry manure in large amounts, may "burn" plants, use them in moderate amounts. Composting with low nitrogen materials may be advisable.

Nitrogen becomes available to plants in the form of nitrate or ammonium by decomposition of organic matter, the resulting nitrate is either taken up by plants (vegetables typically remove 15 to 50 kgs of nitrogen per acre), leached from the soil, or lost to the atmosphere as gaseous nitrogen. Manure containing 0.25 percent nitrogen and 0.15 percent phosphate applied at the rate of 20 tonnes per acre furnishes 50 kgs nitrogen and 30 kgs phosphate. Adding steamed bone meal containing 25 percent phosphate at the rate of 120 kgs per acre will supply the remaining 30 kgs of phosphate. Although rock phosphate and colloidal phosphate often are recommended to organic gardeners, they become available slowly especially in alkaline soils. Phosphate made available by decomposition of organic matter is generally either removed by plants or fixed in slowly available mineral complexes in the soil. Vegetables typically remove 5 to 25 kgs of phosphate per acre per year from the soil. Soil retains excess phosphate, unlike nitrogen, making it available to future crops. As the phosphate level of the soil is built up, the gardener may decrease or stop supplementing the manure with bone meal. When organic matter is added to the soil, the gardener may assume the potassium needs are being met. There is no need to add greensand, wood ashes, granite meal or kainite, which are potassium sources.

Micronutrients: Iron and zinc are the only micronutrients verified as deficient in most of the soils of India. Vegetables most likely to exhibit these deficiencies are corn, potatoes and beans. Many woody plants are also sensitive to lack of iron and zinc. Iron deficiency commonly appears as yellow areas between greener veins of young leaves. In zinc deficient plants, leaves and stems often fail to grow to normal size. There also may be yellowing between the veins of leaves, but usually on older leaves. Decomposing organic matter supplies these necessary nutrients. In addition, organic complexes (chelates) may be formed. These hold the nutrients in a form available to plants and protect them from fixation in the soil in unavailable forms. Since the requirements of plants for iron and zinc are very small, they almost certainly will be met by any program adding a variety of organic materials to the soil. Of course, materials slow to decompose, such as peat and sawdust will be less effective than manures or dried blood.

5.1 Use of Natural Fertilizer

Organic Gardeners has to prefer natural sources of fertilizers and use organic materials (such as manures or plant residues) or naturally occurring inorganic materials as nutrient sources. Natural fertilizers are often materials that release nutrients at moderate or slow rates. Slowly available nutrients are less likely to burn plants or to leach out of the soil. *However, some natural materials do contain highly soluble nutrients and care must be taken not to apply excessive quantities.* Fresh manure, wood ashes and blood meal are all very soluble materials and must be used with care. Frequent application of any fertilizing material, including manures and wood ashes, can cause nutrients to build up to excessive levels in the soil. Be sure to soil test regularly and follow the recommended rates. This section will describe some natural sources that are available locally or as the ingredients of blended fertilizers.

***Animal manures*:** Animal manures add both organic matter and plant nutrients to the soil. Obtain composted or well-rotted manure if possible. Composted manure releases nutrients slowly, but raw manure releases highly soluble nutrients, including ammonia, that can burn plant foliage and leach out of the soil. The nutrients in fresh poultry manure, in particular, are very soluble (approximately 80 per cent of the nitrogen). Do not apply manure to frozen ground because of run-off.

The average nutrient content of animal manures varies from farm to farm. Many factors, such as animal feeds, handling methods and the age of the herd, influence the content. Packaged dried manures have a higher nutrient content when compared by weight with the fresh product, because they contain less water.

***Green manures*:** Green manure crops are grown to improve the soil, and are tilled under rather than harvested. They add organic matter to the soil and hold nutrients. A deep-rooted green manure brings up minerals from the subsoil. As the green manure decomposes, these minerals become available to garden crops. Legumes (for example, cowpea, clover, green/black gram, etc.) should be added to the seed mixture because they increase the amount of available nitrogen in the soil by fixing atmospheric nitrogen. Use annual legumes if possible.

If your garden is large enough, take a section out of production each year to grow a green manure crop. In all your garden is needed for vegetable production, consider inter-planting white clover or cowpea with established vegetable crops to build up nitrogen.

Keep nutrients from leaching out of the soil in winter by planting a cover crop. Deep rooted crops that bring minerals up from the subsoil should be planted. Oats and spring barley are also suitable as cover crops. Plant 6 to 8 weeks before the date of the average first frost in your area so that they will have adequate growth before they are winter-killed.

5.2 Use of Manure

Fresh manure: Some organic materials, such as sawdust or straw, may require special handling in the form of composting before use in the garden. Direct application of high nitrogen materials, such as dried blood or poultry manure in large amounts, may "burn" plants.

Raw manure/Farm yard manure: Raw manure is an excellent resource for organic crop production. It supplies nutrients and organic matter, stimulating the biological processes in the soil that help to build fertility. Still, a number of cautions and restrictions are in order, based on concerns about produce quality, food contamination, soil fertility imbalances, weed problems, and pollution hazards.

Contamination: Some manure may contain contaminants such as residual hormones, antibiotics, pesticides, disease organisms, and other undesirable substances. Since many of these can be eliminated through high-temperature aerobic composting, this practice is recommended where low levels of organic contaminants may be present. Caution is advised, however, as research has demonstrated that *Salmonella* and *E. coli* bacteria appear to survive the composting process much better than previously thought. The possibility of transmitting human diseases discourages the use of fresh manures (and even some composts) as pre-plant or sidedress fertilizers on vegetable crops, especially crops that are commonly eaten raw. The growers can use raw manures as suggested below:

- Apply animal manures at least 60 days prior to harvest of any vegetable that will be eaten without cooking. If possible, avoid manuring after planting.
- Do not use dog, cat, or pig manures (fresh or composted). These species share many parasites with humans.
- Wash all farm produce from manured fields thoroughly before use. Persons especially susceptible to food-borne illnesses (children, the elderly, those with compromised immune systems, etc.) should avoid uncooked produce.

Unlike conventional farmers, who have only safety guidelines regarding manure use, certified organic farmers must follow stringent protocols. As per USDA-NOP, raw manure may NOT be applied to food crops within 120 days of harvest where edible portions have soil contact (i.e., most vegetables, strawberries, etc.); it may NOT be applied to food crops within 90 days of harvest where edible portions do not have soil contact (i.e., grain crops, most tree fruits). Such restrictions do not apply to feed and fibre crops.

Produce quality concerns: It has long been acknowledged that improper use of raw manure can adversely affect the quality of such vegetable crops as potatoes, cucumbers, squash, turnips, cauliflower, cabbage, broccoli, etc. As it breaks down in the soil, manure releases chemical compounds such as skatole, indole, and other phenols. When absorbed by the growing plants, these compounds can impart off-flavours and odours to the vegetables. For this reason, raw manure should not be directly applied to vegetable crops; it should instead be spread on cover crops planted the previous season.

Fertility imbalances: Raw manure use has often been associated with imbalances in soil fertility. There are several causal factors:

- Manure is often rich in specific nutrients like phosphate or potash. While these nutrients are of great benefit to crops, repeated applications of manure can result in their building to detrimental levels. An overuse of broiler litter can put excessive phosphate in the soil and pollute surface waters. Nutrient excesses also tie up other minerals. Excessive phosphate interferes with plant uptake of both copper and zinc; excessive potash can restrict boron, manganese, and even magnesium.
- Continual manure use tends to acidify soil. As manure breaks down it releases various organic acids that assist in making soil minerals available, a benefit of manure that is often unrecognised. Over time, however, this process depletes the soil of calcium and causes pH levels to fall below the optimum for most crops. Manures do supply some calcium, but not enough to counterbalance the tendency toward increased acidity. Possible exceptions include caged-layer manure (when oyster-shell or similar calcium supplements are fed) and manure from dairy operations where barn lime is used.
- When fresh manure containing large amounts of nitrogen and salts is applied to a crop, it can have the same effects as excessive applications of soluble commercial fertilizers it can burn seedling roots, reduce immunity to pests, and shorten produce shelf life. Excessive salinity is often associated with heavy applications of feedlot manure in regions where little leaching naturally occurs as in most western states. For example, growers in southwestern states like Arizona are advised to apply gypsum and leach the soil with about 4 inches of irrigation water following incorporation of dairy or feedlot manures.

To avoid manure-induced imbalances, continually monitor soil fertility, using appropriate soil tests. Then apply lime or other supplementary fertilizers and amendments to ensure soil balance, or restrict application levels if needed. A soil audit that measures cation base saturation is strongly advised.

Understanding the soil's needs is only part of the equation. You must also know the nutrient content of the manure you're applying. Standard fertilizer values of manures should be used only for crude approximations. The precise nutrient content of any manure is dependent not only on livestock species, but also on the ration fed, the kind of bedding used, amount of liquid added, and the kind of capture and handling system employed. Also, some traditional assumptions about manure composition may need to be updated. Because of the abundance of sulphur in rations, manure has long been recognized as a good source of sulphur. However, less sulphur is applied to crops in contemporary high-analysis fertilizers, and atmospheric deposition has been decreased by pollution controls. Sulphur deficiencies are appearing in many soils, and levels in manure may also be diminished. It is advisable to test manure as you would test the soil, in order to assess its fertilizer value.

Weed problems: Use of raw manures has often been associated with increased weed problems. Some manure contains weed seed, often from bedding materials like small-grain straw and old hay. High-temperature aerobic composting can greatly reduce the number of viable weed seeds. In many cases, however, the lush growth of weeds that follows manuring does not result from weed seeds in the manure, but from the stimulating effect manure have on seeds already present in the soil. The flush of weeds may result from enhanced biological activity, the presence of organic acids, an excess of nitrates, or some other change in the fertility status of the soil. Depending on the weed species that emerge, the problem may be related to the sort of fertility imbalances. Excesses of potash and nitrogen in particular can encourage weeds. Monitor the nutrient contents of soil and manure and spread manure evenly to reduce the incidence of weed problems.

Pollution: When the nutrients in raw or composted manure are eroded or leached from farm fields or holding areas, they present a potential pollution problem, in addition to being a resource lost to the farmer. Leached into groundwater, nitrates from manure and fertilizers have been linked to a number of human health problems. Flushed into surface waters, nutrients can cause eutrophication of ponds, lakes, and streams. The manner in which manure is collected and stored prior to field use affects the stabilization and conservation of valuable nutrients and organic matter. Composting is one means of good manure handling and is discussed in more detail below. Reducing manure run-off and leaching losses from fields is a matter of both volume and timing. Manure application far in excess of crop needs greatly increases the chances of nutrient loss, especially in high-rainfall areas. Manure spread on bare, frozen or erodible ground is subject to run-off, especially where heavy winter rains are common. Under some conditions, however, winter-applied manure can actually slow run-off and erosion losses from fields, likely by acting as organic mulch.

Sheet-composting manure (tilling it into the soil shortly after spreading) or applying it to growing cover crops are two advisable means of conserving manure nutrients. Grass cover crops, grown to absorb soluble nutrients from the soil profile to prevent them from leaching. All cover crops function as catch crops to a greater or lesser degree. It is a sound strategy, therefore, to apply manure to growing catch crops or just prior to planting them.

Note that both sheet composting and applying to cover crops have trade-offs. Sheet composting improves the capture of ammonia nitrogen from manure, but requires tillage, which leaves the soil bare and vulnerable to erosion and leaching losses. Surface-applying to cover crops (with no soil-incorporation) eliminates most leaching and erosion losses but increases ammonia losses to the atmosphere.

Composted manure: An effective composting process converts animal wastes, bedding, and other raw products into humus-the relatively stable, nutrient-rich, and chemically active organic fraction found in fertile soil. In stable humus, there is practically no free ammonia or soluble nitrate, but a large amount of nitrogen is tied up as proteins, amino acids, and other biological components. Other nutrients are stabilized in compost as well. Addition of compost has the following effects:

- Increase in cation exchange capacity (CEC), nitrogen (N), sulphur (S) and to a lesser extent P content
- Improved soil structure with improved moisture retention and improved aeration
- Reduction in pests and harmful organisms and neutralization of toxins
- Levelling of extremes of (soil) temperature and pH
- Stimulation of soil life, biota diversity and complexity of soil food
- Increased soil weathering and breakdown of soil minerals.

Composting livestock manure reduces many of the drawbacks associated with raw manure use. Good compost is a 'safe' fertilizer. Low in soluble salts, it doesn't 'burn' plants. It is also less likely to cause nutrient imbalances. It can safely be applied directly to growing crops. Many commercially available organic fertilizers are based on composted animal manures supplemented with rock powders, plant by-products like alfalfa meal, and additional animal by-products like blood, bone, and feather meals.

The quality of compost depends on the feedstuffs used to make it. Unless it is supplemented in some way, composted broiler litter though more stable than raw litter will be abundant in phosphates and low in calcium. Continued applications may lead to imbalanced soil conditions in the long term, as with some raw manures. Soil and compost testing to monitor nutrient levels is strongly advised.

Composting can degrade many organic contaminants, but it cannot eliminate heavy metals. Broiler litter composts have been restricted from certified organic production in some countries largely for this reason. Arsenic once used in chicken feed as an appetite stimulant and antibiotic is a particular concern. Since the precise composition of commercial livestock feeds is proprietary information, arsenic may still be an additive in formulations in some western countries. A more recent concern is the inclusion of additional copper in poultry diets and its accumulation in the excreted manure. While copper is an essential plant nutrient, an excessive level in the soil is toxic. This concern is most relevant to organic horticultural producers, who often apply significant amounts of copper as fungicides and bactericides, increasing the hazard of build up in the soil. Whenever you import large amounts of either composted or raw manure onto the farm, it is wise to inquire about the feeding practices at the source or have the material tested.

The NOP has put no specific restrictions on when farmers can apply composted manure to crops; however, it is very specific about manure composting procedures. According to the USDA-NOP Regulations, compost must meet the following criteria:

- An initial carbon: nitrogen ratio of between 25:1 and 40:1 must exist for the blend of materials in the 'pile', and
- Temperatures between 131° F and 170° F must be sustained for three days using an in-vessel or static aerated pile system or temperatures between 131° F and 170° F must be sustained for 15 days using a windrow composting system, during which period the materials must be turned a minimum of five times.

6. FIELD-APPLYING MANURES AND COMPOSTS

When?

The 90- and 120-day restrictions on manure application are mainly intended to prevent food contamination with manure pathogens. Beyond these timing constraints, however, additional agronomic considerations are involved in scheduling manure applications. According to experienced market gardener, crops like squash, corn, peas, and beans do best when manure is spread and incorporated just prior to planting. The same holds true for leafy greens, though only well-composted manure should be used. Cabbages, tomatoes, potatoes and root crops, on the other hand, tend to do better when the ground has been manured the previous year. Obviously, crop rotations that feature non-manured crops following manured crops would be ideal.

To achieve maximum recovery of the nutrients in spread manure, sheet composting i.e. ploughing or otherwise incorporating the manure into the soil as soon as possible after spreading is the best option. Research has shown that solid raw manure will lose about 21% of its nitrogen to the atmosphere if spread and left for four days; prompt soil incorporation reduces that loss to only 5 per cent.

However, since excessive tillage is discouraged in sustainable systems, options for sheet composting may be limited on some farms. The next-best option appears to be spreading onto growing cover crops. This reduces the chances of loss through surface erosion and cuts leaching significantly. However, it does little to control ammonia losses to the atmosphere.

How?

One of the weakest links in the use of manure as a fertilizer appears to be the actual process of field spreading. According to some researchers, the conventional box spreader is an 'engineering anachronism'- an outdated piece of equipment designed principally to dispose of a waste product, not to manage a nutrient resource. Many machines are built to 'dump' as much material as possible in a short time and are difficult to calibrate if you want to distribute manure accurately and according to crop needs. Still, the basic box spreader is the only technology available and affordable to most farmers.

In all cases, fresh material should be composted to kill harmful pathogens and weed seed. In addition, fresh material can damage plants and be hazardous to the environment through runoff. *Organic materials rich in cellulose and lignin such as crop wastes, straw, etc should be recycled in the first instance as surface mulch and subsequently after one season partially degraded straw should be incorporated in the soil.* This is a cheap method of utilization of surplus crop residues. The results have shown that organic mulch is helpful in increasing the yield of crops significantly when the leftover organic matter is ploughed in the soil.

6.1 How to Apply Organic Soil Amendments

Animal manures

- *Annually:* Spread 16-20 tonnes per hectare and mix in soil to 6 inch depth. Wait 10 to 14 days before planting.
- *Semi-annually*: If a garden is to be planted in a successive season, reapply at the reduced rate of 8-10 tonnes per hectare; mix and wait 10 to 14 days before planting.
- *Side-dressing*: Add the solid form or manure tea as a side-dressing at mid-season on long season crops like tomato. Manure tea is made by mixing up to equal amounts of poultry manure and water.
- *Plant hole/pit application*: For single plant application, thoroughly mix the manure in the planting hole/pit 10 to 14 days before planting to prevent burning the roots. Adjust amount according to size of plant grown; for example, tomatoes - 2 kg per hole; herbs - ½ kg per hole.

Note: Plant injury due to fertilizer root burn may be greatly reduced or eliminated by mixing the animal manure with compost prior to planting (1:1 ratio). In this manner, you may apply and plant immediately.

Composted waste: *Annually* spread and mix well into soil up to 25-30 tonnes per hectare. Supplement with animal manure at rate of 8-10 tonnes per hectare. You may plant immediately or wait a few days.

- *Semi-annually* if a garden is to be planted in a successive season, a second application of composted waste may be applied, but at the reduced amount of 15-20 tonnes per hectare. Supplement with animal manure as before (8-10 tonnes per hectare).
- *Side-dressing* not suggested for adding to a growing crop.

Note: If composted waste is used alone, plant stunting will occur due to low content of primary nutrients. Following 2 to 3 years of 'aging' in the soil, however, plant growth should improve noticeably. For best results, always supplement composted waste with organic fertilizer. The actual amount of manure depends on soil test and requirement of the plant to be grown.

Tree leaves: Certain plant residues may be mixed into the garden soil as a soil improvement practice. Woody residues such as sawdust rot very slowly and are low in nitrogen content. Sawdust should always apply along with animal manure. Tree leaves constitute a fairly large portion of waste. For that reason, the use of tree leaves seasonally may be piled onto the garden site, mixed with the soil, and allowed to decompose. Shredding speeds up breakdown and results in better growth and yields of vegetables. Add animal manure (½kg/sq ft) to hasten decomposition and minimize nitrogen depletion. Even without manure supplementation, good yields of such crops as cucumbers, tomato and greens can be expected after 2 to 3 years of applications of at least 20 tonnes per acre annually.

Mulching: Mulch is any material, usually organic, which is placed on the soil surface around plants. Most commonly used mulches are bedding straw, leaves, pine needles, pine bark, and wood chips. Coarse ground composted waste makes excellent mulch. Mulch is helpful in many ways: 1) it conserves soil moisture and nutrients, 2) it prevents weed growth, 3) it reduces fruit rot; and 4) it moderates soil environment.

Many types of organic mulches may be ploughed into the soil at crops end, thus benefiting as a soil amendment. Keep in mind that mulches are high in carbon content, so need to be supplemented with manure to facilitate decomposition and prevent nitrogen depletion of the plants.

Plastic film (usually black to exclude light) is one of the most popular mulches used in vegetable gardening and farming. However, because it is energy-based, it should be discouraged in an organic garden.

6.2 Soil Science: Medicine of Organic Farming

Potential of organic farming systems to produce plant and animal products appropriate for the human diet. Developing an organic garden by maintaining good soil fertility for plant growth requires some effort. For centuries mankind has been feeding itself through organic gardening. Recent technological advances have made considerable improvements in those primitive ways, greatly enhancing the likelihood for successful gardening endeavours. Non-renewable energy based resources and water resources are disappearing at an alarming rate. Hence a better and more efficient ways to grow foods organically need to be found. Regular additions of organic matter through compost, vermicompost, green manure or crop residues, biofertilizers etc. are necessary.

Voisin (1954) stated: the soil must be kept in good health if the man and animal is to remain in good health. Soil Science is the foundation of protective medicine, the medicine of tomorrow.

Chapter 7

On-farm Organic Manure Production

Plant growth for thousands of years is sustained solely by the recycling of organic matter in nature till the advent of chemical fertilizers on the agricultural scene during the 20th century. However, chemical fertilizers are only complementary to organic manures. Injudicious use of former without the latter would lead to more harm than good in terms of deterioration of soil health, destruction of physical structure of soil and reduction in soil biological activities that are vital for plant growth and survival. Nature is richly endowed with micro and macro organisms that degrade the cellulose and lignin in biomass, fix nitrogen and accelerate decaying and mineralization of organic matter and make nutrients available to plants apart from improving and maintaining the desired physico-chemical and biological properties of soil.

In organic farming, farmer has to manage the farm with coherent diversity by utilising all the on-farm and adjacent resources. There are vast potential of manurial resources and organic wastes available for recycling. Residual wastes from crops and natural biomass hold dependable promise for innovation in nutrient recycling in addition to their own optimum utilisation. Some of the commonly used on farm practices organic manure production and their management in organic farming are discussed below:

1. GREEN MANURE

Green manuring, being indigenous and cost effective is an age old practice in farming system. Incorporation of leaf, stem and other plant parts in to the soil when the plant is green is called as green leaf manure management. Green leaf manures are the cheapest source of nitrogen for the crops. By using green leaf manures, farmer can save on the cost of expensive concentrated organic fertilizers. Further, it will more advantageous if the green leaves are composted and then used.

Use of green leaf manure is advantageous both for crops and soil. They decompose rapidly and improve both physical and chemical properties of the soil by retaining organic matter in the soil. They provide nutrients to the crops and energy to the microbes. It can also be used as mulch to prevent soil and water losses and also control the weeds. There are different green leaf manure crops that can be cultivated in organic farms.

The distinct benefits from green manuring are:

1. Promoting microbial activities, liberating carbon di-oxide, including chemical reactions and nutrient conversion into available forms, especially the phosphates
2. Soil nitrogen build-up through fixation
3. Improvement of soil structure and absorptive capacity due to humus formation
4. Nutrient translocation from the lower to the upper layers for plant use
5. Improvement in environmental quality.

Majority of green manure crops are legumes such as Cowpea, Dhaincha, Sesbania, Sunhemp, etc. These annual green manure crops can be cultivated in all type of soils. About 6-8 tonnes of green leaf can be obtained from one hectare of land from these crops. The harvesting should be done just prior to flowering and green leaves should be incorporated into the soil. Cultivation of Dhaincha in saline and saline-alkali soils continuously for 4-5 years can reclaim the soils.

Table 7.1: Common leguminous green manure crops and nitrogen availability to the succeeding crop

Local Name	Botanical Name	Green Matter ($t\ ha^{-1}$)	Nitrogen (per cent)
Sunhemp	*Crotalaria juncia*	21.2	0.43
Dhaincha	*Sesbania aculeate*	20.0	0.43
Pillipesara	*Phaseolus trilobus*	18.3	1.10
Mungbean	*Phaseolus aureus*	8.0	0.53
Cowpea	*Vigna sinensis*	15.0	0.49
Gaur	*Cyamposis tetragonaloba*	20.0	0.34
Senji	*Melilotus alba*	28.6	0.57
Khesari	*Lathyrus sativas*	12.3	0.54
Berseem	*Trifolium alexandrium*	15.5	0.43

Similarly other perennial crops like *Triposia* and *Indigofera* can be cultivated for green leaf purpose. Bushy plants like *Glyricidia,* hedgerows (*Leucaena leucocephala, Leucaena liliricidia*) can be cultivated on bunds and vacant lands. Glyricidia can be cultivated either by using seeds or stem cuttings. Green leaf can be harvested one year after plantation. During the cropping season, bushes are lopped at about 0.5 m height. These loppings are used as mulch to reduce moisture loss and improve the nutrient status of soil. The leaves of perennial trees like *Pongamia, Neem* and *Calotropis* can also be used for green manuring purposes.

Table 7.2: Common shrubs and trees utilized for green leaf manure

Glyricidia maculate	*Leucaena leucocephala*
Glyricidia sepium	*Leucaena glauca*
Pongamia glabra	*Cassia auriculata*
Azadirachta indica	*Cassia tora*
Deria indica	*Cassia accidentalis*
Ipomoea cornea	*Cassia pistula*
Thespesia populnea	*Calotrophis gigantean*
Delonix regia	*Hibiscus viscose*
Delonix elata	*Vitex negundo*
Jatropha gossypitolia	*Lantana camara*
Tephrosia purpurea	*Echhornia crassipes*
Tephrosia candida	*Chromoleana odorata*
Didonea viscosa	*Salvnia molesta*
Trienthama portulacastrum	

Source: Abrol and Palaniappan, 1987

There are several leguminous plant species commonly found in nature which can be advantageously used as cover crops to serve as shade plants to young plantations, forage and green leaf manure crops, and as soil binders to reduce erosion and minimize the growth of obnoxious weeds. The promising species are *Calopogonium mucumoides, haseolus trilobus* (Pillipesara), *Centrosema pubscens, Macroptilium* (Siratro), *Phaseoloides* (Kudji M.), *Dolichos lab lab* (var Ignosus), etc.

1.1 Raising Green Manure Crops in Different Systems

(i) *Dryland*: The scope for raising green manure crop as catch crop and recycling *in situ* or *ex situ* on dryland is very limited due to inadequate soil moisture for decomposition process. Dryland are the worst hit with respect to supply of organic matter to crops due to various constraints including poor management. Regular recycling of organic matter through compost is an alternate approach to regenerate their production capability.

(ii) *Wetland*: Green manure crops like *Azolla* can be grown conveniently during cropping seasons. Besides, the field bunds can be effectively utilized by raising glyricidia (at 2 cm spacing) which can yield valuable green leaf manure (up to 20 kg/bush/yr). The green leaf manure can be recycled to the field directly 2-3 weeks prior to transplanting of paddy crop or it can be used for compost making.

(iii) *Garden land*: The tree species having compatibility, quick growth, and higher leaf biomass can be selected and planted in the garden. These species may be medium and small canopy plants like *Sesbania* and *Glyricidia* or Perennial shades tree like *Jack, Erythrina,* etc. The leaf litter provides support biomass and plant nutrients apart from shade and nutrient recycling from the lower layers.

2. RURAL COMPOST

Compost is a balanced organic fertilizer and it provides both macro and micronutrients to plants. In addition, it also improves soil fertility, buffering capacity, porosity, aeration, temperature and water holding capacity. Compost is prepared from organic substances obtained from different sources like plant and animal wastes of the organic farm. Composted organic manures provide all the nutrients that are required by plants but in limited quantities. It helps in maintaining C:N ratio in the soil and improves the chemical, physical and biological properties of the soil. Due to increase in the biological activities, the nutrients that are in the lower depths are made available to the plants.

In general, the raw materials may contain 15-60 per cent cellulose, 10-30 per cent hemicelluloses, 5-30 per cent lignin, 5-40 per cent protein, 5-30 percent minerals (ash) besides 2-30 per cent water soluble fractions (sugars, starch, amino acids, urea, ammonium salts) and 1-15 per cent of ether and alcohol soluble fats, oils and waxes. These components of organic wastes undergo decomposition under mesophilic and thermophilic conditions in heaps, pits or windrows and finally yield a dark coloured humified material in about 3-4 months.

Several methods of compost making (worm, heap, sheet, biodynamic etc.) are being researched. A further area of research examines the effects of adding minerals such as rock dust, lime, clay, rock phosphate to the compost heap rather than directly on the land. Various inoculants (starters), including the biodynamic compost preparations are also being researched. Some of the important processes of composting are described below.

Microbial process: Bioconversion of organic materials is carried out by different groups of heterotrophic microorganisms such as bacteria, fungi, actinomycetes protozoa, earthworms, mites, ants, termites, etc. The organism represents both the plant and animal kingdoms. During the course of composting, both qualitative and quantitative changes occur in the active flora, some species multiply rapidly at the first stage, change the environment and then disappear to allow other population to succeed them.

In the second and third weeks, the different physiological groups may be in the following order:

Bacteria (10^6-10^7) > Ammonifying bacteria (10^4) > Proteolytic (10^4) > Pectinolytic (10^3) > Nitrogen fixing bacteria (10^3). The role of cellulytic and lignolytic microorganisms notably fungi is of prime importance.

2.1 Aerobic Composting

In this system, about 2/3rd of carbon is evolved as CO_2 and remaining 1/3rd is combined with nitrogen in the living cells. A great deal of exothermic energy is released during the process generating substantial amounts of heat, thus raising

Table 7.3: Source of composting material

Livestock and human wastes	Cattle shed wastes, livestock excreta, slaughter house waste, piggery, duckery and poultry waste
Crop Wastes, Tree wastes, Aquatic weeds	Cereal residues, sugarcane trash, cotton weeds, aquatic weeds, groundnut shells
Urban and Rural wastes	Solid wastes, sewage and sludge wastes, bio-gas slurry
Agro-industries wastes	Press mud, paper and pulp wastes, oil cakes, paddy bran and husk, coir pith, sawdust, etc.
Marine wastes	Sea weeds, fish meal

the temperature in heaps and creating favourite conditions for thermophilic microorganisms. However, if the temperature exceeds 65°-70°C, the microbial activity is ceased due to thermal killing of organisms.

2.2 Anaerobic Composting

The breakdown of organic matter takes place in the absence of oxygen. Firstly, facultative acid producing bacteria degrade organic matter to fatty acids, aldehydes, etc. then other group of bacteria convert the intermediate products to methane, ammonia, carbon dioxide, and hydrogen. Thus oxygen is also required for anaerobic decomposition but its sources are chemical compounds instead of dissolved oxygen. Under anaerobic composting only about 26 Kcal/mole glucose molecules are released as compared to aerobic process where release of energy is much greater (484-674 Kcal/mole glucose molecules).

2.3 Composting Process Parameters

Several parameters need to be considered for efficient composting. These are temperature, particle size, moisture content, carbon to nitrogen ratio, and nutrient amendments.

During the composting process, the temperature of the composting material passes through mesophilic (10°C - 45°C) and thermophilic (45°C - 70°C) temperature ranges. The temperatures generated are due to the metabolic activity of the microorganisms involved in the composting process. There are two important considerations involving temperature. First, the temperature of the composting material should reach 55°C for at least 3 days to destroy any plant, human or animal pathogens present. This temperature requirement should be reached during the several successive turnings of the material. Secondly, the temperature should not be allowed to exceed 70°C, because at temperatures above this most microorganisms involved in composting die or enter a resting phase which slows decomposition.

Reducing particle size exposes more surface area to microbial attack and enhances decomposition. However, grinding the material too finely can cause compaction and restricted air flow which reduce the rate of composting. Also, the finer the material is ground, the more energy is consumed in the grinding process.

A variety of materials can be used as nutrient amendments. All animal manures are useful as nutrient sources, although they provide differing amounts of nutrients and moisture. Food processing wastes can also be used as an amendment. Seafood waste is a good amendment, although this waste may be seasonal. A steady supply of the chosen amendment is preferred. The use of nutrient amendments enhances the rate of composting, but their use requires extra care. The composting materials should be mixed properly and turned on a regular basis. If they are not, an odour and fly problem may result. The length of time required for composting depends on the composting system used, the initial feed stock, the initial particle size, nutrient balance and moisture content.

Length of time, the temperature of composting, moisture condition and the C/N ratio of the original material are the main factors influencing the outcome. For composting, the moisture content should be between 45-60%. Moisture can be added through watering attachments to the in-vessel or windrow turner or grinder. Microbial activity is affected by the carbon (C) to nitrogen (N) ratio.

Starting with organic matter that has an average C/N ratio of 33/1 makes it theoretically possible to capture all the nitrogen from the organic matter without loss to the air, as sufficient C is available to supply maintenance and growth and reproduction for microorganisms. A C/N ratio lowers than 33/1 increases nitrogen losses to the environment. High moisture contents create anaerobic conditions, stimulating anaerobe microorganisms. The by-products of anaerobic decomposition are (if excessive) toxic to crop plants.

All these contributing factors make composting a science as much as an art. The quality of the compost depends on how the composting process has been managed. Different degrees of maturity of compost fulfil different purposes in the soil. For example, applying mature compost leads to better soil structure and disease suppression.

Table 7.4: Optimum value of composting parameter

Parameter	Optimum Value
Feed particle size	25-40 mm
Moisture content	50-60%
C/N	30/1
Temperature	55-60°C for 4-5 days
Aeration	Periodic turning (2-3 per month)
Heap Size	10 m length x 1.5 m high x 2.5 m wide

2.4 Types of Compost Units

Ideas for constructing your own compost unit can be obtained from various sources. Many of agriculture extension offices have composting demonstration projects with homemade compost units.

Making your own compost unit: There are at least as many different ways to construct a compost unit as there are people who build them. If you decide to build a compost unit, the first step is to decide which type of unit is best for you. The important factors to consider are mentioned below:

1. quantity of material to be composted
2. likelihood of problems with rodents
3. cost and labour requirements
4. skill level required in building
5. skill level required in operating
6. desired time for composting

Compost units can be classified in several ways. They can be classified into 'holding units' and 'turning units'. Holding units include bins which have been constructed from masonry, plastic, wood, wire, or combinations of these materials. Most of the manufactured bin units are plastic or wire. Turning units include barrels which are turned horizontally or end to end.

Holding Bin units: This is the most popular type of home yard compost unit. The simplest and least expensive type of bin compost unit can be constructed from wire fence material. Consider using the vinyl coated fence wire which is now available at building supply stores. The length of wire which is needed for a circular unit will be approximately 3 times the diameter of the bin. The ends of the fence can be wired together or post can be used to permit quick disassembly for removing the compost or for turning the composting material. Wire bins can also be square or rectangular if corner posts and/or other side support are provided. Wood/bamboo posts can be used. Because of the large amount of open area on the sides and top of these units, they may dry out very easily. They may also accumulate excessive water in rainy weather. Because of this, a cover should be added if animals or excessive rainfall is a problem. Bin units can also be constructed from timbers. An inexpensive bin unit can be constructed by fastening four wood pallets together at the corners with wire or posts (wood or metal). Bin compost units can also be constructed from concrete blocks. A non-permanent unit can be made by stacking the blocks and driving wooden post through the holes and into the soil. The blocks can also be mortared or fastened together with surface bonding materials. Gaps can be left between the ends of the blocks (½ to ¾ inch) for aeration, or some of the blocks can be turned on their side to achieve the same effect. The bin composting units should be at least 3 feet wide, 3 feet long and 3 feet high. Larger volumes will work even better because of better heat retention. These units can also be made in two or three bin units to facilitate turning of the composting material, particularly if one bin is not large enough to handle the composting materials.

Turning units: The turning units can produce compost more quickly than a holding unit, if they are intensively managed. Plastic barrels are most commonly used to make turning units. They are available in many places that handle mineral oils, detergents, etc. Caution should be exercised to avoid using barrels that have contained hazardous or toxic materials. Compost units can be turned on either the vertical or horizontal axis. Steel pipe and pipe flanges are convenient ways to mount the barrels on supports made from wood or concrete block. Some barrels will have a large screw on lid for loading and unloading. A door can be made on other barrels by cutting a hole in the side or end and using a piano hinge and a latch. Holes should be drilled in the barrels to provide aeration and drainage of excess water.

Note: Always mix high nitrogen and low nitrogen materials together when building the compost heap. The simplest method of composting is to build a heap by alternating layers of organic material and soil; the layers of soil may be ½ to 2 inches thick and the layers of organic material 6 to 12 inches thick. As the heap is built, water the material until it is moist but not soggy. Two to 4 inches of soil should cover the final heap. After a few weeks, turn the material for aeration and mixed well to move the outer parts to the centre. The whole composting process should be complete in about three months in warm weather.

2.5 Methods of Composting

***Indore method*:** In this method animal dung is used (as catalytic agent) along with all waste materials available on the farm to decompose in a pit having a width of 2 to 2.5 m and depth 0.9 m. Depending on the availability of waste materials, the length may be from 3 to 10 m. The materials that can be used are crop residues, cow dung, cattle shed waste, and ash. In fact, this is the first aerobic method of composting in which manure is ready within four months. Materials needed for composting may be mixed plant residues, weeds, sugarcane leaves, grass, wood ashes, bran etc. Hard woody materials are chopped in to small pieces and are cursed. Animal dung of all the farm animals like cow, buffaloes, bullocks, horses, goats etc. is collected with bedding. The urine soaked mud of the cattle shed can also be utilized for composting. Wood ashes reduce acidity of the compost and add potassium.

These materials are filled in the pit layer by layer. The first layer is of cattle shed waste filled to height of 7 to 8 cms. Then ash and cow dung are spread in layers. These layers are repeated to a height of 1 m. It should not take more than six or seven days to fill the 3/4 length of the pit, leaving about one -fourth of the pit empty to facilitate subsequent turnings. At the end one more layer of bedding material with wood ash and urinated mud should be provided. The sides of the pits should be sloppy and should have sufficient space between them for the carts to move freely. These materials are turned over after 15, 75 and 105 days after filling. To maintain moisture to a level of 40 to 50 per cent it is necessary to sprinkle water often in the evening and morning.

In rainy season the compost may be prepared in heaps above ground. In the areas with average rainfall, the size of the heaps may be 8 x 8 ft broader at the base and 7 x 7 ft at the top and not more than 2 ft high. In regions with heavy rainfall, compost should be prepared at safer places and protected by shed. Compost making should be avoided between the months of June and September.

Bangalore method: This method overcomes many of the disadvantages of Indore Method. This method eliminates the necessary of turning up of the materials as composting is done in trenches of 30 x 6 x 3 ft or in pits of 20 x 6 x 3 ft. Farm wastage are collected and spread in a six inch layer of cattle dung and urinated mud is followed by a layer of 1 to 2 inch thick layer of soil. This process is repeated till the heap is 1.5-2 feet above the ground level. Ultimately the heap is covered with mud plates about one inch thick. Compost will be ready within 8 to 9 months.

NADEP method: This method envisages lot of composting through minimum use of cattle dung. Decomposition process flows through 'aerobic' method and it requires about 90 to 120 days for obtaining the decomposed compost. In this method, tanks are constructed by using bricks or locally available materials. These tanks should be located near cattle shed or on easily approachable farm sites. The dimensions of the tanks are 3 to 4 m length, 2 m width, 1 m height with 9-inch thick wall. Lining of the bricks is made with the help of mud, which saves expenditure on cement. For circulation of air, open spaces of 15 cm width are provided at several places in the walls. About 1400 to 1500 kg of farm residues and, refuses like weeds, grasses, leaves, sugarcane trash, stubbles, stalks, roots, stems, pruning, chaff, pigeon pea stalks etc., about 90 to 100 kg (8-10 baskets) of cattle dung, 1750 kg (120 baskets) dry sieved-soil (urinated earth is more effective), and 1500 to 2000 litres of water (less during rains and in abundance during dry spells) is required for composting.

Materials are filled in the tank layer by layer. Before filling the tank, slurry made of cow dung and water should be sprinkled the floor and the walls. The first layer will be of crop residues, weeds, grass etc. (100-110 kgs) to a height of 15 cm. Over this layer biogas slurry or cow dung slurry (4-5 kg) in 125 litres water is spread in the form of thin layer. Then 4 to 5 baskets of soil are spread over this thin layer. These layers are repeated to a height of 0.6 to 0.75 m (1 and 1/2 feet above the brick level). A hut like shape may be given at the top. The whole tank is to fill within 1 or 2 days. Eleven to twelve layers are required for filling the tank to its capacity.

In case cattle dung is not available in desired quantity, collection of the same is done for 8-10 days under a shade by covering it with a light layer of soil. As an alternative practice, tank can be filled 1/3 or 1/2 of its capacity in parts. Full tank should be covered and sealed by 3-inch layer of soil (300 to 400 kg). It should be pasted with a mixture of dung and soil. Cracks should not be allowed to develop on the heaps, to check gas leakage, for that the pasting can be repeated. After 15 to 20 days the trash contracts and becomes more compact and goes down in the tank by 8-9 inches which can be refilled in the same manner and again sealed and pasted with mud and dung. In order to maintain 15 to 20% moisture, the compost is sprinkled with cattle dung and water. The microorganisms present in the material will utilise the oxygen in the air and convert the organic matter into compost.

Normally from one tank about 160-175 cubic ft. of manure and 40 to 50 cubic feet of undecomposed raw material, weighing about 3 tonnes are obtained. According to an estimate, from annul collection of dung from one cow, 80 tonnes of manure can be prepared which contains 800 kg N, 560 kg P and 1040 kg K. It takes about one thousand rupees to construct a tank. Eight man-days are required for filling the pits to get it emitted and transported to the fields.

VAT Method: VAT method is one of the improved methods of composting. In this method, tanks are constructed using stones slabs and the dimensions of the tanks are 10 m length, 2 m width and 1 to 1.2 m height. Open spaces of 2 to 4 cm are left between two slabs. If stones are not available, locally available materials like bamboo poles, casurina poles or areca trees with coconut fronds can be used for construction of tanks. But this type of tanks will be temporary in nature. The floors of the tank are cemented to prevent infiltration of nutrients in to the soil.

The tanks are filled in layers. The first layer should have slowly degradable materials like coconut fronds, sugarcane thrash, coconut fibre or maize stalks cut to small pieces and filled to a height of 15 cm. Over this cow dung slurry and a thin layer of soil are spread. As these materials are highly fibrous, microorganisms like *Pleurotus, Aspergillus,* or *Trichoderma* have to be mixed with cow dung slurry for early decomposition. The second layer should be of dry grasses, weeds and crop residues and filled to a height of 25 to 30 cm. After these materials are filled, water in sufficient quantity is added to make these materials wet. Over this layer, cow dung slurry with microorganisms is applied as explained earlier. A third layer of green leaves to a height of 25 to 30 cm and a fourth layer of tank silt, bio gas slurry and other animal wastes like poultry manure, sericulture waste etc. are added to a height of 25 to 30 cm. The fifth and final layer of cow dung, cattle shed waste and animal urine is then applied and covered with soil to a thickness of 3 to 4 m. In this method, maintenance of moisture to a level of 50 to 60 per cent is very much necessary, and for this frequent application of water is needed. These materials are turned over 45 days after filling and covered again with 50 to 60 per cent of moisture level.

Compost will be ready in about 90 to 100 days. In composting, green leaves from trees like pongamia, neem and weeds like water hyacinth, ipomoea, calotropis etc. can be used. Well-decomposed compost will have a stable temperature with no bad odour. The colour of the well decomposed compost will be brownish black. The pH will be around 7 and C:N ratio will be less than 15.

The site selected for the pit should be easily approachable for inspection. Pit should be at a comparatively higher level so that neither rainwater gets into it nor the water table rises and causes water stagnation in the pit during monsoon. It should be near the cattle shed and the source of water supply.

2.6 Nutrient Enrichment of Compost

If the organic wastes used for composting are poor in nitrogen they can be enriched by incorporating nitrogen rich organic wastes (fish meal, pongamia cake, non-edible cake, poultry manure, etc.). For phosphorus enrichment, rock phosphate equivalent to 5 per cent P_2O_5 (250 kg of Mussorie rock phosphate per tonne of organic wastes) can be mixed with 10 per cent cowdung, 5 per cent soil and 5 per cent well decomposed manure to serve as inoculum for decomposition. Phosphorus enriched compost reduces the loss of nitrogen which normally occurs during decomposition of nitrogenous rich material due to ammonia volatilisation. Experimentally, it has been found that 2 kg of phosphocompost prepared was found equivalent to 1 kg of superphosphate with respect to crop response studied.

Precautions: Avoid waterlogging of compost pits and washing away of the nutrients from the already decomposed manure. If bad smell is observed during decomposition, it is due to poor aeration. Give 1-2 turnings at monthly intervals to provide more oxygen. Do not enrich more than 5 per cent P_2O_5 equivalent of rock phosphate. Higher levels of rock phosphate incorporation decrease the rate of decomposition.

3. VERMICOMPOST

The term "vermicomposting" means the use of earthworms for composting organic residues. Earthworms encourage the growth of microorganisms in their gut by providing ideal conditions therein. Earthworms can consume practically all kinds of organic matter. Earthworms play a key role in soil biology by serving as bioreactors to effectively harness the beneficial soil microflora and destroy soil pathogens, thus converting organic wastes into valuable bio-fertilizer with bio-pesticides, vitamins, enzymes, antibiotics, growth hormones and protein rich worm mass. This compost is an odourless, clean, and containing adequate quantities of nitrogen, potash, phosphate, calcium, sulphur, magnesium, iron, manganese, zinc, copper, boron, molybdenum, carbon, etc. essential for plant growth. Vermicompost is a preferred balanced nutrient source for organic farming. It is eco-friendly, non-toxic, consumes low energy input for composting and is a recycled biological product. Vermicompost also has immobilised micro flora which continue their function in the soil. Every farmer can do vermicomposting in his own farm using agriculture waste.

Benefits of Vermiculture

First priority benefits	– Additional price gain
	– Better taste of food
	– Better size of farm produce
	– Less irrigation water requirement
	– Better cultivation in saline-alkaline conditions
Second priority benefits	– More tillering, flowering and grain setting
	– Less insects and pests attack on crops
	– Less weed infestation
Third priority benefits	– Better germination
	– Less termite attack
	– Better overall appearance of crops
	– Improved soil texture

Species of earthworms

Only selected species of earthworms which feed exclusively on organic matter can be used for this process of composting. The species of genus *Perionyx* are good composters, though not as efficient as *Eisenia fetida* or *E. eugeniae* which are preferred in western countries. *Perionyx species* can survive on many agricultural wastes, sericulture waste, house waste, etc. Further, their biomass production is directly associated with the nature of culture medium. When earthworms are used for composting of organic waste, the relationship between biomass production and population increase is of utmost importance. Involvement of earthworms in the composting process decreases the time of stabilisation of the waste and produces an efficient organic pool, with energy reserves as vermicompost.

3.1 Production Technology

The vermicompost pit should be located in a place protected by shade and should be on a higher plane and free from water stagnation. The pits size could be 2 metre length, 1 metre width and 1 metre depth. First bottom layer of 15 cm should comprise only small stones and pebbles. Second bottom layer should contain 15 cm. of sand. Third bottom layer of 10 cm. should have soil from the field. Then, the earthworms are placed over the bed @ 30 worms per sq.ft. Then, add topsoil of 5 cm. over the worms. Then, paddy husk or coir pith or saw dust or any other such organic matter mixed with 40% cow-dung is placed in a layer of 10 cm. The top layer of 40 cm. could be filled with a mixture of green leaves, dried leaves, coir pith waste, egg shells or any other organic agriculture waste and covered with jute cloth or coconut leaf. Then the vermi bed should be sprinkled with water on a daily basis to maintain the moisture.

Vermicompost application based on nutrients requirements of different crops: Generally, all agricultural crops have been categorized based on their nutrients consumption for per unit output from an area. For any given location all crops cultivated in that area are expected to give certain optimum yields. Based on experiences, it can be divided into following categories.

- All rainfed crops such as Sesame, Moong, Urd, Cowpea, Moth, Gram, Mustard, Rapeseed, etc. are considered as low nutrients requirement crops. Traditionally no chemical fertilizers are used for their cultivation. In such crops an average of 500-750 kgs of vermicomost per hectare can give excellent results.
- In the second category of rainfed crops such as Bajra, Jawar, Castor, Isabgol, Fenugreek, Arhar, etc. farmers have been traditionally using farmyard manures. In these crops an average of 900-1200 kgs of vermicompost per hectare can give satisfactory yields.
- In third category of crops requiring moderate irrigation such as Sunflower, Barley, Maize, Wheat, etc. farmers generally use a combination of chemical fertilizers and farmyard manures. In most of these cases an average dose of 1700-1900 kgs of vermicompost per hectare is recommended as a substitute for either of the two i.e. chemical fertilizer or farmyard manure as the case may be.
- In fourth category comprising of Tobacco, Beetroot, Onion, Carrot, Sweet Potato, Okra, Coriander, Brinjal, Cucumbers, Ginger, Opium, Mentha, etc. a dose of about 2400 kgs of vermicompost per hectare can be recommended as a substitute for chemical fertilizers. The use of farmyard manure can also be reduced to 50 percent level.
- The crops like Cabbage, Cauliflower, Potato, Chilli, Sugar Beet, Paddy, Tomato, Garlic, Turmeric, Broccoli, etc. use of vermicompost @ 2400-2900 kgs per hectare can be recommended to substitute half the dose of chemical fertilizers. The use of farmyard manure has to be continued at previous levels.
- In high nutrients requirement crops such as Radish, Jute, Sugarcane, Banana etc., vermicompost at the rate of 1500- 2400kgs per hectare should be applied. It has been observed that along with vermicompost, full dose of farmyard manure and the balance being provided other organic fertilizers gives optimum yields.
- In horticulture crops especially fruit orchards depending on the stage of growth, vermicompost use of 1-20 kgs per plant is recommended.

It is advised that individual farmer should make specific assessment for his own conditions by bench-marking nitrogen requirement of crops. Vermicompost can be applied at any stage of crops. As compared to one single dose split doses have been found to give better results.

Table 7.5: Vermicompost application for conversion of organic farming based on previous use of chemical fertilizers

Category	Substitution of Chemical fertilizer by Vermicompost
Very high chemical fertilizers used areas	3-5 crop season
Moderate chemical fertilizers used areas	2-3 crop season
Low chemical fertilizers used areas	1 crop season

3.2 Quality Evaluation

Chemical: The range of chemical composition of compost and vermicompost are given below:

Table 7.6: Range of chemical composition of vermicompost

Parameter	Range of value
pH	: 6.0-7.5
Organic Carbon (%)	: 15 – 30
Nitrogen (%)	: 0.90 – 2.00
Phosphorus (%)	: 1.00 – 1.90
Potassium (%)	: 1.00 – 1.50
Carbon: Nitrogen	: 14-15: 1
Calcium (%)	: 2.0 – 4.0
Magnesium (%)	: 0.4 – 0.7
Sodium (%)	: 0.02 – 0.30
Sulphur (%)	: Traces to 0.40
Iron (%)	: 0.3 – 0.7
Zinc (%)	: 0.02 – 0.03
Manganese (%)	: Traces to 0.40
Copper (%)	: 0.002 – 0.010
Aluminium (%)	: Traces to 0.07
Boron (%)	: 0.003 – 0.007
Cobalt, Molybdenum	: Present in available form

Biological: Total bacter count of 2.5´x10^{6} comprising of Azotobacter, PGPR, PSB (Phosphorus Solublising Bacteria), Actinomycetes. Also contains Gibberalline, Auxins and Cyctokinine in sufficient quantities. It contains sufficient moisture (25-35%).

Enrichment and quality improvement of vermicompost: Enriching vermicompost through the use of naturally occurring minerals and microorganisms ensures higher availability of required nutrients for the plant.

Table 7.7: Possible improvement in chemical and biological composition of vermicompost

Parameter	Possible Improvement
Organic Carbon	35%
Nitrogen	2.5%
Phosphorus	3.5%
Potassium	2.25%
Sulphur	0.5%
Zinc	0.05%
Nitrogen fixing bacteria	$2.4' \times 10^{9}$
Phosphate solubilizing microbes	$2.1' \times 10^{9}$
Beneficial fungi (spores incl.)	$1.9' \times 10^{10}$

3.3 Economic Evaluation

Cost comparisons for different fertility inputs should be carried out on per unit of cultivated area basis. However, the correct methods for comparison should be either the cost per unit of output or the cost per unit of fertility inputs. Here a comparison has been made on the basis of cost per unit of fertility inputs provided through chemical fertilizers vs. organic compost.

***Cost of chemical fertilizer use*:** Presently chemical fertilizers are used for providing nutrients such as nitrogen, phosphorous and potash. The cost of these chemical fertilizers in the context of their plant uptake i.e. fertilizers use efficiency can be estimated as follows:

UREA: Say 100 kg of urea is used per hectare, containing 46% nitrogen. Cost of urea is Rs.5.50 per kg and thus cost of nitrogen is Rs. 11.95 per kg. Plant uptake i.e. use efficiency is 15-40 percent (average 20%) for nitrogen. Therefore, the cost of nitrogen actually used by the plants is Rs.59.75 per kg.

DAP: Say 100 kg of DAP is used per hectare. This contains 18 percent nitrogen and 46 percent phosphorus. Cost of DAP is Rs.11.00 per kg and thus the cost of nutrients (combined for nitrogen plus phosphorus) is Rs.17.18 per kg. Plant uptake i.e. use efficiency is 15-40 percent for nitrogen (average 20%) and 10-25 percent for phosphorus (average 15%). The cost of nutrients actually used by the plants from DAP is Rs.95.45 per kg.

SSP: Say 100 kg of SSP is used per hectare, containing 16% phosphorus. Cost of SSP is Rs.6.00 per kg and thus cost of phosphorus is Rs.37.50 per kg. Plant uptake i.e. use efficiency is 10-25 percent for phosphorus (average 20%). The cost of nutrients used by the plants from SSP is Rs. 187.50 per kg.

CAN: Say 100 kg of CAN is used per hectare, containing 20% nitrogen. Cost of CAN is Rs.6.50 per kg and thus the cost of nitrogen is Rs.32.50 per kg. Plant uptake i.e. use efficiency is 15-40 percent (average 20%). The cost of nitrogen actually used by the plants is Rs.162.50 per kg.

MOP: Say 100 kg of MOP is used per hectare, containing 60% potash. Cost of MOP is Rs. 8.00 per kg and thus cost of potash is Rs. 13.35 per kg. Plant uptake i.e. use efficiency is 40-60 per cent (average 50%). The cost of potash actually used by the plants is Rs. 26.70 per kg.

Cost of organic compost use: In case of organic farming, over 70 percent of the organic compost is produced by the farmer's in-house. For remaining, the farmers are buying from commercial production units at an average price of Rs. 6000/- per metric tonne. The average nutrients contents reported for compost is nitrogen-1.0 to 2.0 percent (average 1.5%), phosphorus-1.25 percent (average 1.00%), and potash-1.00 to 2.00 percent (average 1.5%). In addition compost also contains all micronutrients and trace elements that would also add up to at least one percent equivalent of nutrients.

The compost has active biological life containing Azatobacter, PSB (Phosphorus Solublising Bacteria), PGPR, etc. During 90-100 days of crop duration they also add up to 1.5 to 2.5 percent nutrients (average 2%). Say 1000 kg of compost is used per hectare. The total nutrients provided by 1000 kg of compost will add up to 70.00 per kg (7.00%) and at an average plant uptake i.e. use efficiency of 65 percent will provide 45.50 kgs of nutrients. At an average cost of compost at Rs.6000/- per MT per hectare, the cost of nutrients is Rs.132.00/- per kg.

In addition to nutrients, compost also provides better aeration, water retention capacity and many other benefits. Some of the major advantages of compost use are also the lower cost of labour (saving due to fewer weeds in the field) and saving from the cost of treatment for termites. Compost on subsequent use has been found to provide at least 20-30 percent more nutrients. This ability can continuously reduce the quantities of compost used in the field over long durations.

Table 7.8: Nutrient-wise cost analysis

Fertilizer and Compost	Average Cost of Nutrient per kg	Remarks
Urea+SSP+MOP	59.75+187.50+26.70 = Rs. 273.95	Provide only NPK
DAP+MOP	95.45+26.70 = Rs.122.25	Provide only NPK
CAN+SSP+MOP	162.50+187.50+26.70 = Rs. 376.70	Provide only NPK and Calcium
COMPOST	= Rs. 132.00	Provide NPK, Secondary and Micronutrients

Table 7.9: Advantage of compost over chemical fertilizer

Criteria for Comparison	Compost	Chemical Fertilizers
Macronutrient content	Contains all i.e. nitrogen (N), phosphorus (P) and potassium (K) in sufficient quantities	Mostly contains only one (N in urea, P in SSP and K in MOP) or at the most two (N and P in DAP) nutrients in any one type of chemical fertilizer
Secondary nutrient content	Calcium (Ca), magnesium (Mg) and sulphur (S) is available in required quantities	Not available
Micronutrient content	Zinc (Zn), boron (B), manganese (Mn), iron (Fe), copper (Cu), molybdenum (Mo) and chlorine (Cl) also present	Not available
pH balancing	Helps in the control of soil pH and checks the acidity, salinity and alkalinity in soil	Disturb soil pH to create acidity, salinity and alkalinity conditions
EC correction	Helps in balancing the EC to improve plant nutrient absorption	Creates imbalance in soil EC affecting nutrients assimilation
Organic carbon	Very high organic carbon and humus contents improves soil characteristics	Not available
Moisture retention capacity	Increases moistures retention capacity of the soil	Reduces moisture retention capacity of the soil
Soil Texture	Improves soil texture for better aeration	Damages soil texture to reduce aeration
Beneficial bacteria and fungi	Very high biological life improves the soil fertility and productivity on sustainable basis	Reduces biological activities and thus the fertility is impaired
Plant growth hormones	Sufficient quantity helps in better growth and production	Not available

From the above table, the actual nutrient used and the cost incurred for various chemical fertilizers, it is clear that the chemical fertilizers are more expensive than compost. The use of chemicals has also been promoted by *large amounts of subsidies* that enabled them to be used as nutrients for crops.

Vermicomposting is an economical venture when it is taken up at the vicinity of the waste generating places. With the available resources in the field, vermicompost can be produced by the farmers itself and it has to be part of the organic farming system.

4. RECYCLING OF CROP RESIDUES AND AQUATIC WEEDS

Agricultural wastes like crop residues of vegetables; weeds like parthenium, water hyacinth; dry leaves of trees etc. can be recycled by converting them into compost with little effort or by way of mulch or direct incorporation in the soil.

Cereal straw and residues: On an average, cereal straw and residues on maturity contain about 0.5 per cent nitrogen, 0.6 per cent P_2O_5 and 1.5 per cent K_2O. The quantity of nutrients in legume residues is much higher than in cereal straw. Even if 50 per cent of these crop residues are utilised as animal feed, the rest should be mobilised for recycling for their plant nutrient potential and plants like mulching.

Water hyacinth: Water hyacinth is a free floating weeds plant which grows luxuriantly in ponds, lakes and water reservoirs. Water hyacinth can be used as soil mulch, green manure and compost. The plants can directly be ploughed in the soil and allowed to decompose for a month or so before sowing of a crop.

On an average, the fresh plant contains 95.5 per cent moisture, 3.5 per cent organic matter, 1.0 per cent phosphorus (P_2O_5) and 0.2 per cent potash (K_2O). This is good source of potash. It can be converted in to compost without additional source of nitrogen. Addition of small amount of soil will accelerate the process of composting of water hyacinth. It is good for crops like vegetables, potato, upland rice, maize and jute.

Forest- litter manure: The manurial value of forest litter is as good as farm compost. A portion of the surface litter should be removed in a regular manner without adversely affecting the natural regeneration of the forests.

Hedgerows: The contour-hedgerow-farming system is based on a modification of agroforestry system in which nitrogen-fixing hedgerow species like *Desmodium rensonii, Leucacephala spp., Tephrosia candida,* etc. are planted along contours with desired food crops and other useful plants in the alleys. The technology is environmentally sound and ensures adequate control of soil erosion, helps restore soil structure and fertility, conserves soil moisture and meets food, fodder, fuel and timber needs along with income generation. It can be applied easily by the upland farmers using local resources.

5. EDIBLE AND NON-EDIBLE OILCAKES

The manurial value of edible oilcakes like groundnut, rape-mustard, sesamum, linseed and castor and non-edible cakes such as neem, karanj, mahua, castor, etc. lies mainly in its nitrogen content. These cakes contains small quantities of P_2O_5 and K_2O. The nitrogen content varies from 3 to 9 per cent, depending on the nature of oilcakes. The C/N ratio is low ranging between 3 and 15 for different types of oilcakes. Due to low C/N ratio, its decomposition rate is faster than cereal and legume residues and other bulky organic manures. This nitrifies very quickly and about 60 to 80 per cent of its nitrogen is converted in available form within 2-3 months.

Mahua cake is poor in nitrogen and takes longer time to nitrify. It is better to apply mahua cake about 2 months in advance to soil before sowing of the crop. Oilcakes on economic grounds are not recommended for cereal crops. Neem cake and castor cake is also considered to be good vermicide against some pests. The efficiency of these oilcakes can be increased by moistening it and allowing it to ferment for some time to decompose the toxic substances.

6. BIOFERTILIZER

Microbial inoculants have attained special significance in organic agriculture. Biofertilizers are microbial inoculants consisting of living cells of bacteria, algae and fungi alone or in combination, which may help in increasing crop productivity by biological nitrogen fixation (BNF), solublising insoluble nutrients in soil and minerals, stimulating plant growth or in decomposition of plant residues. Biofertilizers are widely accepted as low cost supplements to chemical fertilizers and have no deleterious effects either on soil health or environment.

6.1 Type of Biofertilizers and their Utility

Depending upon the activity of mobilizing different nutrients, biofertilizers are classified as:

- o Nitrogen fixer
- o Phosphate solubilizer and mobilizer
- o Compost accelerators and enricher

6.1.1 Nitrogen fixer

The nitrogen fixing organisms can be broadly be grouped as symbiotic and non-symbiotic organisms based upon their nature of association with the plant. At present, Rhizobium, Azospirillum and Azotobacter are commonly used nitrogen fixing biofertilizers.

Rhizobium: Biological nitrogen fixation by legumes in association with Rhizobium is one of the most important means of converting large quantities of nitrogen in air and makes it available to plants. The bacteria infect the legume root and form root nodules (also form stem nodules in *Sesbania rostrata*) within which they reduce molecular nitrogen to ammonia which is readily utilised by the host plant in exchange of photosynthates to produce valuable proteins, vitamins and other nitrogen containing compounds. *Rhizobium*-Legume association can fix 50-300 kg N per hectare in one crop season and can leave behind substantial quantity of nitrogen in soil for the succeeding crop. The quantity of nitrogen fixed varies with the kind of legume and agro-ecological conditions available. The important point is the inocultation of efficient *Rhizobium* strain specific to each crop. Crops like groundnut, Bengal-gram, cowpea, green gram, cowpea, Bengal gram and soybean can leave behind 20-30 kg nitrogen per hectare for the succeeding crop, while the crops like lucern, pasture legumes and green manuring crops like sunhemp, sesbania, etc. add about 50-60 kg nitrogen per hectare to the soil. *Rhizobium* can meet more than 80 per cent of the needs of the legume crop and about 10 to 25 per cent increase in grain yield of pulses can be obtained in most of the cases. The response varies depending on soil conditions and effectiveness of native population. In order to derive greater benefit from legumes grown in rotations and as inter-crops with non-legumes, it is necessary that legume seeds are inoculated with specific *Rhizobium.*

Table 7.10: Cross inoculation group of rhizobium

Rhizobium species	Cross inoculation grouping	Legume types
Rhizobium		
R. leguminosarum	Pea group	Pisum, Vicia, Lens
R. phaseoli	Bean group	Phaseolus
R. trifolli	Clover group	Trifollium
R. meliloti	Alfalfa group	Melilotus, Medicago, Trigonella
Bradyrhizobium		
B. lupini	Lupini group	Lupinus, Orinthopus
B. japonicum	Soybean group	Glycine
B. sp	Cowpea group	Vigna, Arachis
Azorhizobium	Stem nodular	Sesbenia rostrata

The main objective of inoculating legume seeds or soil to be planted to legumes is to ensure sufficient viable N_2-fixation *Rhizobia* close to the germinating seed. Large number of viable effective Rhizobia in the rhizosphere of the developing seedling is needed to bring about effective nodulation. Different *Rhizobium spp.* has been placed in 7 cross inoculation groups based on the host plant specificity. These are Alfalfa (*R.meliloti*), Clover (*R. trifolii*), Pea (*R. leguminosarum*), bean (*R. Phaseoli*), Lupine (*R. lupini*), soybean (*R. japonicum*) and cowpea (*R.* spp.). Recently these groups have been placed in 3 main categories represented by *R. leguminosarum, R. meliloti* and *R. loti.* A new genus *Bradyrhizobium* has also been proposed to inciude *R. japonicum* and *R.* spp.

Carrier-based inoculum is prepared for crop inoculation. Rhizobia are first multiplied and then blended with the sterilized carrier materials such as lignite, peat, charcoal etc. A good quality inoculum contains 10^8 - 10^9 cells/g carrier materials and a self-life of over 4 months. Seed inoculation is the easiest way to inoculate the crop. For acid or alkaline soils, an additional coating of the seeds with lime or gypsum respectively is recommended to protect the bacterial cells.

Stem nodulating legumes: There are few legumes like *Sesbania, Aeschynomene and Neptunia* where nodules are found on the stem also and nitrogen fixation occurs. Unlike root nodules, nitrogen fixation in stem nodules is not affected by increased concentration of applied fertilizer nitrogen. Number of nitrogen fixing nodules can also be increased by repeated application of *Rhizobium* culture. *Sesbania rostrata* is known to fix as much as 250 kg nitrogen per hectare in just 50 days time in rice fields. Its high nitrogen fixing ability and succulent nature of the plant tissue at the flowering makes it a very ideal green manure crop for rice fields.

***Azotobacter*:** This is free living N fixing bacterium has been used as a bio-fertilizer for several non-leguminous crops including vegetables. An economy of 10 to 20 kg N per hectare has been obtained with *Azotobacter* application on rice, wheat, cotton, sorghum, brinjal, tomato, cabbage and other crops both under rainfed and irrigated conditions. Better effects are obtained in soils having adequate organic matter. Some of the *Azotobacter* species confer an additional advantage since they are known to produce antibiotics which suppress a variety of root pathogens. The species *A.chroococcum* has been found to be most potential to be used for the production of *Azotobacter* biofertilizer. *Azotobacter* biofertilizer can be applied to the crops like wheat, paddy, cotton, maize, barley and the vegetables. It contributes nearly 10% increase in grain yield and 12-15% increase in vegetative biomass yield.

Table 7.11: Standard Specifications for rhizobium and azotobacter

Parametrs (Base)	Rhizobium inoculant (RI) (Carrier based)	Azotobacter (AI) (Carrier based)
Cell number at the time of manufacture	10^8 per g carrier within 15 days of manufacture	10^7 per g carrier within 15 days of manufacture
Cell number at the time of expiry	10^7 per g carrier within 15 days before expiry date	10^6 per g carrier within 15 days before expiry date
Expiry Period	Six months from the date of manufacture	Six months from the date of manufacture
Permissible contamination	No contamination at 10^8 dilution	No contamination at 10^5 dilution
pH	6.0-7.5	6.0-7.5
Strain	Should be check serologically	Nothing specific, but Chroococcum species is mentioned
Carrier	Should be pass through 150-212 micron (72-100 mesh)	Should be pass through 106 micron
Others	Nodulation test +ve	Minimum amount of N-fixation not less than 10 mg per g of sucrose utilised.

Azospirillum: This is very important and very widely used biofertilizer. This bacteria not only fixes the atmospheric nitrogen but also secretes growth promoting substances like IAA, GA, cytokinins and vitamins which have beneficial effects on seedling vigour and prolonged photo-synthetic activity. The important feature is that it will have intimate association with the roots of cereals and grasses. The wide adaptability of Azospirillum to varied temperature, soil pH and wide host range make this as a widely used biofertilizer throughout the world. The increase in the yield ranges between 10-30%.

Method of using *Azotobacter* and *Azospirillum*:

- Seed treatment
- Seedling root dip treatment
- Vegetative propagate dipping treatment
- Soil treatment

***Acetobacter*:** This biofertilizer is a quite recently introduced biofertilizer for nitrogen mobilization in sugarcane, sweet potato and sweet sorghum. *Acetobacter diazotrophicus* is the most widely used species in preparing the biofertilizer. It has a potential of fixing as high as 300kg N/ha/year. The bacterium is an endophyte and therefore for optimum result the inoculum is applied to the freshly cut seed sets as sets dip treatment.

***Blue Green Algae (BGA)*:** BGA are photosynthetic prokaryotes capable of fixing atmospheric nitrogen particularly under waterlogged rice fields. They are known to contribute about 25 to 30 kg of N per hectare. Although BGA are widely distributed in different ecosystems, waterlogged rice fields provide the ideal conditions for their growth and multiplication. Besides nitrogen fixation, BGA are known to produce growth promoting substances and vitamin (B-12) which benefits rice plants. Algalisation improves the soil aggregation, water holding capacity of the soils and soil aeration. Algae add considerable amount of easily decomposable biomass to the soil and liberate organic acids and other compounds which have chelating properties.

Generally, 4-10 kg of dried BGA inoculum is required to inoculate one acre of paddy field. BGA increase 7-32% grain yield over their corresponding control. It secrets certain growth promoting substances and vitamins. In the main field, algal inoculum is broadcast over standing water at the rate of 10 kg per hectare, one week after transplanting rice seedlings.

Cyanobacteria (Blue Green Algae) like *Nostoc, Anabaena and Tolypothrix* can be multiplied by the farmers in plots of 20m x 1m x 22 cm lined with polythene sheet and pH neutralised by adding lime. Starter culture is inoculated in the plots with water. A thick layer of algal growth which develops in 15 days is allowed to dry, flakes collected and used as inoculum. About 10 kg of algal inoculum can be obtained from a plot of 20 m^2.

Azolla: Azolla is a floating fresh-water fern and is distributed in low land fields and fresh water bodies. It is a genus of small, delicate, heterosporous water fern harbours a nitrogen fixing algae *Anabaena azollae,* at all the stages of its growth and development. The dorsal lobes of the minute leaves of the plant have closed cavities on their ventral side, which house the N fixing cyanobiont *Anabaena azollae.*

It is capable of growing in shallow water systems and under ideal conditions, it puts on prolific growth producing enormous quantities of biomass. It can contribute 100 to 150 kg N per hectare per year. It also adds organic matter and thus improves the physico-chemical properties of soil.

Azolla is grown in 1m^2 plots, 3 weeks before transplanting rice. Inoculation is done at the rate of 100 g per m^2 (1 q per ha). *Azolla* multiplies 100 folds without additional efforts under specific soil and environmental conditions. *Azolla* is added to the main field at the rate of 10 tonnes fresh material per hectare and incorporated into soil before transplanting rice. The other method includes growing *Azolla* along with standing crop of rice as dual crop. The incorporation method is better.

Phosphate solubilisers: Most of tropical and sub-tropical soils are deficient in phosphorus due to its fixation by calcium, iron or aluminium and no natural channels exist for the return of large annual net losses. Hence the supply is inevitably shrinking and the deposits are limited. In this context, the release of insoluble phosphates in soils and fixed phosphorus in the clay minerals by microorganisms assume significance.

A large number of microorganisms belonging to diverse families and genera and of heterotrophic to autotrophic in nature are known to have the ability to solubalise the fixed or insoluble phosphate present in the soil solution and make it available to plants. Out of these, *Pseudomonas striata, Bacillus polymyxa, Aspergillus awamorii, A. niger* and *Penicillium digitatum* have been known to solubalise unavailable form of P to available form. The solublisation effect is generally due to the production of organic acids by these organisms. They are also known to produce amino acids, vitamins and growth promoting substances like IAA and GA which help in better growth of plants.

This biofertilizer can be applied in the crops like oat, wheat, rice, potato, gram, lentil, mustard, chickpea, barley, soybean, groundnut, sugar beet, cabbage and tomato. Seed or soil inoculations with phosphate solubalising organisms are known to increase the growth and yield to the extent of 10 to 20 per cent. However, concerted efforts are needed to identify more efficient organisms either by selection or increase their efficiency by suitable soil amendments and combined inoculation with other organisms like mycorrhizae and nitrogen fixers. The P solubalising bacteria can be mass multiplied and mixed with the carrier material used.

Phosphate mobilising mycorrhizal fungi: There are certain fungi, which form symbiotic association with the roots of plants and help in the uptake of phosphorus. Ectomycorrhizal fungi like *Pisolithus, Laccaria, Amanita, Scleroderma, Russula, Tricholomu* form association with tree species belonging to family *Pinaceae, Betulacea* and *Fagceae.* These fungi increase the surface area of absorption of the roots and thus help in absorption of nutrients, especially those which are less mobile in soil solution like P. They also help in the uptake of water and protect roots from root pathogens.

Vesiculur arbuscular mycorrhizal (VAM): It is an endo-symbiont fungi, ubiquitous in nature colonizes roots of several crop plants, important in agriculture, horticulture and tropical forestry. They form appresorium inside the intercellular spaces in the roots and other end of the hyphae extended towards soil system and thereby covering more soil spaces available for nutrient absorption. They are zygomycetous fungi belonging to the genera *Gigaspora, Glomus, Acaulospora, Enthrpphopora, Sclerocystis,* etc. These fungi extract plant essential nutrients from the soil and make it available to the roots. They help plant growth through efficient absorption of phosphorus from large volume of soil through their profusely ramified mycelia, in addition to absorption of water and some micronutrients and protect the roots against pathogens.

These fungi are obligate symboints and can not be cultured on laboratory media. They can be multiplied as pot cultures, using a suitable host and the root pieces along with the substrate is used as inoculum. They have currently recommended for use in transplanted and nursery raised crops because of difficulty in inoculum production. They are currently recommended for use in transplanted and nursery raised crops. They improve seedling growth and vigour of many plants important in agriculture and horticulture like tomato, chillies, citrus, cardamom, mango, finger millet, and forestry like *Leucaena, Dalbergia, Acacia, Casurina, Tectonneea,* etc. VAM fungi not only improves plant yield but also saves P fertilizers to the tune of 25 to 30 per cent. Combined inoculation of *Rhizobium* and VAM could meet more than 80 per cent of the N requirement of pulse crops.

Potash mobilising biofertilizer: *Frateuria aurentia* is a free-living bacterium which helps in mobilising potassium in soil. It can tolerate a wide range of pH (3.5-8.5) conditions. The bacterium solublises the potassium found in the crystalline structure of primary minerals in soil. The mobilising power has been reported to be so high that it can fulfil up to 50 to 60 per cent of potassium requirement of the crop. The bacterium has been isolated from banana plants and is recommended for use in other crop plants. The organism has been reported to increase maximum level of potassium availability to the magnitude of 5-60 kg K per hectare.

6.2 Use of Biofertilizer as Compost Accelerator/Enricher

Decomposition or composting is essentially a microbiological process accomplished by the combined activity of bacteria, fungi, actinomycetes and protozoa. Compost accelerators are needed to decompose the lignin and cellulose of the waste materials. Species of *Trichoderma, Penicillium, Aspergillus, Trichurus* and *Paecilliomyces* are the compost accelerators.

To improve the nutritional status of compost, biofertilizers like phosphorus solublisers like *Aspergillus niger* and *A. awamorii* when used alone with rock phosphate, solubilise insoluble P and make the compost rich in available phosphorus. Addition of nitrogen fixer such as *Azotobacter* leads to the fixation of appreciable quantities of atmospheric nitrogen.

Precautions to be taken while purchasing biofertilizers

- Ensure that each purchased packet of biofertilizer is provided with necessary information like name of the product, name of the crop for which intended, date of manufacture, date of expiry and instruction for use
- Use the packet before expiry on the specified crop.
- Store the biofertilizer packets in cool and dry place away from direct sun and heat.
- Never mix biofertilizer treated seeds with other organic fertilizers.

Economics of using biofertilizer

- Saving of 40-50 kg nitrogen fertilizer per hectare.
- One tonne of *Rhizobium* inoculants is equivalent to 100 tonnes of nitrogen considering minimum fixation of 50 kg nitrogen per hectare from 0.5 kg per hectare application dose.
- One tonne of *Azotobacter* and *Azospirillum* each is equivalent to 40 tonnes of nitrogen considering maximum fixation of 20 kg nitrogen per hectare from 0.5 kg per hectare application dose.
- One tonne *Blue Green Algae* (*BGA*) is equivalent to 2 tonnes of nitrogen considering minimum fixation of 20 kg per hectare from 10 kg *BGA* per hectare application dose.
- One tonne of *Phosphate Solublisers* (*PSM*) are equivalent to 24 tonnes phosphorus pentoide considering 30 per cent solublisation at the minimum dose of 40 kg P_2O_5 per hectare (through Rock Phosphate) with 0.5 kg PSM per hectare application dose.
- Biofertilizer can also improve 10 to 40 per cent of grain yield and 15 to 30 per cent in vegetative growth.

Prospects of biofertiliser in organic farming

- Biofertilizers are harmless, improve the soil properties like C/N ratio, soil texture, structure, water holding capacity, etc. and maintain soil fertility.
- Secrete growth promoting substances, hormones and vitamins. They help in better nutrient uptake and increased tolerance towards drought and moisture stress.
- Confer disease resistance in crops with the help of fungistatic and antibiotic like substances.
- Solublise the insoluble phosphate and help in decomposing plant residues in soil.

- In composting processes, improve the quality and nutrient value of the compost.
- Improve seed germination and induces synchronous flowering and fruiting.
- The wasteland and lowland can be enriched and productivity may be improved by the application of biofertilizers.
- The economic benefits to cost ratio of biofertilizer is always higher so their use is acceptable in all places of cultivation.
- Biofertilizers have been found for a wide variety of crops such as cereals, legumes, millets, oilseeds, spices, vegetables, fruits, plantation crops.
- Biofertilizers are very useful for organic farming in dryland and rainfed conditions.

Note: *Biofertiliser must NOT carry Genetic Modified Organism (GMO).*

7. EFFECTIVE MICROORGANISM (EM) TECHNOLOGY

The technology behind the concept of Effective Microorganisms and its practical application was developed by Professor Teruo Higa at the University of the Ryukyus in Okinawa, Japan. He has found microorganisms that can coexist in mixed cultures and are physiologically compatible with one another. When these cultures are introduced into the natural environment, their individual beneficial effects are greatly magnified in a synergistic fashion.

Effective Microorganisms or EM is a mixed culture of beneficial microorganisms (primarily photosynthetic and lactic acid bacteria, yeast, actinomycetes and fermenting fungi) that can be applied as inoculants to increase the microbial diversity of soils. This in turn, can improve soil quality and health, which enhances the growth, yield, and quality of crops. EM cultures do not contain any genetically modified microorganisms.

The following are some of the beneficial influences of EM:

- Promotes germination, flowering, fruiting and ripening in plants.
- Improves physical, chemical and biological environments of the soil and suppresses soil borne pathogens and pests.
- Enhances the photosynthetic capacity of crops.
- Ensures better germination and plant establishment
- Increases the efficacy of organic matter as fertilizers.

7.1 Preparation of EM

EM 1: Original EM 1 is yellow-brown liquid with a pleasant odour and sweet-sour taste and dormant. The pH of EM 1 should be below 3.5. If it has a bad smell or foul odour or pH is more than 4.0, the EM1 has deteriorated. It should not be used.

Original EM1 is dormant, thus it needs to be activated by the provision of 'water' and 'food'. This is done by adding water and molasses.

1. 1 litre (1000 cc) of water
2. 1 cc of EM1
3. 1 cc of Molasses or 1g of any sugars

This solution is left for 2-24 hours and sprayed to plants, soil or organic matter. Use EM diluted solution (0.1%) to apply to crops.

Fermented organic matter (Bokashi): *Bokashi* is a Japanese word which means "Fermented organic matter". It is made by fermenting organic matter (rice bran, oil cake, fish meal etc.) with EM. *Bokashi* is normally found as a powder or as granules. It can be used as soil amendments to increase the microbial diversity of soils and supply nutrients to crops. However, *EM Bokashi* is fermented organic matter using EM instead of forest or mountain soil. Thus, *EM Bokashi* is an important additive to increase effective microorganisms in the soil.

Bokashi is equivalent to compost, but it is prepared by fermenting organic matter with EM. It can be used 3 to 14 days after treatment (fermentation). *Bokashi* can be used for crop production even though the organic matter has not decomposed as in compost. When *Bokashi* is applied to soil, organic matter can be utilized as a feed for effective microorganisms to breed in the soil, as well as supplying nourishment to crops. Bokashi is classified as "Aerobic Bokashi" and "Anaerobic Bokashi" based on the manufacturing process. In Aerobic type, fermentation period is shorter than in the anaerobic type. Energy of organic matter is lost, if temperatures during fermentation are uncontrolled. In Japan, the anaerobic type is popular, but in Thailand the aerobic type is widely used.

The materials which can be used as organic matter in preparing *Bokashi* are rice bran, corn bran, wheat bran, maize flour, rice husk, bean husk, rice straw, oil cake, cotton seed cake, pressmud, bagasse, chopped weeds, sawdust, coconut fibre and husks. Crop residues such as empty fruit bunch in oil palm, fish meal, bone meal, dung of any animals, kitchen garbage, sea weed, crab shells and similar material.

However, rice bran is recommended as an important ingredient of Bokashi, as it contains excellent nutrients for microorganisms. It is desirable to combine organic matter which has low and high C/N ratios. Generally the use of at least three types of organic matter is recommended in order to increase microbial diversity. Adding wood or rice husk charcoal, zeolite, kelp, grass and wood ash to Bokashi is desirable. These porous materials improve soil physical conditions and nutrient holding capacity. They also act as harbouring points for effective microorganisms.

There are many type of *Bokashi*, depending on the organic matter used. The preparation of a typical *Bokashi* is as follows:

1. Rice bran : 100 litres (volume)
2. Oil cake : 25 litres
3. Fish meal or Chicken dung : 25 litres
4. EM1 : 150 cc
5. Molasses : 150 cc
6. Water : 15 litres

Note: If molasses are not available, any kind of sugar can be used. Some materials that can be used are raw cane sugar, juice of any fruits and wastewater of alcohol industries. The quantity of water is a guideline. The quantum of water that needs to be added will depend on the moisture content of the materials used. The ideal quantum of water is that required to moisten the material, without drainage.

Anaerobic-type

1. Mix rice bran, oil cake and fish meal well.
2. Dissolve molasses in the water (1:100). It is easily dissolved in warm water.
3. Add EM into the above prepared molasses solution.
4. Pour the EM mixture gradually onto the organic matter and mix well while checking the moisture content. There should be no drainage of excess water. The moisture content should be about 30-40 per cent.
5. Put the mixture thus made into a bag that does not permit air movement (e.g. paper or polyethylene bag). This is placed within another polyethylene bag (black vinyl) to prevent movement of air. Close the bag tightly to maintain an anaerobic condition. This is placed away from direct sunlight.
6. The fermenting period is:

Temperate zone	• In summer more than 3-4 days. • In winter more than 7-8 days (keep it in warm place).
Tropics	• More than 3-4 days.

7. Anaerobic Bokashi should be used soon after preparation. If storage is required, spread it on a concrete floor, dry well in the shade and then put into vinyl bag. Care should be taken to prevent rodent or other pest attacks.

Aerobic-type

1. Follow the procedure from Sl no. 1 to 5 as described above in *anaerobic type* preparation.
2. Put the mixture made above on a concrete floor, and cover with gunny bag, straw mat or similar material. Avoid exposure of this material to rain.
3. Under aerobic conditions, Bokashi ferments rapidly. Thus the temperature increases. Ideally, the temperature should be kept around 35-45°C. Thus,

please check temperature regularly using a normal thermometer. If the temperature rises beyond 50 °C, mix the Bokashi well to aerate it.

4. The fermenting period is same as given in case of anaerobic type.
5. It is ready for use when it gives a sweet fermented smell and white mould is observed. If it has a sour and rotten smell, it is failure.
6. This *Bokashi* is best used soon after preparation. If storage is required, spread it on a concrete floor, dry well in the shade and then put into vinyl bag. Please prevent rodent or other pest attacks.

The efficacy of *Bokashi* made at temperatures above 50°C is 50 per cent lower than that made at a lower temperature. This is due to the loss of heat energy at high temperatures. The key of preparing good Bokashi is to know suitable moisture content and temperature of Bokashi through practice. A Practice is required for preparation of *Bokashi* several times.

7.2 How to Apply EM?

Basically, EM for plant nutrition can be applied in two ways, namely as EM1 stock solution and EM Bokashi.

EM1 Stock Solution: EM1 stock solution can be applied by:

1) Watering into the soil (by watering cans, sprinklers or irrigation systems).
2) Spray onto plants (foliar spray) by sprayer or watering can.

EM Bokashi (EM fermented organic matters): Apply Bokashi 200g per 1 square meter on the topsoil, when enough organic matters have been applied. It can be applied more (maximum 1 kg per 1 square meter), when soil is poor or has little organic matter.

7.3 What Precautions to be taken?

(1) *EM is a living thing*: EM is completely different from chemical fertilizers or agrochemicals. EM does not work when applied in the same method as chemical fertilizers or agrochemicals. It is important to note that EM increases population of beneficial microbes in the soil.

(2) *Use good quality water*: It is important to use good quality water when watering crops, diluting EM1 and preparing Bokashi. Using polluted water (high BOD, low DO) causes infection of pests and diseases, reduction of yields and crop quality.

(3) *Storage of diluted solution*: It is desirable to utilize diluted EM1 solutions within 3 days.

(4) *Storage information*: Store of EM1 - up to 6 months in a closed container, in a cool and dark place, (Please do not store in refrigerator).

(5) *Check smell if in doubt*: EM1 always has a sweet and sour smell. If smell is foul, do not use it. After the cap of bottle is opened and air comes in, a white

membrane (yeast) may be observed on the surface of EM1.

The desired effects from applying effective microorganisms to soils can be somewhat variable, at least initially. In some soils, a single application (i.e., inoculation) may be enough to produce the expected results, while for other soils even repeated applications may appear to be ineffective. The reason for this is that in some soils it takes longer for the introduced microorganisms to adapt to a new set of ecological and environmental conditions and to become well-established as a stable, effective and predominant part of the indigenous soil microflora. The important consideration here is the careful selection of a mixed culture of compatible, effective microorganisms properly cultured and provided with acceptable organic substrates. Assuming that repeated applications are made at regular intervals during the first cropping season, there is a very high probability that the desired results will be achieved.

There are no meaningful or reliable tests for monitoring the establishment of mixed cultures of beneficial and effective microorganisms after application to a soil. The desired effects appear only after they are established and become dominant, and remain stable and active in the soil. The inoculum densities of the mixed cultures and the frequency of application serve only as guidelines to enhance the probability of early establishment. Repeated applications, especially during the first cropping season, can markedly facilitate early establishment of the introduced effective microorganisms.

Once the "new" microflora is established and stabilized, the desired effects will continue indefinitely and no further applications are necessary unless organic amendments cease to be applied, or the soil is subjected to severe drought or flooding.

Note: *Address for procurement of EM Stock Solution is given in appendices.*

8. LIVING FERTILE FARM

Organic farming is a systems approach where the farm is viewed as a living whole, in which each farm activity affects the others. Both raw and composted manures and biodynamic preparation are useful in organic crop production. An attention to balancing soil fertility, manures can supply all or most needs for purchased fertilizer, especially when combined with a whole-system fertility plan that includes crop rotation and cover cropping with nitrogen-fixing legumes.

Management is based on the farmer's own careful observations, plus the results of tests and analyses. The grower needs to monitor nutrients in the soil via soil testing, and learn the characteristics of the manure and/or compost to be used. The grower can then adjust the rates and select additional fertilizers and amendments accordingly. This leads to a modern approach in which traditional knowledge finds a renewal.

Because organic farming practices uses very limited external inputs, and reuses most on farm waste, it has a low impact on the environment. It provides an economical way of farming in which most of the costs are met at the time they are incurred. It thus offers a solution to conflicts between economics and the environment. A further benefit is the quality of the produce. Flavour and keeping quality of the foods, lustre and comfort of fibres provoke favourable comment from consumers and buyers.

Chapter **8**

Bio-control of Pests and Diseases

Pest and disease control which is not dependent on the use of chemical require a recognition that there is no single factor which is responsible for a pest or disease problem and rely on a range of husbandry practices which promote stability and balance between crops and their pests. In particular, an ecological approach should seek to enhance the activities of natural enemies of crop pests including other insects and animals such as hedgehogs and birds, as well as fungal, bacterial and viral pathogens.

The diversity and stability of insects-pests and their predators can be influenced by:

- o The diversity of plant species and structure within the field, including temporal and spatial arrangements
- o The composition, management and permanence of surrounding plant communities
- o The soil type and surrounding environment
- o The distance of the crop from sources of colonists
- o The permanence of crops and the time available for colonization
- o The complexity of tropic relationships between crop and non-crop plants, herbivores and natural enemies, etc.

Ideally, insecticides should reduce pest populations, be target-specific (kill the pest but not other organisms), break down quickly, and have low toxicity to humans and other mammals. Some synthetic insecticides (e.g., chlorinated hydrocarbons, organophosphates, and pyrethroids) leave unwanted residues in food, water, and the environment. Some are suspected carcinogens, and low doses of many insecticides are toxic to mammals. As a result, many people are looking for less hazardous alternatives to conventional synthetic insecticides.

1. WHAT DO THE ORGANIC STANDARDS SAY?

Use of synthetic/chemical pesticides, fungicides and weedicides is prohibited. Natural enemies should be encouraged and protected (for example, raising trees in the farm attracts birds which kills pests of the crops, nest construction, etc.). Products collected from the local farm, animals, plants and microorganisms and prepared at the farm are allowed for control of pests and diseases. (e.g. NSKE-neem seed kernel extract, cow urine spray). Use of genetically engineered organisms and products are prohibited for controlling pests and diseases. Similarly, use of synthetic growth regulators is not permitted. Weeds under the base of the plants should be cleaned and put as mulch around the plant base. The weeded materials should be applied as mulch in the ground itself.

1.1 Permitted Products for Control of Pest and Diseases

- Neem oil and other neem preparations
- Mechanical traps
- Plant-based repellents
- Soft soap
- Clay
- Physical method of weed control
- Thermic method of noxious weed control

The following products can be used when they are absolutely necessary and taking environmental impact into consideration. The certification body should be consulted before using these inputs.

- Copper salts (maximum 8 kg copper/ha/yr) e.g. Bordeaux mixture
- Plant and animal preparations e.g. Cow urine spray, Garlic extract, Chilli extract
- Light mineral oils e.g. Kerosene
- Release of parasite predators of insect pests e.g. *Trichodramma*
- Sulphur
- Viral, fungal and bacterial preparations

1.2 Contamination Control

It is necessary to take some measures to minimise the contamination from outside and within farm.

- If neighbouring fields are non-organic, a buffer zone must be maintained.
- If the farm is under conversion, equipments used for conventional areas should be well cleaned before using for organic areas.
- Products based on polythene, polypropylene and other polycarbonates are allowed to cover protected structure, insect netting, nursery, drying, etc. subject to the condition that these materials should be removed from the field after use and they should not burnt or put in the soil. Use of polychloride based products like PVC pipe is prohibited.
- Farm equipments from conventional farming should be free from residues and clean.

2. INTEGRATION OF CONTROL MEASURES

Pest-management measures are either preventive or therapeutic. Prevention relies on an intimate understanding of the pest life cycle, behaviour and ecology. It involves natural enemies, host resistance and cultural practices. Therapeutic measures are applied as a correction to the system when necessary. Actual integration involves proper choice of compatible tactics and blending them so that each component complements the other.

The following control methods could be effectively combined according to the need of the situation and the pest species involved.

- Deep summer ploughing
- Selection of tolerant varieties resistant to diseases and pests
- Seed treatment
- Growing intercrop and trap crop
- Pheromone traps to monitor pest build up
- Use of light and sticky traps
- Using bio-pesticides
- Using microbiological NPV
- Releasing and conserving parasites and predators
- Neem seed kernel extraction (NSKE) suspension 5%
- Chilli-garlic extract
- Cattle-dung and urine extract

2.1 Cultural Methods

Cultural control is generally preventive in nature and hence need to be carried out at appropriate time. Cultural control is the cheapest of all control measures. However, this technique is indirect and intangible. The following practices can be gainfully adopted in different cropping systems to suppress certain pest species.

Tillage operation: Ploughing or hoeing helps to expose stages of soil inhibiting insects to sun or to the predatory birds. Earthing up of soil in sugarcane reduces seedling borer infestation. Scraping of bunds in paddy fields destroys eggs of grasshoppers.

Field and plant sanitation: Regular removal of weeds, pest-affected parts of the plants, crop stubbles and their destruction will eliminate the sources of infestation of the pest. Destruction of bored shoots and fruits of brinjal prevents further build up of the pest population. Many virus diseases like leaf curl, bud and stem necrosis in crops like tomato, groundnut, sunflower can be minimised by uprooting the infected plants.

Crop rotation: Growing of a non-host crop after a host crop of the pest will break the breeding cycle of pest species and reduce their populations. Like wise, crop rotation prevents the build up of plant pathogen in soil.

Growing pest resistant varieties: Certain varieties of crops are less damaged or less infested than others by insects. The resistant varieties have physical and physiological features, which enable them to avoid or withstand pest attacks.

Trap cropping: The growing of susceptible or preferred hosts as trap crops along with the main crop attract the pests in large numbers and these can be selectively killed. This minimise the damage on main crop. For example, marigold is grown as trap crop for tomato fruit borer and mustard as trap crop for cabbage pests. In cotton, ladies finger can be grown as a trap crop.

Plants more preferred by the pest for egg-laying and feeding are grown as trap crops on the bunds of the man crop or at random.

Crops	Pest	Trap crop
Cotton, Groundnut	Spodoptera	Castor, Sunflower
Cotton, Chickpea	Helicoverpa	Marigold
Pigeon pea	Helicoverpa	Marigold
Groundnut	Spodoptera	Castor
Cotton	Spotted bollworm	Bhendi

Suggestion to be followed:

- Removal and destruction of egg masses and small caterpillars fram trap crop.
- Spraying NPV on trap crops.

Adjusting the time of sowing: The simultaneous sowing of crops in a locality helps in reducing pest damage. Many a times early sown crops escape pest attack. Jowar crop sown before the end of June month usually escapes attack by shoot fly.

Wider spacing: Wider spacing between rows and plants, and lesser crop density often prevent build up of populations of certain pests and also incidence of certain diseases. Wider spacing reduces incidence of brown plant hopper in rice and woolly aphid in sugarcane. In tomato, the spacing of more than 1m between rows and 75 cm between plants reduces disease incidence and its spread.

Growing of barrier crops: Maize is traditionally used by farmers around many vegetable crops. This acts as a barrier for pest and vector species.

Water management: Draining of water for a few days in paddy fields suppress brown plant hopper population. Flooding of fields wherever possible kills root grubs and termites.

2.2 Mechanical Methods

Hand picking of egg masses, gregarious larvae and sluggish adults, and their destruction helps in reducing pest populations in certain situations. Grown larvae of *Helicoverpa* sp. and Red headed hairy caterpillar can be hand collected and destroyed. Fixing of tin bands over coconut trunks prevents damage by rats. In cabbage, scouting and mechanical destruction of gregarious early instar larvae of tobacco caterpillar and leaf webber is very effective.

Pheromones and light traps: Sex pheromones are complex organic compounds emitted by insects, mostly females to attract the opposite sex for mating. Such chemicals compounds, specific to each species, can be synthesized in the laboratory and made available as lures for use in traps. These pheromones attract the insects in large numbers to the traps where they get trapped and killed. Female sex pheromones are available for all these species of cotton bollworms like *Helicoverpa armigera, Sapodoptera litura, Pectinophora gossypiella, Earias vittella* and *Earias insulana* as also for rice stem borer, diamondback moth, apple codling moth etc.

Sex pheromones are helpful in the early detection of pests and their timely control. Mass trapping of males results in depriving the females of mating opportunities, thereby affecting reproduction. Pheromone is non-toxic and species specific. It is safe to other organisms, plants and environment.

To manage brinjal shoot and fruit borer, 50-60 pheromone traps per hectare can be erected. Lures like methyl eugenol, cue-lure can be used to suppress fruit flies. Red palm weevil on coconut can be controlled by using sex pheromone traps.

Ready-to-use sex pheromone lures and traps are commercially available. These may be installed in the field at 30 to 60 cms above the crop canopy. Generally, at least 5 traps are recommended for monitoring every hectare for each pest species. More traps may be deployed for mass-trapping. The traps may be set up when the crop is 30 to 40 days old. A lure has to be replaced in the field once in 15 days or as recommended by the suppliers. However, in sealed packets the lures can be stored in the refrigerator for about 12 months. The lure, being species specific, is different for each pest species. When the traps fitted with lures are deployed in the field, the moths of the cotton bollworms or concerned pests get attracted to their respective lures and fall into the traps.

Yellow sticky traps: Yellow sticky traps can be used to monitor aphids and whitefly. In cabbage and cauliflower, light traps are useful in reducing diamond back moth incidence. A bulb (100 watts) can be fixed by hanging in front of a thin gunny bag or polyethylene sheet (3'X4') smeared with oil or grease. A bucket with water can also be kept below the bulb. About 5-8 light traps are to be set up per hectare.

Nylon net: Cultivation of vegetables and other crops under nylon net drastically reduces the pest populations.

Preparation of Bordeaux mixture

Dissolve 1 kg of copper sulphate in 10 litres of water. In another vessel, slake 1 kg of quicklime by adding small quantity of water preferably warm water (1.25 kg of lime can be taken if the lime is not of good quality). When slaking is over, add 5 litres of water and stir well to get a uniform suspension of lime. Transfer the lime suspension through a sieve to a vessel containing 85 litres of water and stir well. A small quantity of lime solution may be kept separately.

Add 10 litres of the copper sulphate solution to the 90 litres of lime solution with constant stirring. To test the correctness of the mixture, dip a brightened iron knife for a minute in the mixture. If the knife remains bright, the mixture is correctly prepared. If the knife turns rusty brown or if its brightness is lost, add more lime suspension. Correctly prepared Bordeaux mixture will turn red litmus to blue and turmeric powder to orange red in colour.

Important points:

1. For dissolving copper sulphate or Bordeaux mixture, use copper, wooden or earthenware or plastic pots or drums.
2. Use fresh quicklime.
3. Bordeaux mixture should be passed through a sieve before transferring to the sprayers.
4. Spraying of Bordeaux mixture should be done on the same day of preparation.

3. BIOPESTICIDES AND OTHER NATURAL PRODUCTS

The pesticidal compounds derived from living organisms are biopesticides. These products are more selective, eco-friendly and leave no toxic residues in the environment unlike synthetic insecticides. Many botanical insecticides have been known and used for hundreds of years, but were displaced from the marketplace by synthetic insecticides. Botanical insecticides have different chemical structures and modes of action.

Some general traits of botanicals and other natural products include the following:

Fast breakdown: Botanicals degrade rapidly in sunlight, air, and moisture, and by detoxification enzymes. Rapid breakdown means less persistence and reduced risks to non-target organisms. However, precise timing and/or more frequent applications may be necessary.

Fast action: Although death may not occur for hours or days, insects may be immediately paralysed or stop feeding.

Toxicity: Most botanicals have low to moderate mammalian toxicity, but there are exceptions (e.g., nicotine). Even though botanicals are naturally derived and are relatively safe if used properly, they are nevertheless poisons and should be handled with the same caution as synthetic insecticides. All products must be used according to the label on the product container. They are most effective when used in an integrated pest management (IPM) program, which includes sanitation, cultural practices, mechanical controls, use of resistant plant varieties, and biological control.

Synergism: Some botanicals quickly break down or are metabolised by enzymes inside bodies of their target pests. Breakdown may occur rapidly, so that the insecticide only temporarily stuns the insect, but does not kill it. A synergist may be added to a compound to inhibit certain detoxification enzymes in insects. This enhances the insecticidal action of the product. Synergists are low in toxicity, have little or no inherent insecticidal properties, and have very short residual activity.

Pyrethrins are often mixed with a synergist such as piperonyl butoxide (PBO), MGK 264, rotenone, or ryania to increase their effectiveness. PBO and MGK 264, however, should not be mixed with lime or soap solutions because of accelerated breakdown. PBO has also been implicated as a carcinogen, and may not be used in some organic certification programs.

Selectivity: The rapid break down and fast action make botanicals more selective to certain plant-feeding pests and less harmful to beneficial insects.

Phytotoxicity: Most botanicals are not phytotoxic (toxic to plants). However, insecticidal soaps, sulphur, and nicotine sulphate may be toxic to some vegetables or ornamentals.

3.1 Botanical Insecticides

Limonene and Linalool: Citrus oils are extracted from oranges and other citrus fruit peels and refined to make the compounds d-limonene and linalool. Both compounds are generally regarded as safe, and are used extensively as flavourings and scents in foods, cosmetics, soaps, and perfumes.

Limonene and linalool are contact poisons (nerve toxins) and may have some fumigant activity against fleas. They have low oral and dermal toxicities. Both compounds evaporate readily from treated surfaces and have no residual. They are effective against all external pests of pets, including fleas, lice, mites, and ticks. Commercial products (usually called "d-Limonene") are available as sprays, aerosols, shampoos, and dips for pets.

Neem: Neem or neem oil is extracted from the seeds of the neem tree, *Azadirachta indica,* a native of India. The neem tree supplies at least two compounds with insecticidal activity (azadirachtin and salannin), and other unknown compounds with fungicidal activity. Azadirachtin acts as an insect feeding deterrent and growth regulator. The treated insect usually cannot moult to its next life stage and dies. It acts as a repellent when applied to a plant and does not produce a quick knockdown and kill. It has low mammalian toxicity and does not cause skin irritation in most formulations.

Neem has some systemic activity in plants. Currently registered products for ornamental pest control claim activity against a variety of sucking and chewing insects. Neem is most effective against actively growing immature insects.

Neem has been evaluated against more than 500 species of insects at global level and most of these insects have been found to be susceptible. Neem products give good control of grass hoppers, leaf hoppers, plant hoppers, leaf minors, whiteflies, scales, mealy bugs, caterpillars such as diamondback moth on cole crops and gram pod borer on red gram, cotton, tomato etc. Neem kernel powder at 1 to 2 per cent gives good control of pulse beetle and other storage pests. Neem oil is used to control powdery mildew.

Neem and Pongamia cakes suppress nematodes. Soil application of neem cake (250 kg/ha) at planting and repeated at flowering reduces leaf minor and fruit borer in tomato, shoot and fruit borer in brinjal, leaf minor and fruit fly in cucurbits, pod borers in vegetable pigeon pea, dolichos lab lab, etc. coconut mite infestation can be reduced by tying cloth bags containing neem cake in crown region. Cake should be replaced at 45 days interval.

Pyrethrum/Pyrethrins: Pyrethrins are highly concentrated active compounds which are extracted from the daisy-like flower of *Chrysanthemum cinerariaefolium.* When the flower is ground into a powder, the product is called a pyrethrum. Pyrethrum is the most widely used botanical insecticide. Synthetic insecticides that mimic the action of the pyrethrins are known as pyrethroids (e.g., bifenthrin, cyfluthrin, and permethrin).

Most insects are highly susceptible to low concentrations of pyrethrins. The toxins cause immediate knockdown or paralysis on contact, but insects often metabolise them and recover. They act specifically by disrupting the sodium and potassium ion exchange process in insect nerve fibres and interrupting the normal transmission of nerve impulses. Pyrethrins break down quickly; have a short residual, and low mammalian toxicity, making them among the safest insecticides in use. However, people may have allergic skin reactions and cats are highly susceptible to poisoning (e.g., flea powder).

Pyrethrins may be used against a broad range of pests including ants, aphids, roaches, fleas, flies, and ticks. They are available in dusts, sprays, and aerosol. Trade names include Pyrenone and Pyrellin. The use of Pyrethrum/Pyrethrins should be done in consultation with certification body due to its restricted use in organic farming.

Rotenone: Rotenone is extracted from the roots of two tropical legumes, Lonchocarpus and Derris. It inhibits the conversion of nutrients into energy at the cellular level (cellular respiration). Insects quickly stop feeding. Death occurs several hours to a few days after exposure. Rotenone degrades rapidly when exposed to air and sunlight. It is not phytotoxic, but is extremely toxic to fish, and moderately toxic to mammals. It is more toxic to mammals by inhalation than by ingestion, and skin irritation and inflammation of mucous membranes may result from skin contact. Wear protective clothing and a mask. It may be mixed with pyrethrins or piperonyl butoxide to improve its effectiveness.

Rotenone is a broad-spectrum contact and stomach poison that is effective against leaf-feeding insects, such as aphids, certain beetles (asparagus beetle, bean leaf beetle, Colorado potato beetle, cucumber beetle, flea beetle, strawberry leaf beetle, and others) and caterpillars, as well as fleas and lice on animals. It is commonly sold as a 1% dust or a 5% powder for spraying, and trade names include Bonide Rotenone, Rotenone-Pyrethrins Flea Dip, Rotenone/Pyrethrum spray, etc. The use of this pesticide in organic farming is limited.

Ryania: Ryania is extracted from the stems of a woody South American plant, *Ryania speciosa.* Although a slow-acting stomach poison, it causes insects to stop feeding soon after ingestion. It works well in hot weather. Ryania is moderate in acute or chronic oral toxicity in mammals. It is generally not harmful to most parasites and predators, but may be toxic to certain predatory mites. Ryania has longer residual activity than most other botanicals.

It is used commercially in fruit and vegetable production against caterpillars (corn borer, corn earworm, and others) and thrips.

Sabadilla: Sabadilla comes from the ripe seeds of the tropical lily *Schoenocaulon officinale.* The alkaloids in Sabadilla affect insect nerve cells, causing loss of nerve function, paralysis, and death. The dust formulation of sabadilla is the least toxic of all registered botanical insecticides. However, pure extracts are very toxic if swallowed or absorbed through the skin and mucous membranes. It breaks down rapidly in sunlight and air, leaving no harmful residues.

Sabadilla is a broad-spectrum contact poison, but has some activity as a stomach poison. It is commonly used in organic fruit and vegetable production against squash bugs, harlequin bugs, thrips, caterpillars, leaf hoppers, and stink bugs. It is highly toxic to honeybees, however, and should only be used in the evening, after bees have returned to their hives. Formulations include baits, dusts or sprays.

3.2 Horticultural Oil

Considered effective and safe, horticultural or summer oils (e.g., petroleum oils and vegetable oils) are used to control insects and diseases. Commercial products are highly refined, and formulated as dormant or summer oils. Dormant oils are heavier oils, are more likely to damage plant tissue, and are used on dormant plants to control over wintering insects (e.g., aphids, spider mites, and scales). Summer oils are a lighter version of dormant oil and can be applied to actively growing plants to control aphids, mites, thrips, scales, mealy bugs, and their eggs. Oils coat the insects and suffocate them; therefore, thorough coverage is important.

If the product is not used properly, plant damage can occur. This may happen when too much oil is used, plants are water stressed, temperatures exceed 90°F or when dormancy is mistaken (i.e., spraying too early in the fall). Temperatures must be above 45°F for dormant oil application for proper viscosity and coverage on plants.

3.3 Insecticidal Soap

Insecticidal soaps are made from plant oils (neem, pongamia, cotton seed, olive, palm, or coconut) or animal fat (lard, fish oil), but are generally not considered botanicals. They are made from the salts of fatty acids, which are in the fats and oils of animals and plants.

Soaps are thought to physically disrupt the insect cuticle (outer skin), but additional toxic action is suspected. Soaps act on contact and must be applied directly to the insect to be effective. No residues remain on plants. They are effective against soft-bodied insects like aphids, some scales, psyllids, whiteflies, mealybugs, thrips, and spider mites. Hard-bodied insects (e.g., adult beetles or wasps) are not harmed because of their tough, chitinous bodies.

Some plants may be sensitive to soaps, resulting in leaf burn. Plants that have hairy leaves may be more susceptible to soap injury than smooth-leaved plants. Apply the soap spray on a small area of the plant to check for phytotoxicity.

The potassium soaps (soft soap) of Neem and Pongamia oils at 1 per cent are effective against diamond back moth, leafhopper, aphids, leaf miners, fruit borers and fruit fly in vegetable crops. They are also effective against mango leafhopper and citrus leaf minor. Efficacy may be reduced at high temperatures.

4. MICROBIAL INSECTICIDES

The insecticidal substances obtained from the microorganisms are microbial insecticides. The insecticidal crystal proteins (endotoxins) produced by the bacterium, *Bacillus thuringiensis kurstaki* are effective against lepidopteran pest species. These toxins are very specific in their action, easily biodegradable and being stomach poisons, safer to non-target organisms. Some pests like diamondback moth on Cole crops have developed resistance to certain Bt. Endotoxins.

Other microbial products that have been recently introduced for pest management are the insecticidal toxins produced by soil borne Actinomycetes organisms. The *A. vermectins* (abamectine) derived from *Streptomyces avermitillis* possess potent anthelminthic, acaricidal and insecticidal properties at very low doses. Like wise, Spinosyns (Spinosad), toxins obtained from another Actinomycetes organism, *Saccharopolyspora spinosa* exhibit very good insecticidal property against lepidopteran pests like diamondback moth, gram pod borer, etc.

Fungal pathogens like *Metarrhizium* sp., *Beauveria* sp., *Nomuraea rileyi, Paecilomyces farinosus* can be sprayed against caterpillars and other pests. *Verticillium lecanii* can be used against homopteran pests viz. scales and aphids. Another fungus, *Hirsutella thompsonii* has been found to be effective against citrus mite and coconut mite.

4.1 Antagonistic Organisms

The various organisms employed, as bio-control agents are *Trichoderma* spp., *Aspergillus* niger, *Gliocladium* virens, *Bacillus subtilis, Pseudomonas fluorescence, Paecilomyces lilacinus* and *Verticillium chlamydosporium* etc. among these the most commonly employed and commercially available antagonists are discussed here.

Trichoderma: *Trichoderma harzianum* and *Trichoderma viridae* are commonly used fungal antagonists for disease management in nursery and main field. While *T. harzianum* can be used for managing fungal diseases and nematodes, *T. viridae* is specific against fungi. Hence, it is advantageous to use *T. harzianum*. The minimum requirement of the count of *Trichoderma* is 10^6 spores/1g of formulation. Many firms are manufacturing *Trichoderma* formulations and both neem and talc formulations are available. *Trichoderma* is highly effective in reducing wilt disease of pepper.

Pseudomonas florescenes: The foliar application of this insect pests like leaf minor, rice leaf folder etc. and diseases like bacterial wilt, fungal diseases and nematodes. Regular sprays are advocated to minimize the disease incidence in main field and these are believed to induce resistance in plants. Treating the nursery @ 50 g/sq. meter with this antagonistic bacteria induces disease resistance against bacterial wilt in tomato.

Bacillus thuringensis (B.t.): This is the world largest selling bio-pesticide and contributes to 80-90 per cent of bio-pesticides produced in the world. It is a pathogen of *Lepidopteran* pests and causes gut paralysis on ingestion killing the pest. Both liquid and powder formulations are commercially available. Several strains of *B.t.* are available of which *kutarki* and *israèlensis* are more popular.

The caterpillars ingest *B.t.* while feeding on the treated plants. After about 24 hours of infection, their feeding activity gradually diminishes, the larvae become sluggish, turn black with signs of vomiting and diarrhoea and finally die in about 6 days. *B.t.* is used for the control of lepidopterous pests infecting cabbage and other cruciferous crops, cotton, tomato, gram, groundnut etc. It has no adverse effects on plants or environment.

Verticillium lacanii: This is a pathogen of aphids and white flies and one of the most commercially developed microbial pesticides against *Coccus vividis, Archips termias, Empoasca* sp., *Aphis gossypii* and *Tetranychus* sp.

Beuauveria bassiana: *B. bassiana* has been found to be effective against *Spodoptera* *littura, Helicoverpa armigera, Chilo partellus, Scirpophaga certucas* and *Myllocerus* sp. However, care should be taken while using this organism in sericulture areas as the fungus causes muscardine in silkworms.

Paecilomyces lilacinus: *P. lilacius* is a potent egg parasite of root knot nematode *Meloidogyne incognita* and is being used in many agricultural, horticultural, plantation crops and also in mulberry cultivation.

Vesicular-arbusculur Mycorrihza (VAM): VAM increase the water holding capacity, nutritional status of the soil and also help in promoting growth and disease resistance. This can be cultured in the farmer's field itself. For this solarize the soil for 30 days, cultivate the soil and add VAM culture and grow either ragi or jowar. This facilitates growth of VAM in soil. This is added to soil for growing seedlings and to the main field before sowing or transplanting.

Nuclear Polyhedrosis Viruses (NPV): Heliothis and Spodoptera Nuclear Polyhedrosis Viruses (NPV) are baculoviruses, which are target specific and infect a number of important plant pests of the lepidoptera. They are extensively used in many countries. The most important advantage of these groups of viruses is their harmlessness to non-target organisms including natural enemies of pests. They are compatible with other biopesticides. They can be mass produced very easily.

These NPV can be applied at the rate of 250 larval equivalent (LE) per hectare on chickpea, lab lab, tomato, maize, sorghum, alfalfa, sunflower, cottonne, pigeon pea, etc. NPV should be mixed with water along with 0.5 per cent jaggary (local sugarcane product i.e. gur) and 0.1 per cent Ranipal (Blue) and sprayed on the infested crop preferably in the evenings. NPV should be mixed with 0.25 per cent boric or tannic acid on crops like groundnut, cole crops, beet root, etc.

The caterpillars, while feeding on the plant, ingest the virus. The virus multiplies rapidly within the body of the caterpillars and kills them within 6 days. Such diseased larvae die with their head hanging down, purify, releasing a mass of virus which is infective to fresh caterplillars. Being species-specific, NPV does not affect non-target organisms, including *Trichogramma, Chrysoperla* and other bio-control agents. It is safe to plants and environment.

4.2 Insect Parasitoides and Predators

The egg-parasitoids trichogramma: *Trichogramma* are minute parasitic wasps (8 to 10 adults can sit on a pin head) and use host eggs for development and multiplication. The tiny adult parasitoids search for host eggs and parasitise them i.e. they lay their own eggs within the eggs of the pests. On hatching, the parasitoids larva feed on the embryonic contents of the egg and completes its development within. The parasitised eggs turn uniformly dark in about 4 days later. Thus, instead of a caterpillar, a *Trichogramma* adult comes out of the host egg. A single *Trichogramma,* while multiplying itself, can destroy over 100 eggs of the pest.

Among all parasitoids, *Trichogramma* is the most important and widely used for control of a variety of pests throughout the world. They are extensively used for the control of bollworm, sugarcane borers, rice stem borers, rice leaf folder, apple codling moth etc. *Trichograma* eliminates the pest before it has a chance to damage the crop.

Trichogramma are mass produced and supplied in the form of *Trichogramma*-cards, each containing about 20,000 ready-to-emerge parasitoids. Such cards are safely kept in packets. Each card may be cut into small pieces and distributed all over the field. The pieces may be stapled to the underside of the leaves or they may be placed in small boxes fitted with fine wire mesh and tied to the plant. The parasitoids emerge and disperse in search of host eggs. Alternatively, the adult parasitoids may be allowed to emerge in the packets and distributed in the field on the same day.

Release of 1,00,000 to 2,50,000 *Trichogramma* per hectare is recommended. More parasitoids may be released depending upon the crop and pest density. Since *Trichogramma* is an egg parasitoid, its releases should synchronise with the egg-laying period of the concerned pest. *Trichogramma* kills the pest in the egg stage itself, thus preventing potential damage to the crop by the caterpillars. The parasitoids themselves search the host and destroy them while multiplying themselves. They are self-sustaining and have no adverse effects on plants, environment or other beneficial organisms.

Larval-parasitoids, Goniozus and Bracon: *Goniozus nephantidis* and *Bracon brevicornis* are larval parasitoids of the coconut black-headed caterpillar, *Opisina arenosella*, which is a serious pest of coconut palms. These parasitoids are mass-produced and widely used for control of this notorious pest.

When the parasitoids are released in the field, they go in search of the caterpillars feeding on coconut leaves. On locating a caterpillar, the parasitoids paralyses it by stinging and lays up to 15 eggs on it. The eggs hatch in 1 or 2 days. The parasitoid larvae feed on the paralysed host and complete their development in 3-4 days. They form cocoons on or near the remains of the host. After 7-10 days, adult parasitoids emerge and go in search of fresh caterpillars. A female can lay up to 120 eggs and these are distributed on several caterpillars. Thus the caterpillars are killed while the parasitoids multiply themselves. Releases of 3000 to 4500 parasitoids per hectare/season are recommended.

Pupal-parasitoids, Spalangia spp: Several parasitic wasps like *Spalangia* spp. attack the puparia of housefly and other filth flies in their natural habitat. Such parasitoids are mass-produced and released in fly breeding areas for biological control of filth flies. *Spalangia* spp. Dwell within the manure, never becoming pests themselves.

The female parasitoids are ready to mate and oviposit immediately after emergence. The parasitoids search and lay their eggs within the fly puparia. The eggs hatch and the parasitoids larvae develop by consuming the fly puparia. Finally the fly puparia are destroyed as the parasitoids complete their development. It completes its development from egg to adult emergence in 15 to 18 days. Single parasitoids lays its eggs in about 100 fly puparia in about 15 days, thus causing their death as they multiply themselves.

Spalangia spp. may be released in poultries, dairies, piggeries, garbage dumping yards etc. where the fifth flies breed. They are purely beneficial, never themselves becoming nuisance.

Predacious green lacewings, Chrysoperla species: Predacious green lacewings, *Chrysoperla* spp. is well known predators of soft-bodied sucking pests such as aphids, thrips and whiteflies. Species such as *Chrysoperla carnea* feed on the eggs and newborn larvae of Lepidopterans like cotton bollworms. The adult of *Chrysoperla carnea pale* green, possessing net-like wings. It is popularly known as green lacewing. It lays its characteristic stalked green eggs in small groups on the leaf or other parts of the plant. A female can lay 300 to 400 eggs.

The eggs of *Chrysoperla* hatch in about 4 days. The larvae (or grubs) which possess sickle-shaped mouthparts, wander on the plants in search of prey and crush them dry. The larval period lasts for about 8 days. During this period, each larva consumes several hosts every day. The larva pupates within a round. Silken cocoon on the leaf. The entire life cycle is completed in 17 to 20 days. The adult lives for 30-40 days.

Chrysoperla are mass produced in insectaries and supplied to farmers generally in the form of eggs. The eggs are de-stalked and then packed in boxes using suitable inert medium. These may be distributed on the plant. These predators should be applied twice during the cropping season with an interval of 15 days at the rate of 50,000 per hectare.

Predacious ladybird beetles: *Ladybird Beetles* are well known predators of mealy bugs, scale insects and other soft bodied sucking pests. Three species viz. *Cryptolaemus montrouzieri, Scymnus coccivora* and *Nephus sp.* give effective control of mealy bugs in coffee, citrus, grapevine, mango etc. Two other species *Chilocorus nigrita* and *Pharoscymnus horni* destroy scale insects of sugarcane, coffee, sapota and several other crops. *Menochilus sexmaculatus* is a voracious predator of aphids.

The adult ladybird beetles, being winged, search for mealy bugs or scale insects and feed on them throughout their life. They live for about 45 days. They lay their tiny eggs in the colony of the pest. Each female can lay 100 to 250 eggs or more depending upon the species. Eggs hatch in 5 to 6 days and the grubs voraciously feed on all stages of the host insects. They are particularly fond of host egg sacs and gravid females, andoften completely wipe out the pest colonies. They then enter into the pupal stage, usually in groups, in the crevices of the bark, fallen dried leaves, in leaf folds, infested plant surface, etc. the pupae do not move. Adult beetles emerge from the pupae in 8 to 10 days. They mate and again start feeding and breeding. Thus, the ladybird beetles destroy the pest while building up their own populations.

The ladybird beetles are mass-produced and made commercially available to the farmers in the form of grubs or adults. Release of 1500 to 2500 grubs per hectare at the first sign of infestation is recommended.

5. MINERAL INSECTICIDES

Diatomaceous earth: Diatomaceous earth is a non-toxic insecticide mined from the fossilized silica shell remains of diatoms, (single-celled or colonial algae). It absorbs the waxy layer on insect bodies, abrades the skin, and dries out the insect.

Diatomaceous earth occurs as a dust, and is sometimes combined with pyrethrin. The product may control slugs, millipedes and sow bugs, as well as soft-bodied insects like aphids. It has low mammalian toxicity.

Sulphur: Sulphur is probably the oldest known pesticide in current use. It can be used as a dust, wettable powder, paste or liquid, primarily for disease control (e.g., powdery mildews, rusts, leaf blights, and fruit rots). However, mites, psyllids and thrips also are susceptible to sulphur. Most pesticidal sulphur is labelled for vegetables (e.g., beans, potatoes, tomatoes, and peas) and fruit crops (e.g., grapes, apples, pears, cherries, peaches, plums, and prunes). Sulphur is non-toxic to mammals, but may irritate skin or especially eyes.

Sulphur has the potential to damage plants in hot (90°F and above) and dry weather. It is also incompatible with other pesticides. Do NOT use sulphur within 20 to 30 days on plants where spray oils have been applied; it reacts with the oils to make a more phytotoxic combination. The use of sulphur as insecticides has been restricted. A prior approval is required from the certification agency to use it.

6. DISEASE SUPPRESSION BY COMPOST

Many types of compost, if they are prepared and handled properly, have the ability to suppress and control soil borne plant diseases. Disease control, in this case, does not refer to the process of pathogen destruction that occurs during the composting process itself. Rather, it refers to the ability of mature compost to control plant diseases in soil or potting media when the compost is mixed with media or soil containing pathogens. The incorporation of compost has been reported to suppress diseases caused by *Phytophthora spp., Pythium spp., Fusarium spp.,* and *Rhizoctonneia solani.* The disease suppressive quality of compost appears to be due to the microbial population of mature compost. Several bacterial and fungal species present in finished compost have been shown to be involved in disease control.

These biological control agents recolonize compost during the second mesophilic stage which occurs after the thermophilic stage. For compost to retain its biological control properties, it must be prepared properly and handled properly once it is mature. The alteration of the microbial population of the compost can destroy its ability to control disease. Excessive heat alters the compost and there are indications that storing the material in plastic bags also destroys its ability to control plant disease.

7. EM TECHNOLOGY

7.1 EM5 (Sutochu-Fermented solution)

EM5 is a non-chemical insect repellent and is non-toxic. EM5 is used to prevent disease and pest problems in crop plants. It is usually sprayed onto plants at a dilution of 1/500 -1/1000 in water. It is mainly used to repel insects by creating a sort of barrier. EM5 could also control insect populations. EM5 carried by insects to places of food storage could 'contaminate' the stored food. The process of fermentation that takes place in the food due to EM5 makes it non-edible to insects, thereby diminishing populations.

In making EM5, ingredients may vary. A standard set of ingredients is listed below. However, to make effective EM5 for more persistent pests, more organic materials should be added (organic materials that has a high quantum of antioxidants such as garlic, hot peppers, neem leaf, pruned green fruits, and grass) which are considered to be of medicinal value. When using such materials, they should be chopped or mashed in a mixer. Some or all of the materials may be used in making EM5.

Making EM5: The following is a standard set of ingredients for making EM5

1. Water #1	600 cc
2. Molasses	100 cc
3. Vinegar #2	100 cc
4. Distilled spirit (30-50 %) #3	100 cc
5. EM1#4	100 cc

#1: Well water preferred since tap water is chlorinated.
#2: Natural vinegar is better than artificial acids
#3: Whiskey or Ethyl alcohol could be used.
#4 may be procured from Maple Orgtech (I) Pvt. Ltd, New Alipore Block B, Kolkata-53, India (Supplier in India)

Items needed in making of EM5

A large pot may be used to initially blend all of the ingredients. Plastic containers are required to store the EM5 along with a funnel to pour the EM into the containers.

Preparation

1. Blend the molasses with water and completely dissolve it. Warm water may be used for quick dilution of molasses.
2. Add vinegar and distilled spirit, followed by EM1.
3. Pour the mixed solution into a plastic container which can be shut tightly (A glass container should not be used). Remove excess air in container to maintain anaerobic conditions.
4. Store the bottle in a warm place (20-35 °C), away from direct sunlight.
5. When container is expanded by the fermented gas, loosen the cap of the container to release gas. Shut it tightly again.

The EM5 is ready for use when the production of gas has subsided. The EM5 should have a sweet smell (Ester/alcohol).

Storage: EM5 should be stored in a dark cool place, which has a uniform temperature. Do not store in the refrigerator or in direct sunlight. EM5 should be used within three months after preparation.

***Using EM5*:** Since EM5 is not a pesticide, germicide or a harmful chemical, the application method is different from other agrochemicals. Chemicals are used to solve a problem forcefully and quickly and are applied at specific intervals. EM5, on the other hand, should be applied from the time of planting before the development of any disease or pests. If this is not done and diseases or pests appear, EM5 should be sprayed as instructed below:

- Spray EM5 diluted in water 1/500-1/1000 to wet the crop.
- Start spraying after germination, before pests and diseases appear.
- Spray in the morning or after heavy rains.
- Apply EM5 regularly.

Application can be done once - twice a week with a direct spray onto the plants until the problem disappears. Direct spraying on harmful insects should reduce populations leading to eventual disappearance. A thorough spraying to the plant ensures good results. Continuous or regular sprayings ensure that harmful insects which may have escaped or are recent additions will be affected by the EM5. EM5 works over time. Thus a regular application brings out the best results.

7.2 EM Fermented Plant Extract (EM-F.P.E.)

EM fermented plant extract is prepared by using fresh weeds and EM1. EM-F.P.E. includes organic acids, bioactive substances, minerals and other useful substances from weeds. The production cost of EM-F.P.E. is very low, because of the use of weeds.

Making EM fermented plant extract: The following is a standard set of ingredients for making EM-F.P.E.

1.Chopped fresh weeds#1	14 litres
2. Water #2	14 litres
3. Molasses #3	420 cc
4. EM1 #4	420 cc

#1 Use weeds, which have strong life such as mugwort, artemisia, clover and grass which are considered to be of medicinal value. Pruned green fruits and young shoots could be incorporated. The use of various types of weeds is recommended in order to increase bioactive substances and microbial diversity. The weeds should be cut in the morning.

#2 Well water is preferred since tap water is chlorinated. Adding a little amount of seawater (0.1%) is useful to supply minerals to crops.

#3 & #4 3% of water.

Items needed in making of EM-F.P.E.

Large plastic bucket or drum, weight to press chopped weed, black vinyl bag, and wooden lid.

Preparation

1. Cut weeds and chop well (2-5cm).
2. Put chopped weeds into bucket.
3. Mix EM1 and molasses into water and pour the solution into bucket.
4. Cover the top of bucket with black vinyl bag.
5. Put lid on the vinyl, and then put weight on the lid. At the time, take care not to leave air in the bucket.
6. Store the bucket in a warm place (20 - 35 °C), away from direct sunlight.
7. Fermentation begins and gas is generated within 25 days. (depending on temperature).
8. Stir the weeds in the bucket regularly to release the gas.
9. The EM-F.P.E is ready for use when pH of the solution is below 3.5. Put EM-F.P.E. into plastic bottle after removing the weeds by filtration (use gauze or cloth).

Storage: EM-F.P.E. should be stored in a dark cool place, which has a uniform temperature. Do not store in the refrigerator or in direct sunlight. EM-F.P.E. should be used within one month after preparation.

Using EM-F.P.E.

- Watering into the soil (1:1000) by watering cans, sprinkler or irrigation system.
- Spray EM-F.P.E. diluted solution (1:500-1:1000) to wet the crop.
- Start spraying after germination, before pests and diseases appear.

- Spray in the morning or after heavy rains.
- Apply EM-F.P.E. regularly.
- The combination of EM-F.P.E. and EM5 is more effective.

EM based preparation has no adverse effect even with excessive applications. In contrast, it may enhance the plant's strength through the absorption of EM and therefore increase the level of antioxidation (that is, the ability to suppress disease, pest infestation, and overcome any debilitating factors.).

The upliftment of the economic status occurs as EM based solution can be made easily and cheaply. Over the long term, less EM5 solution is needed since the soil conditions change. This ensures a healthy and strong crop to protect itself from disease and pests. The post-harvest crop residues incorporated back to soil as a pre-treatment before the next season is recommended and, additionally, the use of EM5 would help in the suppression of diseases and pests that would be recycled back into the next crop.

EM5 contains EM l - therefore it contributes to the beneficial effects like increasing yield and quality of the crop. Thus, less expense is incurred on fertilizers and no cost would be expended on agricultural chemicals. Although EM solution may take time to create the best condition depending on soil and type of crops grown, it will benefit the environment, the soil, the plant cultivated, and the economic status of the farmer.

8. NON-PESTICIDAL MANAGEMENT OPTIONS

8.1 Neem Seed Kernel Suspension

Method of preparation

- Take 5 kg of dried neem seed kernels and soak them in small quantity of water.
- Grind neem seed kernel and keep the paste in a cloth bag and soak it for 2 hrs.
- Take a big container; fill it with 10 litres of water by measuring with a standard vessel.
- Keep the bag with neem seed kernel paste in vessel, squeeze it thoroughly for 20 minutes while holding in water.
- Filter the milky white suspension using thin cloth and add 100 gms of soap to it.
- Add water to this suspension to make 100 litres of spray solution.
- Spray the suspension on the crop, by shaking the sprayer.
- This gives NSKE 5%.

Insects controlled: *Spodoptera, Helicoverpa, Semiloopers*, leaf folders, all defoliators and sucking pests, including mites.

Precaution while storing neem seed

To store neem in large quantities add sulphur 1:10@ 0.5 kg per quintal.

Do's	Don'ts
Collect only ripe fruits	Don't use the seed stored for more than one year
Separate the seed from fruit before storing	Don't dry seed under sun
Store need seed in gunny bags only	Don't store them in airtight polythene bags.

Benefits of spraying neem seed kernel suspension

- Spraying at flowering stage can prevent the egg laying by female moths of Spodoptera, Helicoverpa, etc.
- Spraying neem presents early instar larvae feeding on the foliage on the foliage, thus leading to starvation of caterpillars.
- Azadirachtin the active ingredient of neem interrupts life cycle of insects resulting in deformed larvae, pupa and adult.
- Neem is safe to mammals, earthworms and natural enemies of the crop pests.
- Neem is an environment friendly bio-pesticide.

8.2 Chilli-Garlic Extract

Method of preparation

- Take 3 gram of chillies; remove the pedicel and grind thoroughly.
- Soak the chilli paste in 10 litres of water overnight.
- Take 500 g of garlic, grind thoroughly and soak overnight in 250 ml of kerosene.
- Prepare two extracts separately by filtering through thin cloth.
- Prepare third solution by dissolving soft soap/potassium soap @ 75g in one litre.
- Mix the three solution in a container and keep it for 4 hours.
- Filter the mixture using a cloth and dilute to 80 litres and spray.

Insects controlled: *Helicoverpa, Spodoptera*

8.3 NVP Preparations

Method of preparation

- Larvae died due to infection of NVP are seen in the field.
- Collect 400 NPV affected *Helicoverpa* or 200 *Spodoptera* larvae from field.
- Grind the collected larvae.
- Filter the solution obtained using a thin cloth.

- Dilute the NPV solution to 100 litres and add 100 g of Robin Blue to protect from UV light in the field.
- Spray this solution during evening hours.
- Virus of one insect species does not kill the other insect species.
- Virus infected dead larvae are observed hanging head upside-down from top branches 2-5 days after spraying the solution in the case of *Helicoverpa* and split body in case of *Spodoptera.*
- Within 10 days all insects in the field are infected with NPV.

8.4 Cattle-Dung and Urine Extract

Method of preparation

- Take cattle-dung 5 kg and 5 litres of urine and mix them in 5 litres of water.
- Ferment the solution for 4 days by keeping a lid over the container.
- After 4 days filter the solution and add 100 grams of lime.
- Dilute the solution in 80 litres of water and spray in one acre.
- Spraying Cow dung urine solution prevents egg laying by the moth, eg. *Heiothis, Spodoptera* , etc.
- It gives protection against some diseases. Crop looks green and healthy.

9. PRODUCTION OF NPV, TRICHOGRAMMA AND GREEN LACEWING

9.1 Mass Production of NPV

- Nuclear polyhedrosis virus of gram pod borer has to be multiplied on its 3rd or 4th instar larvae individually, as they are cannibalistic.
- The 3rd to 4th instar larvae of *Heliothis* are to be collected and kept separately in 5 ml injection vials containing 3 grains of overnight soaked gram and inoculated with a drop of concentrated NPV stock culture. Only one larva is kept in one vial and plugged with cotton.
- These *Heliothis* larvae are fed for 3 to 4 days on soaked gram.
- After 3 to 6 days the virus infected larvae die. Such larvae are isolated and macerated with the help of pestle and morter after addition of a little amount of water. This mass is kept for 1-2 days in sufficient water for purification and then filtered through clean muslin cloth to obtain polyhedra of virus.
- Two hundred and fifty larval equivalent (L.E.) or 250 ml NPV suspension is diluted in 100 to 400 litres of water along with 0.05% 'Teepol' for one ha.

Precautions

- Spraying is to be done during afternoon hours to avoid strong sunlight.
- NPV spraying has to be done at flowering and then depending on pest incidence at the early pod formation stage.
- Two or three rounds of sprayings have to be done.

9.2 Production of Trichogramma

In India *Trichogramma chilonis, T. japonicum* and *T. achaeae* are key mortality factors for many crop pests. These parasitoids attack eggs of many lepidopterous pests e.g. sorghum stem borer, sugarcane borers, paddy stem borer, tomato fruit borer, cutworms, cottonne bollworms etc.

Rice grain moth, *Corcyra cephalonica,* is used as the laboratory host.

- Bold white grains of sorghum are procured.
- The required quantity of the sorghum is milled to make 3-4 pieces of each grain.
- The sorghum grains are sterilized in oven at 100°C for 30 minutes.
- The sorghum is prayed with formalin. This treatment helps to prevent the growth of moulds as well as increases the moisture of grains to the optimum (15-16%).
- The lot is then air-dried.
- The whole lot is kept in boxes @ 2.5Kg/box.
- In each box 300 cc of Corcyra eggs are added and kept closed (with lids) for about 30 days. The same process is to be repeated on 45th, 90th, 135th, 180th and 225th day.
- On the 40th day the moths start emerging and the emerged continues for two months. Up to 75 moths emerge daily. The peak number of moth emergence takes place between 65th and 75th day.
- The moths are collected daily and transferred to the specially designed oviposition cages. Number of moth emergence reduces after 100 days of initial infestation and the boxes are used again after cleaning.
- The eggs are collected and passed through 15, 30 and 40 mesh sieves and rolled over a slope of paper to eliminate dust particles.
- The eggs are then treated with UV rays (15W UV tube for 45 minutes at a distance of 2 feet) to prevent hatching.
- The eggs are glued to 'Tricho' cards of 15 cm x 10 cm size which is prepunched to obtain 8 pieces of 4cm x 3cm size, leaving uncovered space at one end to facilitate stapling. The eggs are exposed to adult female. *Trichogramma,* in the ratio of 8:1 for 24 hrs. In case the cards are kept in this method the females are allowed to parasite the eggs till they die. After parasitisation, 6-day-old parasitised egg cards are prepared for shipment/field release.

9.3 Production of Green Lacewing (Chrysoperla/Chrysopa)

The green lacewing (*Chrysoperla carnea*) is released in the field for the control of aphids, white flies, mealy bugs and eggs and young larvae of lepidopterous pests. It is being mass-produced primarily on the eggs of rice grain moth (*Corcyra cephalonica*) in India.

Rearing technique

- Three thousand adults are kept in oviposition cage, measuring 75 x 30 x 30 cm. The sides of the cages are lined with smooth nylon net and the sliding top cover is fitted with black cloth for obtaining eggs. The top is replaced everyday from the 4th day onwards. The oviposition cage is kept for 20 days and the dead adults are removed every alternate day.
    - Rearing of *Chrysoperla spp.* requires one room of 6x6m^2 maintained at 27$\pm$ 1.0°C, supplied by fluorescent day.
    - The adults in oviposition cage are fed daily on swabs (kept in plastic plates) of drinking water, 50 % honey, diet consisting of protenex 40 gms + fructose 70 gms dissolved in 250 ml of drinking water and castor pollen.
- Twenty-four-hour old eggs are dislodged from the black cloth top cover of oviposition cage by gently working a piece of sponge.
- In first step of larval rearing, three day-old Chrysopid eggs are mixed with 0.6 cc of *Corcyra* eggs; the embryo of *Corcyra* eggs are inactivated by keeping them at 2 feet distance from 15 watt ultraviolet tube light for 45 minutes in a plastic container (27x18x6 cm). On hatching the larvae start feeding. On the 4th day the larvae are transferred to 2nd step individual rearing in 2.5-cm cubical cells of plastic leauvers. Each leauver can hold 192 larvae. *Corcyra* eggs (0.3 cc) are provided in all the cells of each leauver by sprinkling through modified salt-shaker. Leauver is covered on one side by organdie or brown paper sheet and after transfer to larvae it is covered with acrylic sheet and clamped.
- Subsequently 1.3, 1.3, 2.6 and 2.6 cc eggs on 5th, 7th, 8th, 10th and 12th day are provided for ensuring complete development of the larvae in each of the leauvers. One 2m x 1m x 45 cm angle iron rack can hold 100 leauvers containing 19,200 larvae.
- Cocoons are collected after 24 hours of formation by removing organidie or paper from one side. Adults are sometimes allowed to emerge in leauvers and collected against glass window panes by suction.
- One set of leauvers remains in use for 13-15 days. After utilization leauvers are cleaned, sterilized and reused.
- For field release, 3 day-old-eggs, which are about to hatch, are mixed with *Corcyra* eggs before shipment.

The pest management methods listed above can be integrated in to the organic farming systems depending upon the crops and the pest species. It is possible to keep the pest populations below economic injury levels by conserving existing natural enemies and adoption of suitable cultural, mechanical and biological methods. In addition, we have a large number of ITK (Indigenous Technical

Knowledge) passed on from generation to generation, which can also be adopted along with other methods.

When inputs have a relatively high acute toxicity for non-target organisms, a restriction for their use is needed. When it is not possible to take adequate measures, the use of such inputs must not be allowed. Inputs, which accumulate in organisms or systems of organisms, and input, which are suspected of having mutagenic and carcinogenic properties, must not be used.

Chapter 9

Water Management in Organic Farms

Water is a dearer input and difficult to receive, retain and release whenever it is required and in whatever quantity it is needed. The importance of irrigation is recognized for many crops, because the yields of irrigated crops are comparatively better than dryland crops. The productivity of the irrigated crops is nearly two to three folds more than that of the unirrigated crops.

The water demand varies with the crop and therefore planning for development of small water resources in the every organic farm is necessary for judicious use of improves water use efficiency. This can be done by creating *in-situ* water harvesting ponds and adopting suitable irrigation methods like sprinkler and drip irrigation, mulching of the soil and suitable cropping pattern. For flat land, inter-row water harvesting techniques has been found promising not only for *in situ* water conservation but also for harvesting the excess rainfall in a farm pond for irrigation of dryland crops. This runoff should be collected in a suitably designed and lined farm dugout. Since seepage of water from the farm dugout is very high, lining with polyethylene sheet is necessary.

In addition, the surface flow water could be harvested and stored in dugouts, ponds, etc. for reuse as supplemental irrigation. Water harvesting and runoff recycling serve as a woeful contingency measure against weather aberrations in areas where rainfall is erratic and unpredictable. This technology has four major components.

(i) Runoff collection

(ii) Storage of harvested water

(iii) Water application

(iv) Water reuse on crops

Combination of water conservation during the rainy period and optimisation of irrigation during summer will improve the yield of crops significantly. The various quality measures should also be considered for use of groundwater and surface water.

1. CRITICAL STAGES OF GROWTH FOR IRRIGATION

In majority of the cases, use of growth stages criteria have almost extensively used for scheduling irrigation. The most common recommendation which has emerged from the various studies is to apply five to six irrigation of 4-5 cm depth each at different growth stages of crops. The peak demand period of irrigation water for various crops in a given area may occur at different times in the growing season. Any shortage of water during this peak demand period will have some impact on the final yield of crops.

The visible stages of crop growth are germination, emergence, vegetative growth, flowering, grain formation and fruiting. The optimal moisture requirement for plant and growth varies with the stages of crop growth. During certain stages of growth, plants are most sensitive to shortages of water. These stages are known as moisture sensitive periods or critical periods. The period of stage of development in the life cycle of a crop or a plant at which the crop is most sensitive to the deficiency of a production factor and most responsive to correction of the deficiency. Moisture stress due to restricted supply of water during the moisture sensitive periods will irrevocably reduce the yield and provision of adequate water and fertilizer at other growth stages will not help in recovering the yield lost. Critical stages of water requirement (moisture sensitive periods) of various crops are furnished in Table- 9.1.

Table 9.1: Critical (moisture sensitive) stages (s) of crops

Crops	Critical stage (s)	Changes occur due to soil moisture stress
Cereals		
Rice	Tillering, panicle initiation, flowering and milky	Reproductive structure affected
Sorghum	Primordial initiation booting and flowering	Decrease in grain number and weight
Maize	Knee high, tasseling, silking and maturity	Decrease in grain number and weight
Fingermillet	Primordial initiation and flowering	Decrease in grain number and weight
Pearlmillet	Heading and flowering	Decrease in grain number and weight
Wheat	Crown root initiation tillering to booting	Decrease in grain size number and weight
Oats	Ear emergence to heading	Decrease in grain number and weight
Barley	Tillering, early root and soft dough stage	Reduction in grain number and weight
Oil seeds		
Groundnut	Flowering, peg formation and pod development	Decrease in pod number
Seasamum	Flowering and maturity	Reduction in capsule number

Crops	Critical stage (s)	Changes occur due to soil moisture stress
Sunflower	Two weeks before and after flowering	Head size and weight reduced
Soybean	Blooming to seed formation	Reduction in pod number and weight
Safflower	Rosette to flowering	Reduction in head size and grain weight
Castor	Full growing period	Reduction in seed size and weight
Pulses		
Black gram	Vegetative and flowering	Reduction in pod number and grain weight
Green gram	Vegetative and flowering	Reduction in pod number, grain size and weight
Red gram	Flowering	
Bengal gram	Grain maturation	Reduction in grain weight
Cash Crops		
Cotton	Flowering and boll development	Decrease in boll number and weight
Sugarcane	Formative stage (up to 120 days)	Reduction in cane height
Banana	All stages	Reduction in bunch number
Tobacco	Immediately after transplanting, knee high to full bottom	Reduction in leaf weight
Sugar beet	3 to 4 weeks after emergence	Tuber size reduced
Legumes		
Alfalfa	Immediately after cutting for hay and flowering for seed production	Leaf growth and seed production affected
Beans	Flowering and pod setting	Pod number, size and grain weight reduced
Peas	Flowering and pod development	Pod number and seed weight reduced
Vegetables		
Onion	Bulb development to maturity	Bulb size reduced
Tomato	Flowering to maturity	Reduction in fruit number and size
Chillies	Transplanting to tipping	Reduction in fruit number
Cabbage	Head formation	Head size reduced
Carrot	Enlargement of roots	Root size reduced (Tuber size) reduced
Potato	Tuber initiation to tuber maturity	Tuber size reduced
Cauliflower	Planting to harvest	Curd size reduced
Radish	Root enlargement	Growth affected
Broccoli	Head formation and development	Head size reduced
Turnips	Root development	Reduction in root development
Watermelon	Blossom to harvest	Reduction in fruit size

Crops	Critical stage (s)	Changes occur due to soil moisture stress
Fruits and Plantation Crop		
Citrus	Flowering, fruit setting and enlargement of fruits	Fruit size reduced
Mango	Flowering	Fruiting is reduced
Apricots	Flowering and bud development	Fruits numbers reduced
Cherries	Fruit growth prior to maturity	Fruit size reduced
Peaches	Fruit growth prior to maturity	Fruit size reduced
Straw berries	Fruit development	Fruit number and size
Coffee	To ripening flowering and fruit development	Reduction in nut size

In general, water stress during germination and early seedling may have deleterious effect because at this stage plants have only a small root system. Water storage during vegetative period usually has little effect on subsequent production unless it is so severe as to drastically reduce leaf area. Moisture stress during heading and flowering reduces grain formation. Moistures stress during grain development reduces the number of filled grains and grain filling capacity by affecting the translocation of photo-assimilates to the developing grains.

2. QUALITY IRRIGATION WATER

Any appraisal of water resources for agriculture would be incomplete without any examination of the quality of water. The quality of irrigation water must be good for the soil and the crop. There are large seasonal and yearly changes in irrigation water quality due to land clearing, rainfall dilution or geological strata through which the water flows. Rivers/canals which flow through the agricultural fields are easily contaminated with pesticide residues present in the soil under the influence of rain and irrigation water by the processes of surface run-off, sediment transport and movement of groundwater from aquifer to river. Whatever the case may be, it is obvious that a single water analysis is entirely insufficient. Water has to samples at regular interval-at least seasonally. The safe limit of irrigation water for agricultural and horticultural crops is given in Table 9.2.

2.1 Salinity Hazard

The total salt content or electrical conductivity (EC) is the most important criteria for evaluation of irrigation water quality. Generally, an increase in irrigation water salt content will result in an increase in the salinity of the soil solution. The rate and extent of increase will depend on the leaching fraction, i.e., the amount of water supplied by irrigation or rainfall in excess of that required to satisfy consumptive use of crops. Leaching efficiency, on the other hand, depends on soil physical characteristics.

Table 9.2: Safe level of irrigation water

Parameter	Agronomic and Horticultural Crops (Humid land)
pH	5.5 - 8.5
Temperature °C	35
Total Salts (mg/l)	1000 (Non-saline-alkali area) 2000 (Saline-alkali area)
Sodium Absorption Ratio (SAR) (meq/l)	1.25
Nitrogen (mg/l)	12 - 30
Total -P (mg/l)	5 - 10
Sulphate (mg/l)	1
Chloride(mg/l)	250
Boron (mg/l)	1-3
Total Mercury (mg/l)	0.001
Total Cadmium (mg/l)	0.005
Chromium (mg/l)	0.1
Total Lead (mg/l)	0.1
Total Copper (mg/l)	1.0
Total Zinc (mg/l)	2.0
Total Selenium (mg/l)	0.02
Total Arsenic (mg/l)	0.01 - 0.1
Fluoride (mg/l)	2.0 (high F-area) 3.0 (normal area)
CN (mg/l)	0.5
Petrolium (mg/l)	1 - 10
Phenols (mg/l)	1
Benzene	2.5
Tri-colour aldehyde (mg/l)	0.5
BOD (mg/l)	80 - 150
COD (mg/l)	150 - 300
Total Coliform Organism (mg/l)	10000
Roundworm (mg/l)	2

2.2 Sodicity Hazard

Among the soluble constituents of irrigation water, sodium is considered most hazardous. The effect of sodium is two-fold. It may affect the permeability of the soil by causing swelling and dispersion of clay particles, and clogging of soil pores, and it may cause injury to crops specifically sensitive to sodium such as fruit crops.

2.3 Toxicity Hazard

The higher metal levels in soil may cause negative impacts on crops, inhibiting the growth in one or other way. One of the most important factors is the pH of the soils. Alkaline pH of the soil usually restricts the mobilization of the metal in the soil matrix and consequently, the metal uptake by the crops can be controlled thereby reducing the risk of metal toxicity.

Boron is very toxic to most plants at low concentrations in the soil solution. Sulphate also creates a specific nutritional problem by precipitating with calcium or by promoting the uptake of sodium. Under some condition (natural or polluted by industrial wastes) mercury, cadmium, chromium, lead, copper, zinc, selenium, iron, arsenic and fluoride also present in the irrigation water in toxic concentration.

Fluorides (above safe level) in irrigation water may restrict the plant growth due to its adsorption by the soil. Vegetables irrigated with waters of high fluoride levels are likely to contain more fluoride. Excessive fluoride intake beyond permissible limit through water or food crop also causes dental, skeletal *fluorosis* and non-skeletal manifestations.

Arsenic accumulation in soils through irrigation of groundwater can injure the crop and plant growth appears to be restricted. In spite of the capacity of most soils to tie up arsenates, long-term additions of arsenic through irrigation water may result in decided toxicities to some sensitive plants like maize, beans, potatoes etc and may cause *arsenocosis* manifestation to consumer.

Iron present in irrigation water gets converted into ferric oxide when it comes in contact with the atmosphere. As a result, the iron gets precipitated and is deposited on the soil mass. Soil, being a porous media, allows the deposited iron to move down to the root zone of the crop and is likely to affect the balance of plant nutrients and their availability to plants.

Hydrocarbons or petroleum can have considerable impact on soil environment because of its extreme stability and subsequent entry into the food chain. The presence of hydrocarbon in water may also affect the respiration and movement of anaerobic soil microbes by forming a thin film on the soil and water surface.

Special monitoring needs to be implemented to systematically monitor water pollution.

3. IRRIGATION METHOD

An effective irrigation is the controlled and uniform application of water to crop land in the required time, with minimum cost to produce optimum yields without the waste of water and any adverse effect on the soil in the form of soil salinity/ acidity and waterlogging problems that create wet desert. More efficient use of available water through, amongst others, curbing over-irrigation of crops will not only save water but also help in minimizing the loss of nutrients from fields through leaching and run-off.

The evapotranspiration losses of water are the lowest during the winter months and are the highest during the hot months. During the hot season, therefore, irrigation practices are less economic than in the rest of the year. It is therefore, advantageous to bring as much area as possible under irrigation during winter months. However, hot months are very much crucial for survivability of perennial types of crops such as sugarcane, all the fruits and forest plants.

The method of application may profoundly affect the suitability of water for irrigation. Three basic methods are available:

- Surface irrigation system
- Sprinkler irrigation system
- Drip or trickle irrigation system

3.1 Surface Irrigation System

In this system, water is directly applied to the surface of the soil from the conveyance system, channel or pipe located at the upper reaches of the field and spread by gravity flow incidental to the slope of the land. There are several methods in this system, the commonest being hand watering, free flooding from a ditch, check flooding basin, border strip and furrow.

For irrigation with the surface system, fields are laid out every time before the crops are sown, since these layouts are destroyed during preparatory tillage. In some cases, the same layout may be used for irrigating the subsequent crop. However, the *field must be levelled well to achieve higher water application efficiency.*

Hand watering: It is done through bucket, plastic pipe or sprinkler container to individual plants and effective in small scale garden (home garden, kitchen garden, composite garden) or during transplantation of seedlings, saplings, cutting of vegetables, flowers, perennial fodders, etc. More time is wasted but intensive care is taken.

Check flooding or Check basin: This method is well suited to all irrigable soils with smooth gentle and uniform land slopes and to a variety of crops. The method is especially adapted to irrigation of grain and fodder crops in heavy soils where water is absorbed very slowly and is required to stand for a relatively long time to ensure adequate irrigation. The basins are filled to the desired depth and water is retained until it infiltrates into the soil. When irrigation rice or ponding water for leaching salts from the soil, depth of water may be maintained for considerable periods of time by allowing water to continue to flow into basins. Crops, susceptible to complete saturation of the root zone such as potato, maize, chillies cannot be irrigated by this method.

Basins, rectangular or square, with sizes varying from about 10 to 100 sq m or even more are used. The basins are levelled in both directions. Water is conveyed to the field by a system of supply channels and lateral field channels. A supply channel is aligned along the highest contour. Individual flat basins are connected one after another with this channel by breaching or by setting potable siphons. After irrigation, the breached bank is blocked and patched up or the siphons are removed.

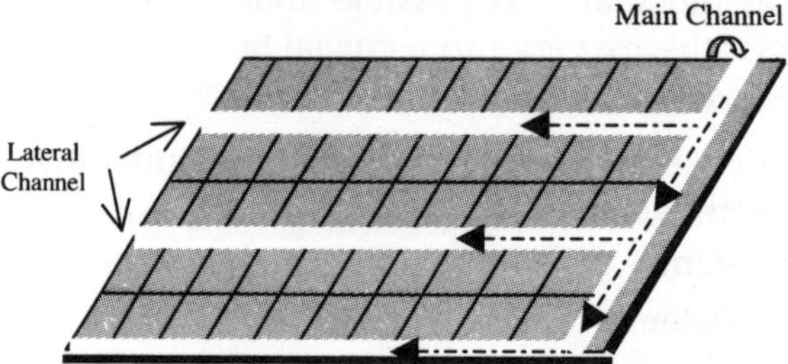

Figure 9.1: Typical layout of check basin method of irrigation

Slopes up to two to three percent can be irrigated by using this method with a good control on irrigation water and high application efficiency. On steeper slopes, this method can be used after terracing. In rolling topography, the ridges follow the contours of the land surface. The contour ridges are connected by cross ridges at intervals. The vertical interval between contour ridges usually varies 6 to 12 cms.

Limitations of this method are that it has too many ridges which occupy a larger area of land, construction and repair of the ridges, careful supervision during irrigation, time consuming and costly.

Basin or Ring basin: This method is commonly used for widely spaced crops to irrigate an individual plant or plants grown in pits or pockets. A ring or flat basin is made around the plant and a number of plants or pockets are connected with a ditch passing between rows of plants. The soil around the plant is soaked with irrigation water and not the entire field. Crops such as sweet guard, pumpkin, ash guard, fruit orchards, etc. are irrigated by this method. Basin irrigation is most effective for leaching purposes. This method considerably economises water.

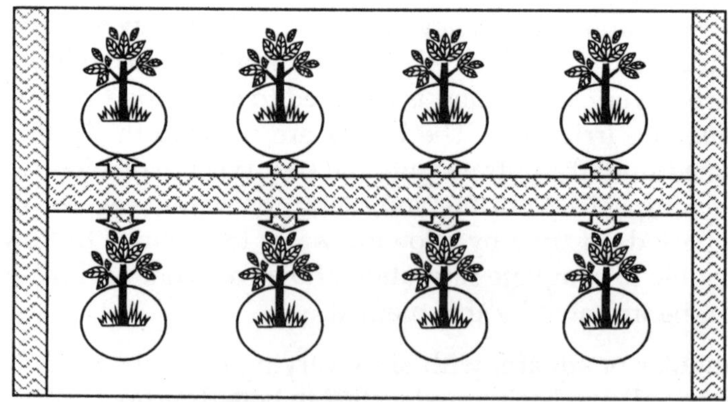

Figure 9.2: Ring basin type irrigation

Border strip: In this method, the field is divided into a number of long narrow parallel strips called boarders that are separated by low ridges. The strips are 2 to 10 m wide. The length of a strip ranges from 10 to 300 m or more depending up on the soil type, slope and flow size.

The following boarder length and slope are suggested:

Soil Type	Boarder length	Boarder slope
Sandy and Sandy loam soil	60-120 metres	0.25 % to 0.60 %
Medium loam soils	100-180 metres	0.20 % to 0.40 %
Clay to clay loam soils	150- 300 metres	0.05 % to 0.20 %

The boarder strip has little or no cross slope but has a uniform gentle slope in the direction of irrigation. Individual strips are levelled perfectly and connected with the supply ditch laid out at the upper elevation. Each strip is irrigated independently by turning in a stream of water at the upper end. After irrigation these are disconnected.

This method is suited well to all irrigable soils and to closely spaced row crops and even to pasture crops which can tolerate saturated soil conditions for an hour or so. Wheat, barley, rapeseed and mustard, peas, beans, grams, nurseries of rice and onions are irrigated by this method.

Slopes up to seven per cent can be irrigated under this system. On steep slopes, proper terracing or trenching along with contours is required to control the shifting of soil. Boarder may be laid along the general slope of the field (straight or down the slope boarders) or may be laid across the general slope of the field (contour boarders).

For laying out the boarder strip, the land needs to be graded uniformly to achieve a high application efficiency of water. Repair of ridges and supervision during irrigation are needed.

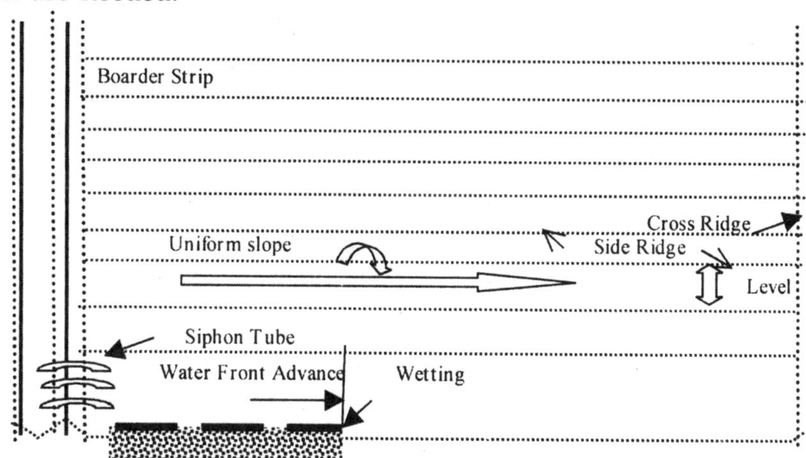

Figure 9.3: Layout of boarder type irrigation system

Furrow method: In row crops, furrows are made in between the two crop ridges. Water is applied to the furrows and the top of the ridge is not directly wetted. The furrows can be made along the slope when the level of the land is sloping gently up to three per cent. When the slope exceeds three per cent and is about 15 per cent, the furrow are laid out on graded contours or as cross-slope furrows. Water distribution can be controlled well to achieve uniform application and the consequent high efficiency. The length of the furrows varies with the soil type, the slope and the quantity of water to be applied and may vary from 10 to 1000 m in different situations. The depth of the furrow should be such that water movement within the soil is predominantly horizontal into the plant root zone.

Irrigation furrows may be classified into two general types. They are (a) straight furrow and (b) contour furrows. Straight furrows are best suited to sites where the land slope does not exceed 0.75 per cent. However, the furrow grade should not exceed 0.5 per cent so as to minimise the erosion hazard. Contour furrows are curved to fit the topography of the land. Soils can be irrigated successfully across slopes up to 5 per cent. Where soils are stable and will not be cultivated (as in orchards), slopes up to 8 to 10 percent can be irrigated. Contour furrows for row crops are constructed annually. But in orchards or other permanent contour plantings, the same furrows and ditches can be used every year after proper cleaning.

Figure 9.4: Furrow method of irrigation

Crops, sensitive to the saturated soil conditions is the root zone and having subterranean storage tissue are generally irrigated by this method as a considerable number of roots and a part of stem remain above the saturated soil after the application of water in the furrows. Crops with underground storage organs such as potato, sugar beet, turmeric, ginger, colocasia, sweet potato, radish and beet get sufficient space for the development of such organs above the wetting zone in the soil. Wide spread crops such as sugarcane, maize and cotton are also irrigated by this method.

3.2 Sprinkler Irrigation Method

In this method, water is conveyed through pipelines under high pressure and sprayed over the crops through nozzles. The spray of water somewhat resembles the rainfall. It is especially designed in order to achieve high efficiency in its performance and economy.

Figure 9.5: Sprinkler method of irrigation

A typical sprinkler system consists of a pump to lift and convey water under pressure, pipes or tubing of the conveyance of water, sprinkler heads or nozzles and risers which connect the sprinkler heads with a pipe line. Based on the equipments with which spraying is done, the sprinkler systems has been classified as the rotating head type and the perforated head type. There are self-propelled sprinkler systems which move laterally or radially around a central pivot feeding-line. Portable type system is designed in such a way that it can cover any area ranging from 3 to 4 hectares to 50 to 60 hectares.

This method is advantageous, as water can be applied at a controlled rate and a uniform distribution. This method can be adopted in the case of almost all crops and is very popular in the case of cash and some orchard crops and in all types of nurseries. This system is especially suited to shallow sandy soils of uneven topography, where levelling is not practicable, and in areas where water and labour are scarce. The sprinkler system also helps in cooling the crops during high temperature and for frost control during freezing temperatures. On some soils with salinity problems, this system is advocated for the leaching of salt more effectively, for emergence and to secure the quicker and better growth of plants.

The planning and design of a sprinkler irrigation system require following information:

Map of the area	A map of the area concerned is prepared and drawn to scale with sufficient accuracy to show all dimensions so that lengths of main and laterals can be scaled therefrom. It should be a contour map or, at least, should show all relevant elevations with respect to water supply, pump location, and critical elevations in the fields to be irrigated.
Water source	The source of water is usually a well or a tank storing rainfall runoff. The water should be relatively clean, free from suspended impurities, chemically suitable for irrigating the crops and soils of the area.

Climatic records	The climatic records will show when and how often irrigation is needed in various seasons of the year. Information on temperature, radiation intensity, humidity and wind velocity helps in designing the system.
Depth of irrigation	It is calculated on the basis of available moisture capacity of the soil in its different layers and the soil moisture extraction pattern of the crop in its root zone depth.
Irrigation interval	This is length of time allowable between successive irrigations during the peak consumptive use of the crop.
Types of crops	The general layout of the system depends on the type of crops due to different plant spacing and irrigation requirements.
Soils	The soil infiltration rate, water holding capacity, texture, and bulk density must be known for determining spacing, and setting up irrigation schedules.

3.3 Drip or Trickle Irrigation System

This system involves the slow application of water, drop by drop, to the root-zone of a crop. There is considerable saving in water by adopting this method. The equipments consists of a pumping unit to create a pressure of about 2.5 kg/sq cm, pipelines which may be of PVC tubing with drip type nozzles or emitters, and a filter unit to remove the suspended impurities in the water. The amount of water dripping from the nozzles can be regulated, as desired, by varying the pressure at the nozzles, and the size of the orifice of the nozzles. Water may be lifted and distributed through overhead pipe lines which are fitted with nipples to drop or trickle water to the desirable site or water may be held at a certain height from where it passes to the orifice which is impregnated into the root-zone.

The initial cost of drip irrigation equipment is considered to be its limitation for large-scale adoption. The cost of the unit per hectare depends mainly on the spacing of the crop. For widely spaced crops like fruit trees the system may be even more economical than sprinklers. The main item of expenditure is the cost of the lateral lines. The crops like grapes, sugarcane, papaya, banana, guava and most other types of fruit trees and vegetables have been found to respond well to drip irrigation.

Water supply may be continuous or intermittent. In this method water is used very economically, since losses due to deep percolation and surface evaporation are reduced to the minimum. The successful growing of orchards even on saline soils has been made possible by the drip system of irrigation.

The following information is required for designing a drip irrigation system:

Water source	The source of water is usually a well or a tank storing rainfall runoff.
Types of crops	The general layout of the system and especially emitters depends on the type of crops due to different plant spacing and irrigation requirements.
Topographic condition	It is necessary to know the general land slope to determine the size and location of main and sub-main lines.
Soils	The soil infiltration rate, water-holding capacity, texture, and bulk density must be known when selecting emitter type, determining spacing, and setting up irrigation schedules.
Climatic records	The climatic records will show when and how often irrigation is needed in various seasons of the year.

The drip irrigation method proved to be of great advantage as regards salinity. By applying water from a point source rather than with aerial uniformity, very heavy leaching is provided near the source. An additional advantage is the higher frequency of water application. Consequently, total soil moisture stress in the limited root zone of a crop irrigated by drip irrigation is considerably lower than for other irrigation methods using water of similar salt contents.

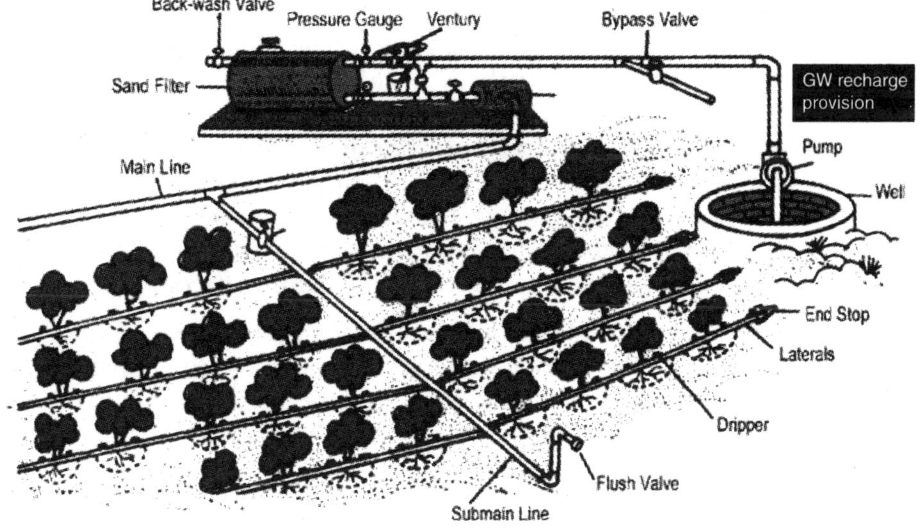

Figure 9.6: Layout of drip irrigation

Figure 9.7: Drip irrigation system

4. LAND SHAPING FOR IRRIGATION

Due to ploughing and other normal cultivation operations, the smoothness of a levelled or graded land is disturbed. Localised high and low patches are developed. When irrigation water is released in the field, it moves forward as well as downward. If the field does not have proper smooth shape the forward movement of water is restricted in case of surface irrigation. More water is accumulated in lower patches and water may not reach the higher patches. As a result, the crop growth becomes non-uniform and a poor irrigation distribution efficiency is obtained. To ameliorate these, land shaping becomes necessary.

Land shaping is usually done by simple equipment as buck scraper, float wooden plank etc. which are tractor/animal-drawn or manually by using spade. A skilled worker and well-trained animals should be used for land shaping. Land shaping requirements can be seen in cropped area by identifying the patches of bad crop growth.

4.1 Drainage

The removal of excess water (free or gravitational/standing or stagnant water) from the surface or below the surface of the soil so as to create favourable soil condition for plant growth. The excess water which is not available to crop plants should be drained to facilitate crop growth and yield. The excess water may originate from high intensity and prolonged pluviometric conditions exceeding the rate of infiltration, more than adequate irrigation water, phreatic and fluxial water from adjoining areas or back water and high tides, sudden breaches in the embankments of reservoirs.

Land conditions for drainage: Land under the following conditions requires drainage provided the land after draining is capable of providing expected production.

- High water table (within or near to the root zone, i.e. 1.5 m.
- When water stands on the land surface for long periods (about 2 hours for potato, grams, peas, mustard or few weeks for rice, para grass, *sesbania*).
- When temperature of the standing water in the crop field exceeds 45°C for a prolonged period or less than cardinal point with respect to crop or the water is turbid or turbulent or when soil application of fertilizer are needed or when good physical, chemical, biological and workable soil conditions are desired.
- In arid and semi-arid regions.
- In humid to sub-humid regions with continuous or intermittent heavy precipitation.
- Areas under saline, saline-alkali and alkali soils.
- Areas with irrigation facilities.
- Areas with level land or land with mild slope.
- Areas prone to be flooded or remain saturated with interflow.
- Where natural drainage is impeded due to construction of embankments, canals, aerodromes, new townships, etc. or where there is impervious hard pan near the root zone.
- Areas that is physically suitable for drainage with required number and size of outlets.
- An area less prone to be reversion and the size of the drainage system is a manageable unit.

There are different methods of drainage under surface drainage system. These are:

a) *Lift drainage*: Draining low-laying areas having stagnant water. Water to be drained is lifted by manually-operated devices like buckets, swing scoop or by pumps run by electricity, wind mill, solar power, etc.

b) *Gravity drainage*: Water is allowed to drain from the areas under higher elevation to lower reaches through the regulated gravity flow through the outlets of various types (grassed or mud compacted water ways). This system is practiced in wet rice areas with gentle to moderate slope.

c) *Field surface drainage*: The irrigated basins or furrows are connected with the drainage ditches used for irrigation to fields under lower elevation or connected to the main outlet through laterals or the farm pond used for water harvesting. If the slope of the land is sufficient to drain excess water from the individual plots this drained water may be collected and stored locally in reservoirs for life saving irrigation.

d) *Ditch drainage:* Ditches of different dimensions are constructed at different distances to drain the excess water accumulated on the surface and inside the soil up to the depth of the ditch. The cross section of field ditches may be 'V' shaped, trapezoidal or parabolic. Such ditches may be-

- interceptors (which are located almost perpendicular to the direction of the interflow as seepage and overflow as run-off).
- relief drains (that lower the water table in a low lying area and relieve the land from waterlogging. These are installed parallel to the direction of the inter and over flow.).

Interceptors may be installed at the upper sides of the fields to intercept seepage or run-off water from fields located at the upper elevation. This is the most useful preventive drain in undulated, sloppy, terrace and hillsides that enter flat lands. These drains isolate the problem area from the source of excess water. This method is adopted in nurseries, seed beds and rainfed crops. Sloping lands are drained by the cross-slope ditch method. Ditches are constructed across the slope to reduce the hazard of soil erosion and overflow.

In some instances, complete drainage is not desirable particularly under erratic and uncertain rainfall and irrigation conditions. However, the excess water should be drained out. Aquatic and semi-aquatic crop plants may be allowed to grow under moderate drainage conditions. Intermittent draining of wet crop soils is essential to control pests and diseases, algal weeds, expel trapped gases, increase availability of plant nutrients and to provide better environment to the crop plants.

Faulty drainage may induce erosion and degradation of productive soils. In higher slopes, drainage should be with check bunds with settling and silting ditches. Inducing quick drainage may reduce percolation and recharge of root-zone and groundwater. In some crops and soil conditions, both drainage and drying are essential. For those elevated/raised beds, corrugated field surface, diversion of inlet of flood water or installation of interceptors are helpful. All drainage ditches should be deeper than the field surface and with non-erosive gradient. Width should be according to the flow rate. Sides of the ditch should be erosion tolerant and for this the special measures namely, growing grass, compacting, caging, masonary work, lining with various cheap and durable materials may be done.

Chapter **10**

Planning Organic Garden

Organic gardening is not just a middle-class hobby. It is part of the wider environmental movement and part of a sustainable future. The truth, so effectively suppressed that it is now almost impossible to believe, is that organic farming is the key to feeding the world. Organic farming could produce enough food to feed large populations. It sounds like an environmentalist's dream.

Planting an organic garden requires planning. A properly planned and planted garden will naturally resist disease, deter insect pests, and be healthy and hardy. With the spring planting season fast approaching, winter is the ideal time to get started. It is important to understand the magnitude of your project before you begin. Getting the background information necessary to fulfil your goals may take an hour or a week, depending upon your level of experience and how involved you plan to get. Consulting this guidebook is a great way to begin and investigate their resources or contact your local organic growers for their suggestions. As you research, write down how long each project will take, what tools you will need, and the approximate cost of everything you will need. This information will be invaluable when you make up your shopping list and schedule of activities.

1. BEGIN PLANNING

1.1 Set Goals

What do you want to do with your plot of agriculture farm this season? Begin planning by setting goals. Grab your garden map, a pencil, your gardening guide, and your thinking cap. List the areas of your main plots, orchard and garden separately (i.e. cereals and pulses, fruits and plantation, vegetable, medicinal garden, etc.), and keeping in mind the size and conditions of your site, brainstorm! Are you planning a garden for the first time? Do you want to expand your existing garden? Did you have pest or disease problems previous year that you're hoping to prevent this year?

1.2 Prepare Map

To create a map of your organic garden, measure the dimensions of your site as a whole, and then the individual dimensions of main plots, orchards, vegetable plots, etc. It's easiest to draw your map to scale on a sheet of graph paper. These measurements will be necessary later for input requirements, certification and documentation of your farms. Once the map is drawn, write in any information you know about soil characteristics, drainage, environmental conditions (sunny, shady, windy), and the names of trees and perennial plants that already exist. Your map will let you know exactly what you have to work with, and will give you a realistic idea of problems that need attention or features you'd like to change or add.

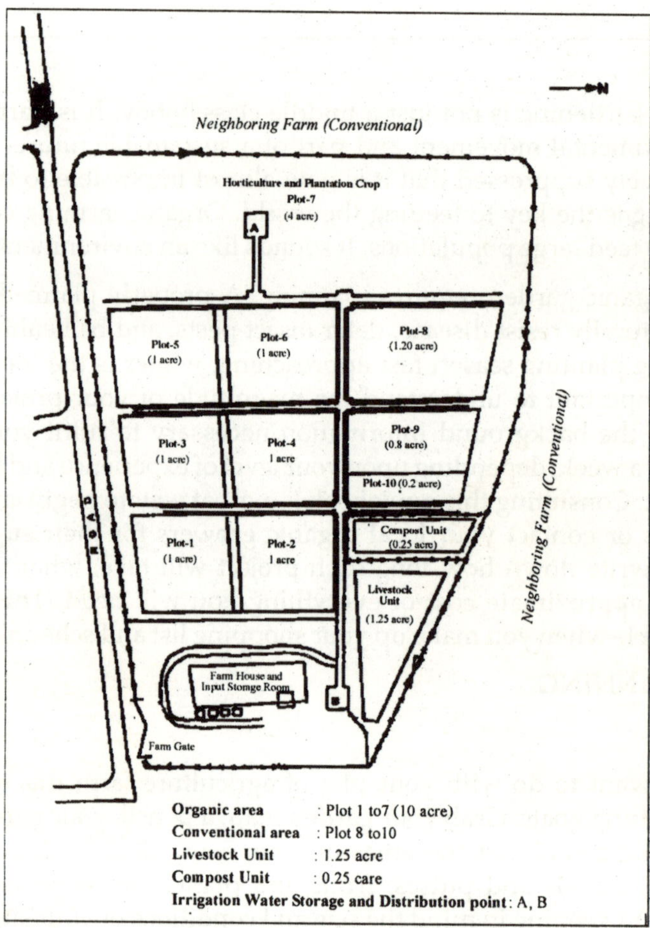

Figure 10.1. Farm map for organic production

1.3 Scheduling of Farm Activities

A schedule of activities lists what you hope to accomplish in what time-frame. It will help keep you on track. It is important to be realistic about what you are capable of. Staggering your major tasks over time will make easier to accomplish and save you the frustration of looming unfinished projects. Planning for the long term will aid in your farm. You can create a year-by- year schedule that maps out a time frame in which to achieve your big goals. Obviously, the schedule can change as time goes by, you learn new methods and you rethink your objectives, but maintaining focus on what you hope to create in the long term can keep you motivated on what you are doing now.

2. LOCATION

Plants are solar collectors. They take energy from the sun, nutrients from the soil and transform that energy and nutrients into protein, carbohydrates and vitamins that are used to sustain human life.

It is important to make the most of the plants ability to change suns energy into a form which can be utilized for our benefit. Therefore, when considering a location for a garden the following important points should be considered:

1. Drainage
2. Protection from frost
3. Shading
4. Surrounding vegetation
5. Pest protection
6. Fertility

2.1 Surface and Subsurface Drainage

Too much moisture will limit the plants ability to produce. There are two types of drainage systems used to correct excessive moisture.

Surface drainage: It is not always possible to select a site that does not suffer from excess moisture resulting from excess rain or runoff. The solution to this problem may be the addition of surface drainage to the garden. The cross-slope or diversion system should work for most situations. The cross slope system is used to drain sloping land that is wet because of:

a. Slowly permeable soils

b. Water accumulation from higher slopes

c. Surface concentrations in shallow depressions within the garden.

In the cross slope system, a diversion ditch is used at the head of the garden to intercept surface water and to a lesser extent subsurface water before it can reach the garden. A second ditch is made at the bottom of the garden and parallel to the interceptor ditch. The garden is crowned so that the centre of the garden is a foot higher than the edges. Raised beds can be used to help raise plants out of a poor drainage area in a temporary garden.

Sub-surface drainage: Not all soils are suited for sub-surface drainage. But yields in soils suited to sub-surface drainage can improve dramatically with tile installed. Capillary water is used by plants and beneficial organisms to grow. Free water, the water that occupies soil pores, is detrimental to both. Tile removes free water but does not have an effect on capillary water. The elimination of free water will mean that the garden can be planted earlier and harvested earlier with high yields.

Tile should be installed to a minimum depth of 3 ft. If the tile is to go across a driveway it should be protected or buried to a minimum of four feet. The area a tile can drain is directly related to the depth of the tile. For most situations a single tile 6 ft deep and centred in the garden will drain a garden 50 ft wide.

2.2 Protection from Frost

Protection from frost should be a major consideration when locating a garden. Frost has a tendency to travel down a slope and collect in pockets in low-lying areas. These areas should be avoided.

Irrigation can be used to offer a degree of protection to susceptible plants. If a source of water is not near then plants should be covered during suspected periods of frost. Covers can be commercially available materials or common household materials. Additional protection is available using plastic row covers and plastic mulch.

2.3 Shading

Shading within the garden should also be avoided. Usually there should be 3 trees per acre around the farm. If any tree cause unwanted shading, tall growing plants should be grouped in a manner so that they do not shade low growing plants. Rows should be planted north and south to take advantage of the movement of the sun and to minimize shading.

Shading can be put to advantage to help control weeds. Planting in wide or mat rows instead of single rows will give greater volume per unit of area. Once the crop has established it will be harder for weeds to establish. Rows should not be so wide as to make it difficult to reach across to weed. Wide rows can be raised beds allowing for faster drainage. Rows must not run the length of the terrace and interfere with the natural drainage. Proper row orientation would be north and south.

2.4 Surrounding Vegetation

In addition to shading from tall growing plants, roots from trees and other vegetation will rob the garden of nutrients. Hedges used to keep out grazing animals should be far enough back from the garden so as not to cause a problem. If vegetation is a problem then steps should be taken to rectify the situation.

- Arrange the garden to a location not affected by trees or hedges
- Prune the offending trees or shrubs
- Trim roots and branches that are causing the problem.
- Edging material placed around the garden will help slow the spread of roots and help control weeds in the garden.

Sufficient diversity should be obtained in time or place in a manner that takes into account pressure from insects, weeds, disease and other pests. To ensure a clear separation between organic and conventional production, a buffer zone or a natural barrier should be maintained. Land under conversion should not get switched back and forth between organic and conventional management.

2.5 Controlling Pest and Disease

Edging material installed to stop the spread of rhizomes will also help to stop the invasion of harmful worm and grubs into the garden. The first year should be green manure or fallow. The second year should be non-root, non-tuber crops.

Use of Compansion planting for plant protection:

Trap cropping: Sometimes, a neighbouring crop may be selected because it is more attractive to pests and serves to distract them from the main crop.

Symbiotic nitrogen fixation: Legumes such as peas, beans, and clover have the ability to fix atmospheric nitrogen for their own use and for the benefit of neighbouring plants via symbiotic relationship with *Rhizobium* bacteria. Forage legumes, for example, are commonly seeded with grasses to reduce the need for nitrogen fertilizer. Likewise, beans are sometimes interplanted with corn.

Biochemical pest suppression: Some plants exude chemicals from roots or aerial parts that suppress or repel pests and protect neighbouring plants. The African marigold, for example, releases thiopene a nematode repellent, making it a good companion for a number of garden crops. The manufacture and release of certain biochemical is also a factor in plant antagonism. Allelochemicals such as juglone found in black walnut suppress the growth of a wide range of other plants, which often creates a problem in home horticulture. A positive use of plant allelopathy is the use of mow killed grain rye as mulch. The allelochemicals that leach from rye residue prevent weed germination but do not harm transplanted tomatoes, broccoli, or many other vegetables.

Physical spatial interactions: For example, tall-growing, sun-loving plants may share space with lower-growing, shade-tolerant species, resulting in higher total yields from the land. Spatial interaction can also yield pest control benefits. The diverse canopy resulting when corn is companion-planted with squash or pumpkins is believed to disorient the adult squash vine borer and protect the vine crop from this damaging pest. In turn, the presence of the prickly vines is said to discourage raccoons from ravaging the sweet corn.

Nurse cropping: Tall or dense-canopied plants may protect more vulnerable species through shading or by providing a windbreak. Nurse crops such as oats have long been used to help establish alfalfa and other forages by supplanting the more competitive weeds that would otherwise grow in their place. In many instances, nurse cropping is simply another form of physical-spatial interaction.

Beneficial habitats: Beneficial habitats, sometimes called refugia, are another type of companion plant interaction that has drawn considerable attention in recent years. The benefit is derived when companion plants provide a desirable environment for beneficial insects and other arthropods, especially those predatory and parasitic species which help to keep pest populations in check. Predators include ladybird beetles, lacewings, hover flies, mantids, robber flies, and non-insects such as spiders and predatory mites. Parasites include a wide range of fly and wasp species including tachinid flies and Trichogramma. Agroecologists believe that by developing systems to include habitats that draw and sustain beneficial insects, the twin objectives of reducing both pest damage and pesticide use can be attained.

Security through diversity: A more general mixing of various crops and varieties provides a degree of security to the grower. If pests or adverse conditions reduce or destroy a single crop or cultivar, others remain to produce some level of yield.

Controlling pests is usually a full season chore. The alternative to pesticide when resistant plants are not available is natural predator introduction. For more information on natural predators see chapter "Bio-control of Pests and Diseases" in this book.

2.6 Fertility

Balanced nutrition: Whether or not it has been customary in the past, organic growers are encouraged to have periodic soil testing done on their fields. How the results of a soil test are used, however, can vary considerably among farmers, depending on their personal philosophy and management skill. While there is certainly a segment of the organic farming community that has no faith in soil audits, most growers use them as a means to monitor progress in building their soils, to identify nutrient deficits, and to guide supplementary fertilization. While there are many ideas about fertilization guidelines, there are two schools of thought that dominate.

The first is commonly known as the *sufficiency* approach or model. Though somewhat oversimplified, the following are among its significant characteristics:

- annual fertilizer additions of organic P and K are based in good part on how much the crop is expected to remove from the soil at harvest
- additional amounts of organic P and K are recommended based on keeping the soil nutrient reserves at a particular level
- lime is added to the acidic soil based on pH
- little to no attention is paid to nutrient balance or to the levels of secondary nutrients Ca and Mg

The second approach is referred to variously as *cation nutrient balancing system*, the *CEC, or* the *base saturation* approach. It differs from the sufficiency approach in that fertilizer and lime recommendations are made based on an idealised ratio of nutrients in the soil and its capacity to hold those nutrients against leaching. Cation nutrient balancing is more popular among practitioners of organic farming and sustainable agriculture in general than it is among conventional growers. However, there is no universal agreement on which approach is most appropriate for organic management.

Green manuring: Green manure, unless it is a legume, does not add nutrients to the soil. A legume with the help of nitrogen fixing bacteria takes nitrogen from the air and stores nitrogen in the plant and nodules. When the plant dies, the breakdown of nodules and plant tissue releases nitrogen to a subsequent crop.

The legume is not turned under until late in the season before freeze-up. To control weeds the legume is cut when the first flower on the companion weeds die or when full bloom occurs on the legume. The cut tops are either left or removed to the compost bin.

Legumes can be annual such as clover, dhaincha, sesbania etc. or perennial such as sweet clover, alfalfa, etc. are also used. The proper live, nitrogen-fixing inoculum (biofertilizer) must be added to legume seed before planting or crop failure may result. Annuals are usually preferred to perennials. Annuals can be turned under in the fall or left for erosion protection and turned under in the spring with less chance of becoming a weed in the garden. An abundant crop of clover can add as much as 170 kg/ha of nitrogen.

Soil testing: The soil should be tested every two years to determine the pH and available nutrients in the soil. pH is the measure of acidity or alkalinity of the soil. A garden for most crops should have a pH between 6.0 and 6.5. Too high a pH will tie up micronutrient. Too low a pH will allow the plant greater access to aluminium, restricting uptake of other necessary macro-nutrients, severely decreasing yields. When the pH falls below 6.0, lime is used to increase pH.

Liming in acidic soil: There are two types of lime: Calcite and Dolomite. Calcite contains calcium. Dolomite contains both calcium and magnesium. Dolomite should be used when a soil test shows the soil to be low in magnesium. Lime also comes in pelleted and powdered form. Lime moves in the soil about one inch per year.

Equal amounts of pelleted or powered lime will raise the pH equally. Powdered lime when mixed with the soil will contact the soil particles more uniformly and create a more uniform pH throughout the soil. Pelleted lime, although easier to apply will create areas of high and low pH within the soil. Lime should not be banded alone or with manure. Bands will be of very high pH and tie up micronutrient within the band rendering the manure useless in the year of application.

Use of manure and compost: Manure, fresh or composted, makes an excellent source of nutrients for the garden. Manure should be covered immediately upon applying to the garden. As much as 30% of the nitrogen can be lost in as little as a day and a half when left exposed and not covered. Compost is easy to make and can be fortified with rock phosphate or other ingredients as needed to make a complete fertilizer. In making good quality compost the following rules should be used:

- Use equal amounts of fresh material and dry material
- Keep material moist but not wet
- Do not put mature weeds in the mix. Use only weeds without seed. Do not add diseased material to the compost, as this will only spread the disease
- Have more than one compost bin. To help lessen disease carry over, group refuse from different crop groups. As an example in bin one place, refuse from corn and root crops. In bin two places, refuse from Legumes, *Brassica* and *Solanaceae*
- Do not apply compost from the same group back to the garden where that group is to be grown. Example: apply bean refuse to corn and not back to the bean crop
- Compact each layer if it contains a lot of browns, but do not over-compact. Do not compact under conditions of adequate moisture and small particle size
- Do not compost dairy or meat products as they may attract rodents. Turn each bin at least once a month
- Do not add lime as adding lime will cause N loss.

What can I use in organic crop production?

One of the greater difficulties that organic producers face on a regular basis is determining whether or not a particular product or material can be used in organic production. All natural or non-synthetic materials can be assumed to be acceptable in organic production. There are a few exceptions. Most organic producers and prospective producers have heard about the National Organic Program Regulations. The synthetic materials that are allowed in organic crop production- for example, sulphur, insecticidal soap, etc. contains natural, or non-synthetic materials. The materials those are prohibited- for example, ash from manure burning, nicotine sulphate, etc. When considering commercial products, the grower must be aware of all ingredients to determine that none are prohibited. If a full disclosure of ingredients is not found on the label, details should be obtained from the distributor or manufacturer and kept in the grower's files. Note that such details must extend to inert ingredients. When in doubt about the acceptability of any material or product for certified organic production, contact your certifying body.

3. GARDEN DESIGN

Switching to chemical free organic gardening will not only mean changing your gardening practices, but also your gardening design. Gardening in beds, as opposed to rows, provides for better weed, disease and pest management. Beds should also more attractive and easier to maintain. Diversity in a garden bed also has many advantages. A variety of plants in a mixed bed provide some natural pest protection by making it difficult for pests to find and eat their target plants, or helping to attract insects that are beneficial to your garden and prey on pest insects. It also reduces the chances that pests and disease organisms will build to epidemic levels, as they won't be able to hop from tasty host to tasty host, as they would if you had planted in rows. Your soil will also reap the benefits of your diverse planting techniques. A good example is planting nitrogen-gobbling corn with nitrogen-giving beans. Pairing up particular plants or planting in variety can help the soil maintain its nutrient balance, ensuring happier plants and a better crop yield.

3.1 Making the Organic Garden Bed

Making your bed can be as simple as marking off 3-by-5-foot sections of garden with pathways left between them. However, to optimise the advantages of planting in garden beds, raise your beds. Raised beds provide lighter, deeper, more nutrient-rich, water absorbent soil. Raised beds, however, must be regarded as permanent in order to maintain their splendour. They cannot be walked on or broken down at the end of the season. You can build sides on your bed with bricks, rocks, planks to maintain the shape instead of raking and reshaping the bed every year. Stay away from pressure-treated wood, as it is treated with wood preservatives that are harmful to you and the environment.

Double-digging raised beds

1. Dig out the top one-foot of soil along one end of the bed. Keep the soil separately.
2. Loosen the exposed subsoil by a spading fork. For extra benefit, add a small amount of organic matter.
3. Once the subsoil is loosened, move over and begin removing the topsoil from the next strip of garden bed. This time, instead of keeping the topsoil that you are removing, shovel it over the subsoil to which you have just added the organic matter. You can add a little more organic matter to the topsoil as you shovel.
4. Repeat step 3.

Options for system design: Agronomists use the term "intercropping" to describe the spatial arrangements of companion planting systems. Intercropping systems range from mixed intercropping to large-scale strip intercropping. Mixed intercropping is commonly seen in traditional gardens where two or more crops are grown together without a distinct row formation. Strip intercropping is designed with two or more crops grown together in distinct rows to allow for mechanical crop production. No-till planting or transplanting into standing cover crops can be considered another form of intercropping.

3.2 Suitable Planting Stock

One of the challenges faced by organic growers is getting suitable planting stock. Certified production requires that seed not be treated with pesticides and be organically grown. Transplants must be purchased from a certified organic source or otherwise be grown using organic methods. Most transplants sold in garden centres have been treated with chemical fertilizers or pesticides. Seeds themselves bought at garden centres may be coated in fungicides, so be very careful about what you buy or buy from an organic seed supplier.

Generally, planting stock for perennial crops like tree fruits and berries is available from conventional sources, even if treated with pesticides. However, no production can be sold as organic for a minimum of 12 months following transplanting to an organic yield.

In all instances, organic growers are not permitted to use varieties of crops that have been developed through genetic engineering (GE). At first glance, this might seem perplexing. GE crops promise further means of non-chemical pest control and the possibility of nutritionally enhanced foods. However, the organic community is concerned about environmental, economic, and social impacts from this new technology which, they feel, have not been adequately studied. Natural foods consumers- a large segment of the organic market does not want them.

To start plants from seed in small vegetable or flower garden, you need sterile soil, sterile planting containers, and labels. It is better to grow each seedling in a separate container to avoid the damage incurred by ripping roots apart, and to make for a less shocking transplant. If you purchase soil mix, be sure that it is sterile to avoid spreading disease to your seedlings. To make your own mix, use vermiculite (a mica-based mineral that has been heated to make it expand to many times its original size) and sphagnum (moss that has been collected while still alive, dried, and then finely ground). Add 1 tablespoon of lime for each 2 quarts of sphagnum that you use to counteract its acidity. Seeds actually need heat, not light, to germinate. The heat from a grow light or sunny window may be enough for some. Keep your seeds moist by planting them in moist mix and covering them with plastic wrap. As soon as you see the first sign of life, remove the wrap and place them someplace where they will receive 6-8 hours of sunlight per day. Water them carefully with a spray mister, careful not to knock the seedlings over or wash away the soil.

Before you transplant your seedlings outdoors, they need to be acclimated to the different climate. Bring them outside and place them in a sheltered, somewhat shady spot for a few hours each day, gradually increasing their exposure to the elements over a week or two. Plants have a hardiness zone, an area based on the average annual low temperatures where a plant is most likely to withstand the region's annual low temperature. Growing plants that are outside your hardiness zone is not impossible, but they will need special attention.

3.3 Plan Crop Rotation

Crop Rotation is defined as the practice of alternating the annual crops grown on a specific field in a planned pattern or sequence in successive crop years so that crops of the same species or family are not grown repeatedly without interruption on the same field. Using different crops with varying nutrient requirements assist in breaking weed cycles and in building soil fertility. Organic farmer should not plant the same row crop 2 years in a row unless the rotation is clearly broken by a cover crop. The grower should take pictures of the green matter to be ploughed down from the cover crop.

Agronomic operations are especially dependent on crop rotations that include forage legumes. These provide the vast majority of the nitrogen required by subsequent crops like corn, wheat, rice which is a heavy consumer of that nutrient. Even when livestock are present to generate manure, the animals are largely recycling the nitrogen originally fed by legumes in the system. In horticultural orchard, perennial cropping systems such as alley cropping, intercropping, and hedgerows are generally introduce for biological diversity in lieu of crop rotation.

Ideally an organic crop rotation system would contain crops from different plant groups, that are seeded at different times and that have different nutrient demands. An organic field crop rotation plan should include row crops, legumes/ sod crops and small grains. Production of annual vegetable crops should also include crop rotations.

Crops should be grouped and a three to four year rotation within the garden developed. Crop grouping should include members of the same family and other crops that are attacked by the same pathogen.

For example:

- Group 1: (Brassica) cabbage, cauliflower, broccoli
- Group 2: (Legumes) beans, peas, gram
- Group 3: (Root crops) carrots, beets, turnips
- Group 4: (Solanaceae) potato, eggplant, tomato
- Group 5: (Gramineae) corn, millet

A typical rotation would be:

1st yr.	Buckwheat	Legume	Group 1 2 3 4 5
2nd yr.	Group 1 2 3 4 5	Legume	Buckwheat
3rd yr.	Legume	Buckwheat	Group 5 4 1 2 3
4th yr.	Group 5 4 1 2 3	Buckwheat	Legume

Table 10.1: The rationale for Coleman's eight-year rotation

Crop Rotation	Beneficial effects
Potatoes follow Sweet Corn	...because research has shown corn to be one of the preceding crops that most benefit the yield of potatoes.
Sweet Corn follows the Cabbage family	...because, in contrast to many other crops, corn shows no yield decline when following a crop of brassicas. Secondly, the cabbage family can be under-sown to a leguminous green manure which, when turned under the following spring, provides the most ideal growing conditions for sweet corn.
The Cabbage family follows peas	...because the pea crop is finished and the ground is cleared (early) allowing a vigorous green manure crop to be established.
Peas follow tomatoes	...because they need an early seed bed, and tomatoes can be under-sown to a non-winter hardy green manure crop that provides soil protection over winter with no decomposition and re-growth problems in the spring.
Tomatoes follow beans	... because this places them 4 years away from their close cousin, the potato.
Beans follow root crops	...because they are not known to be subject to the detrimental effect that certain root crops such as carrots and beets may exert in the following year.
Root Crops follow squash (and potatoes)	...because those two are good 'cleaning' crops (they can be kept weed-free relatively easily), thus there are fewer weeds to contend with in the root crops, which are among the most difficult to keep cleanly cultivated. Second, squash has been shown to be a beneficial preceding crop for roots.

The garden should be big enough so that half of the garden can be fallowed using a green manure crop. Green manure crops are grown to control weeds, add nitrogen for the following year crop, provide erosion control and stop organic matter loss.

Buckwheat is planted after the threat of killing frost is over. When the first flowers on the weeds growing among the buckwheat die the buckwheat and weeds are turned under. Before the buckwheat is turned under the buckwheat is checked for seed formation. If seed has not formed then the buckwheat will have to be replanted. Buckwheat is a smother crop, it is used to control weeds and does not add nutrients to the soil.

An excellent substitute for buckwheat, if available is a combination smother crop and fertility crop such as a mixture of 50 % annual ryegrass and 50 % clover. This mixture can be cut with a ride on lawn mower all summer to help control weeds, maintain organic matter levels and add nitrogen for a subsequent crop.

For crop rotation to be effective, you need to know the genetic relationship between common garden crops. For example, you should not follow tomatoes with peppers since these plants are closely related and many of the same diseases and insects attack both.

3.4 Companion Planting

Much of the science of companion planting is figuring out what works for you. Many books can give you guidelines about what plants work well together. Some plants are attractants, some repellents, some can be inter-planted with your crops and flowers, and some compete too vigorously and should be planted in separate borders or hedgerows. For example, sunflowers are a good border plant, attracting lacewings and parasitic wasps; radishes are good to interplant because they repel the striped cucumber beetle; and marigolds are good to both uses as a border and interplant, as they attract hover flies and repel root nematodes, bean beetles, aphids, and potato beetles. It can be confusing, and not all plants work well together. Your best bet is to start simple, determine what pests you encounter, and work from there, altering the plants in your garden bed as needed from year to year. Often, a mixture of flowers, vegetables and herbs work well together in a single bed.

Agricultural Tools: Basic tools which is required for organic farm are diggers, weeders, standard short-handled cultivating tools, tillers, sowers etc.

Table 10.2: Companion planting chart for home and market gardening

CROP	COMPANIONS	INCOMPATIBLE
Beans	Most Vegetables & Herbs	
Beans, Bush	Potato, Cucumber, Corn, Strawberry, Celery	Onion
Beans	Corn, Radish	Onion, Beets, Sunflower
Cabbage Family	Aromatic Herbs, Celery, Beets, Onion Family, Spinach	Strawberries, Beans, Tomato
Carrots	Pea, Lettuce, Onion family, Tomato	
Celery	Onion and Cole families, Tomato, Bush Beans	
Corn	Potato, Beans, Pea, Pumpkin, Cucumber, Squash	Tomato
Cucumber	Beans, Corn, Pea, Sunflowers, Radish	Potato, Aromatic Herbs
Eggplant (Brinjal)	Beans, Marigold	
Lettuce	Carrot, Radish, Strawberry, Cucumber	
Onion Family	Beets, Carrot, Lettuce, Cole family	Beans, Peas
Pea	Carrots, Radish, Turnip, Cucumber, Corn, Beans	Onion Family, Gladiolus, Potato

CROP	COMPANIONS	INCOMPATIBLE
Potato	Beans, Corn, Cole family, Marigolds	Pumpkin, Squash, Tomato, Cucumber, Sunflower
Pumpkin	Corn, Marigold	Potato
Radish	Pea, Lettuce, Cucumber	
Spinach	Strawberry, Bean	
Squash	Corn, Marigold	Potato
Tomato	Onion Family, Marigold, Asparagus, Carrot, Cucumber	Potato, Fennel, Cole family
Turnip	Pea	Potato

4. HANDLING TRANSITION PERIOD

During the transition from conventional to organic farming practices nutrients from organic fertilizers often are mineralised before or after the time when plants can best use them for their growth. When this happens, farmers obtain poor yields from their crops, while nutrients intended for plant growth may be leached into the groundwater or carried into rivers by runoff. This lack of synchrony between nutrient release and nutrient uptake occurs primarily because the soil undergoes chemical and biological changes during the transition from synthetic to organic inputs. Synthetic inputs kill or inhibit the growth of many soil organisms while organic inputs stimulate their growth. As organic matter is added to the soil and populations of soil organisms increase, these organisms become more effective in breaking down added organic matter and making its nutrients available to plants. They also hold within their own bodies many nutrients not used by plants. This process allows for greater nutrient availability during times of plant growth, while conserving excess nutrients against environmental loss.

Unfortunately, this balance does not exist in the soils of many transitional organic farms. To mineralise sufficient nutrients for plant growth, the relatively small populations of soil organisms in transitional farms need to have access to high amounts of organic matter. However, these soil organisms will continue to decompose the added organic matter even after plants are done growing. These excess mineralised nutrients may then contaminate groundwater or streams. The potential for runoff and erosion is higher if synthetic inputs used previously created a surface crust, poor soil aggregation, or other conditions that hinder water infiltration.

4.1 Proper Nutrient Management Planning

Good nutrient management planning is critical for any farming operation in order to optimise plant growth and decrease risks of nutrient leaching and runoff. Nutrient release from organic matter is on going process and most plant uptake of nutrients occurs during the growth stage. This is especially true when 'high yielding' or hybrid crop varieties are being raised. Organic farmers can use a variety of management practices to simultaneously conserve nutrients and enhance soil quality. These practices include:

- Using annual soil tests to calculate appropriate amounts of organic fertilizers to add
- Composting animal manure and other organic residues to form a more uniform and chemically stable fertilizer material
- Using crop rotations to trap and recycle nutrients in the soil profile, increase soil tilth, and provide a diversity of crop residues
- Applying manure and incorporating cover crops in the spring prior to plant production and nutrient uptake.
- Avoiding surface application of manure prior to rainstorms, irrigating, or when the ground is frozen or saturated
- Using practices that enhance soil quality and reduce the potential for water runoff and wind and water erosion
- Providing buffers or filter areas between cropping areas and water bodies to protect against nutrient and sediment movement into lakes and streams

4.2 Proper Storage of Manure or Compost Materials

Many organic farmers make their own compost or store manure to be used in crop production. These materials are a concentrated source of nutrients that can easily leach through the soil into the groundwater or be washed into streams through runoff. To minimize the potential for these environmental losses, manure or compost piles should be set up on soils that have been cemented or compacted to minimize leaching under the piles. Diversions should be established upslope from the manure or compost pile to prevent contamination of clean rainwater and filter areas should be constructed and maintained down slope to treat any contaminated water running off the pile. Setting up manure or compost piles in a roofed enclosure will decrease the potential for runoff while protecting the quality of these valuable nutrient sources.

Pathogens. *E. coli*, *Cryptosporidium*, and *Giardia* are disease-causing microorganisms often found in manure. Crops can be contaminated with pathogens if fresh manure is applied to growing crops or shortly before planting or if contaminated water is used to irrigate crops. Poor sanitary practices by farm workers during crop production and harvesting can also cause produce to become contaminated with pathogens. People who consume contaminated food or water can suffer gastrointestinal problems. These problems pose the greatest threat to young children, elderly people, and persons who are immuno-compromised.

Heavy metals. The term heavy metal refers mainly to lead, cadmium, arsenic, nickel, copper, zinc, and iron. While the last three elements are required for plant growth in small amounts, build-up of these elements in the soil environment can be phytotoxic as well as damaging to the growth of soil organisms. While the National Organic Standards prohibit the use of sewage sludge or biosolids because these products tend to have high concentrations of heavy metals, they do not regulate the source of animal manures used as fertilizers. However, animal manure may be contaminated with heavy metals if these materials were used as feed supplements. To avoid a build-up of heavy metals in your soil, analyse manure used as fertilizer and check your soil regularly for changes in the accumulation of these contaminates.

Plastics. Plastics have revolutionised horticulture production by serving as mulches and row covers. Unfortunately, the advantages of plastic use come with an environmental price. While plastics reduce leaching and waterlogging of covered soils, they also concentrate water that cannot soak through the plastic into the soil. Concentrated water flowing off from plastics forms erosive streams, which has been shown to cause up to four times more water runoff and 15 time more soil erosion as compared to fields mulched with organic materials. Disposal of plastic mulches poses an additional environmental problem. The National Organic Standards require removal of plastic mulches from beds at the end of the production season, and once removed, good methods for plastic disposal or recycling are lacking. Since soiled plastics cannot be recycled economically, plastic users are forced to dispose of this material through incineration, burying on the farm, or land filling. All of these methods are damaging to the environment and ultimately unsustainable.

5.0 LIVESTOCK ON ORGANIC FARMS

An organic farm cannot achieve its full potential or ecological balance without livestock manure; that it is essential to nutrient cycling and to the inner aspects of soil building.

Excellent soil fertility can be built in the absence of farm livestock and livestock manures by using vegetation-based composts and by harnessing the livestock in the soil-earthworms and other soil organisms. However, it is clearly easier to design a contemporary, low input organic farm when traditional livestock are integrated. The biological and enterprise diversity that livestock can bring contributes enormously to stability and sustainability.

6.0 QUALITY MANAGEMENT

Like in every organic market the credibility of organic products is very important. Without consumers trust, organic farming is lost. There will be no success in **Indian Organic Market** without quality assurance on a high level. The principles and standards of organic agriculture has to be known to all stakeholders, such as, farmers, processors, traders, exporters, government etc. and the last but not the least to the consumers.

Chapter 11

Organic Vegetable Gardening

Organic gardening differs from 'conventional' gardening mainly in the areas of fertilization, diversity and pest control. The organic gardener prefers to use natural and organic materials and methods, and avoids using practices and synthetic chemicals that may be detrimental to soil and human health or environment.

Successful vegetable gardens require good planning, constant care, and the will to make things grow. Among the many things, a vegetable garden may offer toward a satisfying experience are fresh air, sunshine, supplemental income, and fresh food, rich in vitamins and minerals, harvested at the best stage of maturity.

The information in this chapter should be beneficial to all gardeners regardless of methods of culture used; however, it is primarily intended to aid the organic gardener to employ workable methods acceptable and compatible with the philosophy of 'organic gardening'.

1. WHAT DO THE ORGANIC GUIDELINES SAY?

1.1 Soil and Potting Earth Substrates

- Growing of vegetables is permitted only in soil. Growing in earth-less media (e.g. stone wall), hydroponics culture, nutrient-rich plastic films and similar methods and techniques are not allowed. Chicory buds may be grown in water, but without added fertilizer.
- The addition of peat to enrich the soil with organic matter is prohibited. The same applies to the use of styropore (Styromull) and other synthetic substances in the soil or in growth media.
- Purchase of organic fertilizer (compost) serves only as a supplementary measure to cultivation methods.

1.2 Seedlings

- Seedlings must either be from the own organic farm itself or purchased from organically-approved operations. When unforeseeable shortages occur, the Certifying body will decide on the procedure within the bounds of legal prerequisites.
- When rearing seedlings are cultivated, the amount of peat used is to be kept to a minimum.

1.3 Steam Treatment of Small Areas and Soil

- When growing under plastic covers and in the production of seedlings, a superficial steaming for control of competing plants is authorised. This applies to soil and other growth media, but the steam treatment is to be kept to a minimum level.
- Deep steam treatment to eliminate soil epidemics requires a special permit.

1.4 Growing under Glass and Plastic

- In winter (December to February) the cropping areas may simply be kept free of frost (about 5 degrees C). An exception is made in the case of seedling production.
- In selecting a heating system and the fuel to be used, compatibility with the environment must be respected. Care must be taken to have housing with good insulation.
- The use of plastic and fibre sheet etc. on the soil is to be kept to a minimum. Used sheeting is to be taken away for recycling and disposed at safe place.

2. SOME EARLY PLANS

Choosing a location: Select a plot of good, well-drained soil near a water supply. The vegetable growing area should not be shaded by tall buildings or trees. Enclosing the garden spot with a fence is usually profitable.

The garden design: Many gardeners find it helpful to draw out on paper the location of each row and the crop or succession of crops to be planted.

2.1 Climate

Each vegetable crop has specific climatic requirements and the season selected for the cultivation should be the one during which period the crop growth, yields and produce quality are best. Some disease and insect problems are especially prevalent at specific, predictable times of the year and vegetable crops can be planted at dates that protect them from these high-risk periods. The following sowing schedules (in India) can be followed:

Rainy Season: Bhendi, Brinjal, Chilli, Onion

Winter Season: Carrot, Cauliflower, French Beans, Onion, Peas, Tomato

Summer Season: Bottle gourd, Cucumber, Pumpkin, Ridge gourd

Avoid tomato cultivation during rainy period, as *alternaria* and *fruit rot* diseases are severe. During high temperature period, virus diseases are severe in Okra (bhendi), Chilli, Tomato and Beans. In capsicum, fruit set is a problem.

2.2 Crop and Variety Selection

The vegetable crop(s) selected to be grown is an economic and/or personal decision.

The potential market should be determined even before deciding what crops to grow, acreage, and the time of year best to be in production. Next comes the matching of soil type to crop and which variety to select.

Organic vegetable growers have fewer choices. Disease and insect resistance must be considered when selecting the variety. Variety trials results will provide information on adaptability of variety under a range of environmental conditions. Disease resistance is the most economical and effective means of pest management. Yield, horticultural quality, and market acceptability also have to be considered in making variety selection.

As per experiments and trials tested at different locations, following Indian Varieties of vegetables resistant to diseases and insects may be used for organic vegetable production:

Vegetable	Varieties
Tomato	Arka Abhijit (F1), Arka Shrestha (F1), Arka Alok, Arka Abha, Arka Vardan (F1), H-24
French Beans	Arka Jay, Arka Vijay
Okra (Bhendi)	Arka Anamika, Arka Abhay, Prabhani Kranti
Cauliflower	K1 (IARI), Pusa Shubra
Cabbage	Pusa Mukta
Brinjal	Arka Keshav, Arka Neelkant (F1), Arka Nidhi, Arka Sheel, Swetha, Surya
Onion	PBR-5
Cucumber	PCUCH-1
Cowpea	Arka Garima, Arka Suman
Cluster bean	Pusa Navbahar, Gauri
Dolichos bean	Arka Jay and ArkaVijay
Muskmelon	Pusa Madhuras , Arka Rajahans
Watermelon	Arka Manik
Garden Pea	VL Ageti Matar-7

Note: *Bacterial wilt resistant varieties in tomato and photo-thermo insensitive hybrids in cabbage should be used. The resistant (disease-pest) vegetable varieties developed under local conditions should be preferred.*

2.3 Soil

To become a successful organic vegetable grower, soil fertility must be enhanced. It takes three or more years of adding cover or green manure crop, crop residue, and compost to sandy soils before good production results can be expected. Helpful practices include crop rotation, use of cover crops, green manure crops, use of mulch, animal compost, plant material compost, etc.

The ideal soils for growing vegetables are well drained, fairly deep, and relatively high in organic matter (2-2.5%) with C:N ratio between 15-20:1 and gentle slope of 2 to 3 per cent. Deep sands may be too well drained and very low in organic matter for vegetable production. Healthy and productive soil helps crop to develop good root systems and reduces crop stress caused by drought or excess rainfall. If drainage is a problem, plant on raised beds to promote drainage and faster warming of soil.

Most of the vegetable crops grow well in pH range of 5.5-7.5. pH below 5.5, liming is required. Salinity problems can be solved by frequent irrigations before planting to leach out salts or effect can be reduced by adopting drip irrigation, while the sodicity can be checked by the application of gypsum.

In order to prevent over-or undersupply, the soil's or substratum's nutrient and humus content must be analysed at least every third year. A soil analysis for harmful substances (heavy metals, organic compounds) has to be carried out and presented at the beginning of the conversion.

2.4 Organic Manure and Fertilizer

A major basis for organic gardening is the use of sufficient quantities of organic material applied to the soil. Usually, it is in the form of animal manures, plant manures, cover crops, compost, or mixed organic fertilizer.

Benefits of adding organic matter

- Improves tilth, condition, and structure of soil
- Improves ability of soil to hold water
- Improves ability of soil to hold nutrients
- Improves "buffering" capacity of soil
- Supports the soil's microbiological activity
- Contributes nutrients, both minor and major
- Releases nutrients slowly
- Acids arising from the decomposition of the organic matter help to convert insoluble natural additives such as ground rock into plant-usable forms.
- Helps vegetables survive stress, as from nematodes.
- Helps dispose of organic waste products.

What happens to organic matter applied to the soil or compost pile?

Under suitable conditions, the organic matter is decomposed by microorganisms such as fungi, algae, bacteria, moulds, and earthworms. In the process, insoluble and unavailable (to plants) nutrients, such as nitrogen, are gradually changed into simple usable products.

For example, *nitrogen* is converted from the unusable organic forms to a usable inorganic form through the process called nitrification. Thus, nitrification is the breakdown of protein (organic nitrogen) into ammonia and then nitrate. Some of the organic matter becomes part of the soil humus.

Proper conditions for nitrification

First, materials containing nitrogen must be present. There is a great variation in the amount of nitrogen the different organic materials contain. Then certain soil or compost conditions are necessary:

- Proper soil acidity (pH) – should be about 7.0; in acid situation below 5.5 it ceases.
- Proper temperature of soil – above 10°C.
- Good aeration (does not occur with wet, soggy soil or compost).
- Adequate lime for use by microorganisms and to keep the soil from being acid.

Animal manures: Where animal manures are available, they are probably the best source of fertilizer and organic matter for the organic gardener. Use manure which has been aged for at least 30 days, or composted. Manures vary greatly in their content of fertilizing nutrients. The composition varies according to type, age, and condition of animal; the kind of feed used; the age and degree of rotting of the manure; the moisture content of the manure; and the kind and amount of litter or bedding mixed in the manure. The average minimal amounts of nutrients may be as high as 4.5% N; 2% P; and 2% K in some cases. Animal manures also provide most of the micronutrients needed. Some manure products are composted, rehydrated or mixed with plant litter to enhance their fertility.

How much to apply as broadcast: A minimum of about 5 tonnes of manure per acre of garden soil. For best results, supplement recommended dose of ground rock phosphate or raw bone meal.

How to apply: Broadcast evenly over plot and spade, roto-till or otherwise work into topsoil. Apply three or more weeks before planting. A small amount may be mixed well in the planting hole; however, plant injury may occur with 850 g or more placed in the hole.

How to apply as a side-dressing: Scatter a band of manure down each side of the row. Place each band at the edge of the root zone and work lightly into the soil surface. For individual plants, open a furrow encircling the plant and fill with manure, then cover.

If mulch is present, rake it back at the edge of the root zone in order to apply the band of manure, then re-cover with the mulch.

Composts: Acceptable manure-like organic fertilizer (artificial manure) may be obtained through the process of composting. Compost is made by alternating layers of organic materials, such as leaves and kitchen table refuse, with manure, topsoil, lime, organic fertilizer, water, and air, in such a manner that it decomposes, combines, and yields artificial manure. Details of different type of composting is given in the Chapter- *On farm Organic Manure Production.*

Use of Compost in the Garden: Broadcast it over the entire garden three weeks or more before planting. Or if you have only a small quantity of compost, it may be mixed into the soil along each planting furrow or at each hill site.

CAUTION: If your compost is made from mostly woody materials, it may temporarily deplete the nitrogen from the soil and plants. Be sure to mix manure with it when applying.

Natural and organic fertilizers: Natural and organic materials which yield plant nutrients upon decomposition are often applied to the garden separately or combined, used in the compost pile, or mixed with manure.

Natural Deposits (Rocks, Sands, Shells, etc.): Such naturally occurring materials are usually not easily obtained in today's modern agriculture; however, where available they represent sources of mainly potash, phosphorus, and lime (calcium and magnesium) for organic gardeners.

Phosphorus: Rock phosphates are natural deposits of phosphate in combination with calcium. The material as dug from the earth is very hard and yields its phosphorus very slowly. When finely ground and with impurities removed, the powdery material is only slightly soluble in water, but may be beneficial to plants in subsequent seasons following application. The reaction of phosphate rock with acids from decaying organic matter in the garden or compost tend to make the phosphorus available to garden plants. Colloidal phosphate is also available and widely used.

Broadcast the material over the soil surface and work into the topsoil at least three weeks before planting. Manure or other organic fertilizer should be added at this time. Since the materials are so slowly decomposed, side-dressings are seldom beneficial.

Potash: Potassium is widely distributed in nature, occurring in rocks, soils, tissues of plants and animals, and water of seas and lakes.

In gardening practice, materials such as wood ashes, seaweed, potash salts, greensand, and ground rock potash are used alone, in combinations with other materials yielding other nutrients, mixed with manure, or in compost piles.

Since the potash bearing materials vary so much in composition and rate of decomposition, specific application rates must be determined for each material and its combinations.

Micronutrients: An advantage for using organic materials as fertilizers is that they contain many of the elements also needed by the plants in addition to N, P, and

K (for example, manganese in manure).

Besides the general amounts of micronutrients found in most organic materials, certain ones are concentrated into such naturally occurring materials as gypsum (calcium and sulphur), marl (calcium), dolomite (calcium and magnesium), limestone (calcium), basic slag (iron, calcium, manganese and magnesium), and finely ground borosilicate.

Since organic fertilizer and soil conditioning materials are slow working in general, they should be mixed into the soil at least three weeks ahead of planting and the soil thoroughly prepared for the seed or transplants. Clumps of unrotted organic materials not only interfere with the seeding operation, but also may result in nutrient deficiency and possible soil-borne diseases problems such as "damping-off" of young seedlings.

3. SOIL CONDITIONING AND MANURING

3.1 Lime Application

Reducing the acidity of the soil is the primary purpose for using lime in the garden. However, liming materials also provide nutrients for plant use. Calcium and magnesium are the two elements most commonly provided by lime. Gypsum is used where more calcium is needed without raising the pH.

Natural deposits of lime which are an organic gardener might use are limestone, dolomite, shell, and marl. All these forms must be finely ground to provide maximum benefit to the soil and plants. Dolomite is preferred due to its content of both calcium and magnesium.

Lime Requirement (LR) for different soils: The lime requirement depends on soil texture (clay content), CEC and sesquioxide content of soil. In the absence of more accurate recommendation the ready reckoner may be followed. Requirement of different quantities to raise the soil pH to 6.5 from different degradation is given in table 11.1 and 11.2.

Table 11.1. Requirement of lime for reclamation of acid soil

Lime required to raise pH to 6.5 (kg/ha)

pH Range	Soil Type				
	Sandy	Sandy loam	Loam	Silt loam	Clay & loamy clay
4.5 to 5.0 (For pure $CaCO_3$)	4250	7250	10750	15000	20000
5.1 to 5.5 (Pure $CaCO_3$)	2500	4250	6250	8500	11300
5.6 to 6.0 (pure $CaCO_3$)	1000	1750	2500	3500	5000

Table 11.2. Requirement of lime at different pH of acid soil

Soil pH	Lime Requirement kg/ha		
	Sandy loam	Loam	Clay loam
5.0	1,262	1,892	2,944
5.2	1,093	1,639	2,551
5.4	925	1,387	2,159
5.6	757	1,135	1,766
5.8	589	883	1,374
6.0	421	630	981

Apply lime well in advance of the planting date, preferably 2 to 3 months before the garden is planted. Mix well with the soil and keep moist for best reaction. Application closer to planting time is permissible, but its benefits are delayed.

3.2 Manure Application

When growing vegetables in the open, the nitrogen fertilization must not exceed 110 kg/ha and year in the average of the crop rotation on horticultural area. Due to higher degree of nutrient decomposition in the soil on account of a higher degree of cultivation intensity in greenhouses, more concentrated manure application (over 110 kg/ha in a year) may be permissible in some cases after consultation with Certifying body.

The quantity of farm manures and approved organic and mineral fertilizers should be based on the results of the soil analysis and the data about the nutrient requirements of the crop rotation. All fertilizer sources have to be accounted for. An analysis of the nitrogen level on the farm has to be calculated annually. The supply expected from harvesting residues, green manure and humus should be considered when applying nitrogen fertilizers. Fields that are expected to lie fallow for more than 12 weeks during the vegetation period (April to November) grow green manure crop.

Manure application for organic vegetable gardening may be as follows:

- Add 20-30 tonnes of fully decomposed farm yard manure (FYM) and 1.5 tonnes of vermin-compost and 250 kg Neem cake with 8 % oil to the soil. Also apply *Trichoderma* treated FYM (*Trichoderma harzianum* at the rate of 1 kg per tonne of FYM) at the rate of 2-3 tonne per hectare to reduce the nematodes and soil borne pathogens infestation.
- Prepare ridges for transplanting at recommended row spacing according to the crop to be grown.
- For supply of nitrogen, green manuring crop can be grown and incorporated at least 3 weeks before transplanting as N source or in addition to organic manure use groundnut or cotton cake or meal (6-7 % N).
- At 30 days after transplanting, repeat vermin-compost (1.5 tonnes per hec-

tare) and neemcake (250 kg per hectare) application.

- As a source of phosphorus, use rock phosphate in combination with phosphate solubalising bacteria (PSB, commercially available as phosphor-bacteria) or bone meal can be used.
- As source of Potash, wood ash (1.5 % K_2O) or sheep manure (3-4 % K_2O) can be used.
- At 10 days after transplantation, spray bio-dynamic solution (e.g. Panchagavya) over the plants and repeat 4-5 times at 10 days interval to supplement nutrient and as tonic.

4. NURSERY RAISING

Disease-free seeds and seedlings

Clean seeds and seedlings are must in organic vegetable production. To avoid problems, do not save seeds from fields or areas where disease are prevalent. Examples of seed-borne diseases are anthracnose of beans, mosaic virus of peas, bacterial blight of beans, black rot of cabbage, leaf spots of turnip, mustard, etc.

Raising seedlings in farm area

- Prepare raised nursery beds of 10-15 cm height providing necessary drainage.
- During summer season solarisation of beds is recommended.
- Mix FYM, vermin-compost and *trichoderma* powder in the ratio of 0:1:0.1 and apply 0.3 kg of this mixture and 1 kg neem cake per sq. meter and prepare the nursery bed.
- Seeds are sown in lines with a spacing of 7.5cm x 5cm and covered by vermicompost or straw.
- It is good to protect nursery beds with 40-50 mesh nylon net cover of 1.5 m height to prevent insect vectors.
- Biofertilizers like Azotobactor/Azospirillum and Phosphate Solublising Bacteria (PSB) need to be applied in nursery stage itself. Apply biofertilizer @50 g per sq. meter of nursery area. While transplanting, dip the root of the seedlings in a slurry made up of these biofertilizer for 30 minutes.
- The seedling will be ready in 25 to 45 days depending on the vegetable crop.

Raising seedlings in plastic pro-trays

- Each flat or pro-tray having 98 cells is used for raising seedlings of tomato, cole crop, etc.
- Use commercially available phosphor rich coco-peat (coir pith powder) available in bags (fully decomposed and sterilized) to fill the trays. About 1.0-1.25 kg coco-peat is required to fill one tray.

- Apply 10g each of Azospirillum/Azatobacter, PSB and *Trichoderma* per kg of coco-peat and mix well with the media.
- Make a small depression in the centre of the coco-peat layer by hand and put one seed per cell and again cover by coco-peat just enough to cover the seeds.
- Keep about 10-12 trays one over the other until germination (3-4 days). After germination, keep them singly over plastic mulched beds prepared under protected net cover or house.
- Irrigate daily by garden can. The seedling will be ready in 20-25 days.

Transplanting/direct sowing

- Spray *Bacillus thuringiensis* formulation (1g or 1ml) to the vegetable (cabbage, cauliflower, etc.) seedlings.
- At the time of transplantation, dip the root portion of seedlings in 1per cent Bordeaux mixture solution to drench the trays with it.
- Similarly biofertilizers such as Azotobacter/Azospirillum and Phosphate Solublising Bacterial slurry can be used.
- When seeds are used for direct sowing the biofertilizers (Azotobacter/ Azospirillum and Phosphate Solublising Bacteria) can be used for seed coating for non-leguminous vegetables. While Rhizobium and PSB are recommended for leguminous vegetables.

5. IRRIGATION

In irrigating the garden, it is advisable to thoroughly wet the soil once a week unless sufficient rainfalls. Thus, the soil will be moistened throughout the root zone. Light sprinklings every day merely tend to wet the surface and encourage shallow root growth. Foliage wetting should be minimized in order to reduce foliar disease incidence. Drip or trickle irrigation with plastic mulching is encouraged to maintain better soil moisture regimes and also reduces pest and disease incidence in addition to saving of irrigation water through reduced evaporation losses as well as preventing weed growth. Use of organic materials as soil conditioners and fertilizers tends to improve the ability of the soil to retain moisture.

6. MULCHING

Mulch is any material, usually organic, which is placed on the soil surface around the plants. Organic materials most commonly used for mulching are leaves, grass clippings, pine straw, sawdust, and wood shavings. Synthetic materials, mostly plastic sheeting, have been used quite often in recent years.

Among the benefits of a mulch are (a) conserves soil moisture, (b) conserves nutrients, (c) reduces soil erosion, (d) reduces crop loss due to nematodes, (e) reduces weed growth, (f) provides barrier between fruit and soil, thus reducing soil rot on fruit, and (g) moderates the soil temperature.

Apply mulch before or after seeding or transplanting. Roll back the mulch with a rake in order to wet the soil beneath when irrigating, for best results.

At the end of the garden season, the mulch (except plastic) may be removed and composted, or cut into the garden soil. Most mulch is woody and should have manure or other rich organic fertilizer applied with it when cutting into the soil.

7. WEED CONTROL

Hand removal (pulling and hoeing) is the most widely practiced method of eliminating weeds. Thermic weed control (for noxious weeds) are permitted. All equipment from conventional farming system must be properly cleaned and free from residues before being used for weed management in organically-managed area.

A garden mulch, such as dry grass, straw, bark, leaves, banana sheath, composted sawdust or other similar material, will help to keep weeds from growing if the mulch is thick enough to exclude light. Plastic mulch is a widely used ground cover in organic vegetable production especially for widely spaced crops such as cucumber, green house capsicum and tomato, etc.

8. INSECT AND DISEASE CONTROL

During periods when infestations of various garden pests are high, control by natural means becomes very difficult. However, the following practices will help to reduce losses without use of chemical pesticides.

- Plant resistant varieties and select disease and pest free plants.
- Rotate garden areas
- For cutworms, place a cardboard of tinfoil collar around plant stems at ground level.
- Spade garden early so vegetation has time to rot before planting.
- Use mulch, vegetables touching the soil may rot. A good garden mulch tends to reduce damage caused by nematodes.
- Handpick insects and clean up crop refuse early.
- Keep out weeds which harbour insects and diseases.
- Summer fallowing (clean cultivation) helps control nematodes.
- Summer flooding, where soil type permits, helps control nematodes.
- Water in morning so plants is not wet at night.
- Dispose of severely diseased plants before they contaminate others.
- Some insects, like cabbage worms, may be killed by spraying with natural preparations such as *Bacillus thuringiensis*.

- Solarize transplanting soil at least 7-10 days.
- *Crotolaria spectabalis* and marigolds, when planted as cover crops, tend to reduce some kinds of nematodes. The use of marigolds to *repel* nematodes from interplanted vegetables is not effective control.
- Many organic gardeners approve of and use sprays and other preparations containing naturally-occurring materials. Pyrethrin, rotenone, and ryania are examples of natural poisons from plant parts. These give some control to some insects under certain conditions.
- Natural predators should be encouraged wherever possible; however, predators raised in captivity, then released into the garden area are usually ineffective.
- Insecticidal soaps, made from fatty acids tend to work well for some insects under average conditions.
- Insect traps, baited with phermone lures, work well in some instances. Many of these have sticky adhesives to catch insects.
- Solar fumigation is effective in reducing some soil-borne problems such as nematodes.

8.1 Pest Control of Some Vegetable Crop

Cabbage, Cauliflower and Knol-Khol (aphids, diamond back moth)

- Removal and destruction of all the stalks and stubbles sown after harvest and deep ploughing of field prevent the carry over of pest specially diamond back moth.
- Hand-picking and destruction of caterpillars, webbed leaves and larvae in early stage of attack helps the pest infestation.
- *Use of trap crop*: Paired mustard rows (bold seeded) should be grown in beginning and after every 25 rows of cabbage to obtain continuous foliage until maturity or cabbage. First row of mustard should be grown 15 days prior to cabbage transplanting, while second row 25 days after transplanting. Mustard (a preferred host) attracts 80-90% diamond back moth (DBM). Aphids are also attracted during flowering of mustard. The limited surviving pest population on inter-cropped cabbage may effectively be controlled with 4% need seed kernel extract.
- A planting pattern of one row of cabbage and one row of tomato (cabbage planted 30 days later than tomato) affords maximum reduction of DBM and leaf webber larvae on cabbage.
- Spreading Blue or Green cloth pieces in evening hours in crop field attract spodoptera larvae, which they use for hiding beneath the cloth, can be collected and destroyed.
- Yellow sticky traps @15-20 per hectare for aphid management.

- *Use of antifeedant*: Triphenyltin acetate (0.1-0.2%) can be used against grubs of epilachana and mustard sawfly and spodoptra litura, where as bitter gourd seed oil emulsion is effective only against sawfly. Consult certifying body before using antifeedant.
- *Use of biocontrol*: *Apanteles plutellae* is an effective parasitiod of diamond back moth and Trichoplusia sp. An entomopathogen, *Fusarium semitectum* can be effectively utilized for control of aphids infesting cauliflower/cabbage. *Menochilus sexmaculatus* (coccinellid predator) and *Aphelinu spp*. (hymenopterous parasite) are effective against several species of aphidsin vegetable crops. Stethorus pauperculus is noted to be highly specialized spider mite pests infesting vegetable crops.
- Bt.@ 0.5 to 1 kg/ha effectively used to control DBM and shoot borer (Hundalis). Adjuvant like whole egg homogenate 1% and whole milk 0.5% increase the efficacy of Bt, NPV@ 250 LE/ha is effectively used for the control of Spodoptera. NSKE 4% at primordial or head initiation stage; 2-3 post heading stage at 10-15 days interval effective.
- *Insect growth regulator*: Moult inhibitor Dimiline (Dflubenjuron) 25 wp @ 0.05% effective against spodoptera more so in case of early instars. Prior approval from certification body is required for use of such growth regulator.

Brinjal (jassids, whitefly, shoot and fruit borer, hadda beetle)

- Ratooning must be avoided to minimise stem borer and shoot and fruit borer infestation.
- Deep summer ploughing and clean cultivation is helpful in keeping borers and hairy caterpillar population under control.
- Pusa Kranti, Pusa purple cluster H-4, A-II Arkakusmark, Chakalasi Doli, Doli-5, SM-67, and SM-68 varieties are found resistant/tolerant against shoot and fruit borer, jassids, whiteflies and *Epilachna spp*. (Haddabeetle).
- Clipping of dropped and dead shoots and fruits and collection and destruction of egg masses of epilachna beetle found useful to suppress the pest population.
- At fruiting stage, neem formulation (4-4.5 lit/ha) or Bt. (1-1.5 kg/ha) may be used to check the infestation of fruit borer.

Tomato (whitefly, jassids and fruit borer)

- BT-1, T-32 and T-27 are found resistant/tolerant against tomato fruit borer *(Heliothis armigera)*.
- Simultaneously planting of 40 and 25 days old seedlings of marigold (African marigold) and tomato, respectively in a pattern of 16 rows of tomato alternated with one row of marigold. Marigold nursery is raised 15 days earlier to tomato. Both the crops will flower simultaneously and buds of

marigold will attract 90% of resident and immigrant moth of Heliothis (fruit borer) for oviposition and colonization. It has added attraction potential for serpentine leaf miner, and conserves natural enemies. Thus the larvae developing on marigold could be killed by spraying little amount of (16 times less) of any safer insecticides.

- Hand-picking and destruction of infested fruits to suppress the fruit borer population.
- 2-3 releases of *Trichogramma brassiliensis* @ 60000 (3-Tricho cards each containing 20000 parasitiods, may be cut in small pieces and distributed in one acre area) at 15-days interval are found effective against fruit borer.
- NPV (Nuclear Polihedrosis Virus) @ 250 LE effectively control *H. armigera.*
- Spray with bacterial formulation (Biolep or Dipel @ 0.5-1kg/ha) at fruit initiation followed by second spray 15 days later.

Okra (Ladies Finger)

- Deep summer ploughing and clean cultivation help minimise infestation of spotted bollworm, red hairy caterpillar and whitefly.
- Collection and destruction of infested plants/plant parts to check infestation of shoot and fruit borer.
- Padmini Kranti and P-7 are resistant against whitefly and kalyanpur Boni, moderately resistant against shoot and fruit borer and Erias sp. AE-57, PM-8 Perkins long green and PKX-9275 are resistant against shoot and fruit borer.
- Intercropping of leguminous crop (Cowpea) prevents pest appearance on Okra (bhendi).
- Spraying of neem-based formulation during fruiting stage to control fruit borer infestation is advisable. The Bt. (biolep or Depel8L) formulation can also be used to reduce insecticidal pressure to suppress the population of spotted bollworm.

Cucurbits (Gourds, Cucumbers)

- Deep ploughing of fields just after harvest to kill grubs of red pumpkin beetle and pupae of fruit fly.
- Frequent racking of soil under cucurbit vines to kill eggs of red pumpkin beetle and pupae of fruit fly.
- Collection and destruction of infested fruits and adult beetles during early morning when they are sluggish.
- Selection of early maturing and resistant varieties Arka Tudia of round gourd, Arka Surya Mukhia of pumpkin against fruit fly.
- The vines may be dusted with ash which acts as repellent against red pumpkin beetle.

Table11.3: Use of potential agents for pest control of the vegetable crops

Crop/pest	Biotic agents	Dosages/ ha	Frequency of application	Application method
Cabbage				
Plutella xylostella	Bacillus thuringiensis (several formulations, including Bt oligosporogen mutant (300g)	500 g	Need-based or at weekly intervals	Spraying with knapsack sprayer
Tomato				
Helicoverpa armigera	Trichogramma brasiliense/ T. pretiosum/ T. schilonis $BioH_1$	50,000 per ha	Weekly intervals 6 times from 25 day after transplanting or during egg laying period	Releasing adults for stapling parasitized eggs cards
	HaNPV	250 LE/ha	Thrice during the growth period	Spraying with knapsack sprayer
Pythium spp.	Trichoderma viride/T. harzianum	4g/L	Before planting	Seedling root dip
Fruit borer	Chrysoperla, Chilonus, Bracon Trichogramma	40,000 per ha	at weekly interval NPV (125 LE/ha)	---
Potato				
Agrotis spp.	Stienernema carpocapsae	5 billion infective juveniles/ha	on detection of cutworm larvae	Apply in the soil through irrigation
Leaf, miner, Aphid, Jassid, White flies Thrips	Chelonus blackburni, Copidosoma koehleri, T. chilonis	50,000 per ha	any one four times at 10 days interval	---
Tuber moth	Bacillus thuringiensis.	0.05%	Spray	---

HaNPV, Helicoverpa armigera nucelopolyhedro virus
LE, Larval equivalent (=6 x 10^9 POB)

Potato (cutworm, aphid, tuber moth)

- B1A 21-29 is resistant/tolerant against potato tuber moth
- Deep ploughing to expose hibernating stages of cutworms for natural enemies.
- 'Shakti' less susceptible to aphid, whitefly and cutworm and 'Jyoti' least susceptible to leaf hopper may be taken advantage.
- Early sowing (end of September to beginning of October) reduces the infestation of aphid.

- When aphid reaches 20 aphids/100 leaves, the crop may be sprayed with dimethoate or methyl-o-demeton @ 1lit/ha
- Cover potato with 2.5 cm thick layer of Banana leaf. Use sex pheromone for trapping moth both in field and in stores.

Root crops (radish, turnip and sugar beet)

- Early sown radish crop severely with mustard sawfly, therefore late sowing is suggested.
- Collection and destruction of sawfly grubs in early sunny hours of the day.
- The crop may be sprayed with one of the neem formulations (4-4.5 lit/ha) to control sawfly and bacterial formulation (Biolep @ 0.5 –1.0 kg/ha) to control diamond back moth and cabbage butterfly.

Garlic and Onion (thrips, onion fly, maggot)

- Regular irrigation and periodical hoeing are effective for reducing the pest. White variety of Onion is least susceptible to thrips.

Leafy vegetables (spinach, lettuce, amaranthus, fenugreek)

- Collection and destruction of egg masses of leaf feeding caterpillars, infested shoots and leaves regularly.
- The crop may be sprayed with bacterial formulations (Biolep 1 @ 1-1.5 kg/ha) to control leaf feeding caterpillars.

9. HARVEST AND POST-HARVEST MANAGEMENT

- Schedule harvest timings during cool hours of the day.
- Pick the produce at right stage to retain firmness so that transport shocks are reduced.
- Handling of the produce should be done by using plastic containers, as the contamination of the produce should be totally prevented.

In the developed countries, organically grown produce fetch premium price of 2-4 fold than conventionally grown produce. In coming years as the purchasing power of the Indian consumers increases, farmers will certainly get higher price for the organically grown vegetables. Farmers can also earn more in the coming years by supplying the organically grown produce for the export market.

Chapter 12

Organic Spices Production

The world demand for organically-produced foods is growing rapidly in developed countries like Europe, USA, Japan and Australia. The current estimated share of organic foods in these countries is approximately 1 to 1.5 per cent. According to the ITC and UNCTAD/GAT, more than 130 countries produce certified organic foods. Internationally, there's a shift towards traditional ethnic system of medicines. Since spices form part of many of these medicines, the demand for originally-produced spices is increasing considerably.

India grows over 50 different varieties of spices. Total production is around 2.7 million tonnes. The country at present exports around 50 tonnes of different varieties of organic spices. Exports will get a significant boost in the coming years as more farmers switch to organic methods.

To promote the production of organic spices in India, Spices Board had started several research programmes on organic cultivation of important spices. The work is carried out at the Spices Board's Indian Cardamom Research Institute at ldukki District in Kerala. Besides organising demonstrations to educate and motivate prospective organic spice growers, the Board is simultaneously involved in training programmes to existing spices growers on organic principles and practices.

1. WHAT DO THE ORGANIC GUIDELINES SAY?

1.1 Duration of Conversion Period

For existing plantation, a minimum of three years is required as conversion period for organic cultivation. For a newly planted or replanted area raised through organic cultivation practices, the first yield itself can be considered as organic provided chemicals have not been used in the previous cropping. In case of cultivation on virgin land and for farms where records are available that no chemicals were used previously, the conversion period can be relaxed. It is desirable that organic method of production is followed in the entire farm, but in large plantations the transition can be phased out for which a conversion plan is to be prepared.

Buffer zone: An isolation distance of at least 25-30 ft width is to be left from all around the conventional plantation in order to avoid contamination of organically cultivated plots from neighbouring farms/conventional farms. In sloppy lands, adequate precaution should be taken to avoid the entry of run off water and drift from the neighbouring farms.

Organic cultivation of some of the important spices as suggested by Spices Board of India and other concerned Research Institute and Universities are discussed below:

2. PRODUCTION TECHNOLOGY

2.1 GINGER

Ginger (*Zingiber officinale*) is one of the important spices grown in India. It is marketed in different forms such as raw ginger, dry ginger, bleached dry ginger, ginger powder, ginger oil, ginger oleoresin, gingerale, ginger candy, ginger beer, brined ginger, ginger wine, ginger squash, ginger flakes etc.

2.1.1 Climate and Soil

Ginger is cultivated in almost all states in India. Kerala is the major ginger growing state. Other major ginger growing states are Orissa, Karnataka, Meghalaya, Himachal Pradesh, Arunachal Pradesh and Manipur.

Ginger grows in warm and humid climate. It is mainly cultivated in the tropics from sea level to an altitude of above 1500 MSL and it can be grown both under rainfed and irrigated conditions. For successful cultivation of the crop, a moderate rainfall at the sowing time till the rhizomes sprout, fairly heavy and well distributed showers during the growing period and dry weather for about a month before harvesting are necessary. Ginger thrives the best in well-drained soils like sandy or clay loam, red loam or lateritic loam. A friable loam rich in humus is ideal. However, being an exhausting crop it may not be desirable to grow ginger in the same site year after year.

2.1.2 Maintenance of Buffer Zone

In order to cultivate ginger organically, a buffer zone of 25 to 50 feet is to be left all around from the conventional farm, depending upon the location of the farm. The produce from this buffer zone belt should not be treated as organic. Being an annual crop, the conversion period required will be two years. Ginger can be cultivated organically as inter or mixed crop provided all the other crops are grown following organic methods. It is desirable to include a leguminous crop in rotation with ginger. Ginger-banana legume or ginger-vegetable-legume can be adopted.

2.1.3 Sources of Planting Material

Carefully preserved seed rhizomes free from pests and diseases which are collected from organically cultivated farms can be used for planting. However, to begin with seed material from high-yielding, local varieties may be used in the absence of organically produced seed materials. Seed rhizomes should not be treated with any chemicals. The seed rate varies from region to region and with method of cultivation adopted. The seed rate varies from 1500- 2500 kg per hectare.

2.1.4 Preparation of Land and Planting

While preparing the land, minimum tillage operations may be adopted. Beds of 15 cm height, 1m width and of convenient length may be prepared giving at least 50 cm spacing between beds. Solarisation of the beds is beneficial in checking the multiplication of pests and disease-causing organisms. Solarisation is a technique by which polythene sheets are covered over moist beds of the field, reaching all the sides and exposing to sun for a period of 20-30 days. The polythene sheets used for soil solarisation should be kept away safely after the work is completed.

At the time of planting, apply 25g powdered neem cake and mix well with the soil in each pit taken at a spacing of 20-25 cm within and between rows. Seed rhizomes may be put in pits and mixed with well rotten cattle manure or compost mixed with Trichoderma, an antagonistic (Parasitic) fungi (10g compost inoculated with Trichoderma). The best time for planting is during May-June with the receipt of monsoon showers. Under irrigated conditions, it can be planted well in advance during February-March.

2.1.5 Cultural Practices

Mulching the ginger beds with green leaves is an essential operation to enhance germination of seed rhizomes and to prevent washing off of soil due to heavy rain. This also helps to add organic matter to the soil and conserve moisture during the later part of the cropping seasons. The first mulching is to be done with green leaves @ 10-12 tonne/ha at the time of planting. It is to be repeated @ 5 tonne/ha at 40th and 90th day after planting. Use of *Lantana camara* and *Vitex negundo* as mulch may reduce the infestation of shoot borer. Cow dung slurry or liquid manure may be poured on the bed after each mulching to enhance microbial activity and nutrient availability. Weeding may be carried out depending on the intensity of weed growth. Such materials may be used for mulching. Proper drainage channels are to be provided in the inter rows to drain off stagnant water.

2.1.6 Manuring

Application of well rotten cow dung or compost @ 5-6 tonne/ha may be made as a basal dose while planting the rhizomes in the pits. In addition, application of neem cake @ 2 tonne/ha is also desirable.

2.1.7 Plant Protection

Pests: Shoot borer is the major pest infesting ginger. Regular field surveillance and adoption of phytosanitary measures are necessary for pest management. It appears during July -October period. Spot out the shoots infested by the borer and cut open the shoot and pick out the caterpillar and destroy it. Spray neem oil (0.5%) at fortnightly intervals if found necessary. Light traps will be useful in attracting and collecting the adult moths.

Diseases: Soft rot or rhizome rot is a major disease of ginger. While selecting the area for ginger cultivation, care should be taken to see that the area is well drained as water stagnation pre- disposes the plants to infection. Hence provide adequate drainage. Select seed rhizomes from disease-free areas since this disease is seed borne. Solarisation of soil done at the time of bed preparation can reduce the fungus inoculum. However, if the disease is noticed, the affected clumps are to be removed carefully along with the soil surrounding the rhizome to reduce the spread. *Trichoderma* may be applied at the time of planting and subsequently if necessary. *Restricted use of Bordeaux mixture (1%) in disease prone areas may be made to control it as spot application.*

2.1.8 Harvesting and Post Harvest Operations

The crop is ready to harvest in about eight to ten months depending upon the maturity of the variety. Harvesting should be done when fully mature leaves turn yellow and start drying up gradually. Clumps are lifted carefully with a spade or digging fork and rhizomes are separated from dried leaves, roots and adhering soil. The average yield of fresh ginger per hectare varies with varieties ranging from 15 to 25 tonnes.

For making vegetable ginger, harvesting is done from the 6th month onwards. The rhizomes are thoroughly washed in water twice or thrice after harvest and sun-dried for a day. For preparing dry ginger the produce is kept soaked in water overnight. Rhizomes are then rubbed well to clean them. After cleaning, rhizomes are removed from the water and the outer skin is removed with a bamboo splinter or wooden knife having pointed ends. *Iron knife is not recommended, as colour will be faded.* In order to get rid of the last bit of the skin or dirt, the dry rhizomes are rubbed together. The peeled rhizomes are washed and dried in the sun uniformly for one week. Rhizomes are to be dried to a moisture level of 11% and they are stored properly to avoid infestation by storage pests. Storage of dry ginger for longer periods is not desirable. The yield of dry ginger is 16-25 percent of the fresh ginger depending upon the variety and location where the crop is grown. *Burning of sulphur for processing ginger is not allowed.*

2.1.9 Preservation of Seed Rhizomes

The rhizomes to be used as seed material should be preserved carefully. The indigenous practices like spreading layers of leaves of *Glycosmis pentaphylla* being followed by farmers can be adopted for this purpose. In order to get good germination, the seed rhizomes are to be stored properly in pits under shade. For seed materials, big and healthy rhizomes from disease-free plants are selected immediately after harvest. For this purpose, healthy and disease-free clumps are marked in the field when the crop is 68 months old and still green. Seed rhizomes are stored in pits of convenient size made inside the shed to protect from the sun and rain. Walls of the pits may be coated with cow dung paste. Seed rhizomes are stored in these pits in layers along with well-dried sand/sawdust (i.e. put one layer of seed rhizomes, then put 2 cm thick layer of sand/sawdust). Sufficient gap is to be left at the top of the pits for adequate aeration. Seed rhizomes in pits need inspection once in twenty days to remove shrivelled and disease-affected rhizomes. Seed rhizomes can also be stored in pits dug in the ground under the shade of a tree provided there is no chance for water to enter the pits. In some areas, the rhizomes are loosely heaped over a layer of sand or paddy husk and covered with dry leaves in a thatched shed.

2.2 TURMERIC

Turmeric (*Curcuma longa*), the ancient and sacred spice of India is a major spice produced and exported from India. Turmeric is used as condiment, dye and cosmetic in addition to its use in religious ceremonies. India is the leading producer, consumer and exporter of turmeric in the world.

2.2.1 Climate and Soil

Turmeric can be grown in diverse tropical conditions from sea level to 1500m above MSL at temperature range of 20-30°C with a rainfall of 1500 mm or more per annum or under irrigated conditions. It is grown on different types of soils from light black, loam and red soils to clay loams. However, it thrives the best in a well-drained sandy or clayey loam.

2.2.2 Maintenance of Buffer Zone

In order to cultivate turmeric organically a buffer zone of 25 to 50 feet should be maintained if the neighbouring farms are non-organic. The produce from this zone should not be treated as organic. Turmeric being an annual crop, the conversion period required will be of two years. Turmeric can be cultivated organically as an intercrop with other crops provided organic methods of cultivation is followed for all the companion crops.

2.2.3 Sources of Planting Material

Carefully preserved seed rhizomes free from pests and diseases which are collected from organically cultivated farms should be used for planting. However, to begin with seed material from high yielding, local varieties may be used in the absence of organically produced seeds. A seed rate of 2500 kg rhizomes is required for planting one hectare.

2.2.4 Preparation of Land and Planting

While preparing the land, minimum tillage operations may be adopted. Beds of 15 cm height, 1 m width and of convenient length may be prepared giving at least 50 cm spacing between beds. Solarisation of such beds is beneficial in checking the multiplication of pests and diseases causing organisms. The polythene sheets used for soil solarisation should be kept away safely after the work is completed.

At the time of planting apply 25 g powdered neem cake and mix well with the soil in each pit taken at a spacing of 20-25 cm within and between rows. Seed rhizomes may be put in shouldow pits and covered with well rotten cattle manure or compost mixed with *Trichoderma* (10 gm compost inoculated with *Trichoderma*). Turmeric can be planted during April-May with the receipt of pre-monsoon showers.

2.2.5 Cultural Practices

Mulching the turmeric beds with green leaves is an essential operation to enhance germination of seed rhizomes and to prevent washing off of soil due to heavy rain. This also helps to add organic matter to the soil and conserve moisture during the later part of the cropping season. Judicious mix of leguminous leaves with high nitrogen content and leaves rich in phosphorous like *acalypha* weed and leaves rich in potassium like *Calatropis* can be used according to availability. The first mulching is to be done at the time of planting with green leaves @ 10-12 tonnes per ha. It is to be repeated again @ 5 tonnes/ha at 50th day after planting. Cow dung slurry may be poured on the bed after each mulching to enhance microbial activity and nutrient availability. Weeding may be carried out depending on the intensity of weed growth. Such materials may be used for mulching. Proper drainage channels are to be provided in the inter rows to drain off stagnant water.

2.2.6 Manuring

Application of well rotten cowdung or own compost from own form @ 5-6 tonnes/ ha may be made as a basal dose while planting rhizomes in the pits. In addition, application of neem cake @ 2 tonnes/ha is also desirable.

2.2.7 Plant Protection

Pests: Regular field surveillance and adoption of phytosanitary measures are required for pest management. If shoot borer incidence is noticed, such shoots may be cut open and pick out larvae and destroy them. Spray neem oil 0.5% at fortnightly intervals if necessary.

Diseases: No major disease is noticed in the crop. Leaf spot and leaf blotch can be controlled by restricted use of Bordeaux mixture 1%. Application of *Trichoderma* at the time of planting can check the incidence of rhizome rot.

2.2.8 Harvesting and Post Harvest Operations

Turmeric is to be harvested at correct maturity. Depending upon the variety, the crop becomes ready for harvest in 7-9 months, medium varieties in 8-9 months and late varieties after 9 months.

Usually the land is ploughed and the rhizomes are gathered by hand-picking or the clumps are carefully lifted with a spade. Harvested rhizomes are cleaned of mud and other extraneous matter adhering to them. The average yield per hectare comes to 20-25 tonnes of green turmeric.

Fingers are separated from mother rhizomes. Mother rhizomes are usually kept as seed material. The fresh turmeric is cured for obtaining dry turmeric. Curing involves boiling of rhizomes in fresh water and the drying in the sun.

No chemical should be used for processing. The cleaned rhizomes are boiled in copper or galvanized iron or earthen vessels, with water just enough to soak them. Boil till the fingers/mother rhizomes become soft. The cooked turmeric is taken out of the pan and the same hot water in the pan can be used for boiling next set of raw turmeric which is already filled in troughs. The cooking of turmeric is to be done within 2-3 days after harvest. Rhizomes may also be cooked using baskets with perforated bottom and sides. The mother rhizomes and the fingers are cured separately.

The cooked fingers/mother rhizomes are dried in the sun by spreading in 5-7 cm thick layers on bamboo mats or cement floor. A thinner layer is not desirable as the colour of the dried product may be adversely affected. During night time the material should be heaped or covered. It may take 10-15 days for the rhizomes to become completely dry. Artificial drying using cross-flow hot air at a maximum temperature of 60°C is also found to give satisfactory product. In the case of sliced turmeric, artificial drying has clear advantages in giving brighter coloured product than sun drying which tends to suffer due to surface bleaching. The recovery of dry product varies from 20-30% depending upon the variety and the location where the crop is grown.

Dried turmeric has a poor appearance and rough dull colour outside the surface with scales and root bits. Smoothening and polishing the outer surface by manual or mechanical rubbing improve the appearance.

Manual polishing consists of rubbing the dried turmeric fingers on a hard surface. The improved method is by using hand-operated barrel or drum mounted on a central axis, the sides of which are made of expanded metal mesh. When the drum filled with turmeric is rotated, polishing is effected by abrasion of the surface against the mesh as well as by mutual rubbing against each other as they roll inside the drum. The turmeric is also polished in power-operated drums. The yield of polished turmeric from the raw material varies between 15-25%.

The colour of the turmeric always attracts the buyers. In order to impart attractive yellow colour, turmeric suspension in water is added to the polishing drum in the last 10 minutes. When the rhizomes are uniformly coated with suspension they may be dried in the sun.

Rhizomes for seed purpose are generally stored after heaping under shade of tree or in well ventilated shade and covered with turmeric leaves. Sometimes the heap is plastered with earth mixed with cowdung. The seed rhizomes can also be stored in pits with sawdust. The pits can be covered with wooden planks with one or two holes for aeration.

2.3 CHILLIES

The native home of chilli is considered to be Mexico with secondary origin of Gautemala. It is also called as hot pepper, cayenne pepper, sweet pepper etc. Chilli belongs to the genus Capsicum under Solanaceae family. Five species of Capsicum are under cultivation, though number of wild species have been identified recently. In India, only two species viz. *Capsicum annum* and *Capsicum frutescens* are known and most of the cultivated varieties belong to the species *Capsicum annum.* Chilli was introduced in India by the Portugese in Goa in the middle of 17th century and since then it had rapidly spread throughout the country.

Chilli besides imparting pungency and red colour to the dishes is a rich source of vitamin A, C and E and assist digestion. Recently Russian scientists have identified Vitamin P in green chillies which is considered to be important as it protects from secondary irradiation injury. The pungency in chillies is due to an alkaloid capsicin which has high medicinal value. It also prevents the heart diseases by dilating blood vessels. Chilli is an important ingredient in day-to-day curries, pickles and chutnies. Oleoresins, sauce and essence are prepared from chillies.

2.3.1 Climate and Soil

Chilli requires a warm and humid climate for its best growth and dry weather during the maturation of fruits. Chilli crop comes up well in tropical and sub-tropical regions, but it has a wide range of adaptability and can withstand heat and moderate cold to some extent. The crop can be grown over a wide range of altitudes from sea level up to nearly 2100 meters. It is generally a cold weather crop, but can be grown throughout the year under irrigation. Black soils which retain moisture for long periods are suitable for rainfed crop whereas well drained chalka soils, deltaic soils and sandy loams are good under irrigated condition.

2.3.2 Maintenance of Buffer Zone

In order to cultivate chillies organically, a buffer zone of 25 to 50 feet is to be left all around from the conventional farm, depending upon the location of the farm. The produce from this buffer zone belt should not be treated as organic. Chilli can be cultivated organically as inter or mixed crop provided all the other crops are grown following organic methods. It is desirable to include a leguminous crop in rotation with chilli.

2.3.3 Sources of Planting Material

For raising nurseries, seeds of high-yielding varieties with tolerance to pests and diseases may be used. They should be carefully selected from certified organic farms or from own seed plot which is raised organically. To start with, chemically untreated seeds from local high yielding varieties could also be used, in the absence of organically produced seeds. Seeds should not be treated with any chemical fungicides or pesticides. However, it is always beneficial to adopt indigenous practices for seed treatment, wherever possible. The seeds may be treated with *Trichoderma* to prevent incidence of seedling rot in the nursery.

2.3.4 Cultural Practices

Normally chilli is grown under rainfed condition. However, under irrigated condition, care should be taken to avoid using water contaminated with fertilizers, pesticides and fungicides. Irrigation should be done judiciously. Stagnation of water should not be allowed in nursery beds and fields in order to avoid fungal infection. Weeds which attract pests should be allowed to grow in the field to act as trap and removed before flowering.

2.3.5 Manuring

Organic manures such as farmyard manure is applied @ 4-5 tonnes/ha. However, it is always advisable to use compost/farmyard manure from own farm rather than from outside the farm. It is desirable to give sheep manure @ 3-5 quintals and neem cake @ 3-4 quintals per hectare at the time of land preparation. Restricted use of permitted mineral fertilizers like rock phosphate under organic system can be done depending on requirement, on the basis of soil analysis. Application of soil amendments such as tank silt, basic slag and gypsum is also allowed in a limited manner. Use of bio-fertilizers can also be resorted to the combination with organic inputs.

2.3.6 Plant Protection

Pests: To avoid infestation of root grub, only well rotten farmyard manure should be applied in the field. Application of neem cake @ 250 kg/ha is also advisable for control of root grubs. Change in the agronomic practices to disturb the life cycle of the grub is also found useful. Application of neem seed kernel extract (NSKE 3%) can be done for control of thrips, aphids and mites. Release of larvae of *Chrysoperla cornea,* a biocontrol agent, once in 15 days is also helpful in controlling thrips and mites. Fruit (pod) borers are the major pests which cause considerable damage to the crop. They can be managed to a certain extent by adoption of biocontrol measures. Restricted installation of pheromone traps in the field @ 5 nos. per acre helps to monitor the adult moths. Ten days after spotting the moths in the traps, spraying with Nuclear Polyhedrosis Virus (NPV) @ 500 LE (larval equivalent)/ha. About 4-5 rounds is beneficial to control the early larval stage of the pod borers. The egg mosses of Spocloptera borer can be mechanically collected and destroyed. *Trichogramma* an egg parasite may be released two days after appearance of moths. Spraying of neem products like neem oil, neem seed kernel extract and restricted use of *Bacillus thuringenisis* @ 1kg/ha are beneficial. All the shed fruits and inflorescence parts should be collected and destroyed at regular intervals.

Diseases: Rot and Die back caused by *Colletotrichurn capsici* and bacterial wilt are the two major diseases of chillies. Careful seed selection and adoption of phytosanitary measures will check the diseases of chillies. Early removal of affected plants will control the spread of the diseases. Seed treatment with *Trichoderma* takes care of seedling rot in nursery. Varieties tolerant to diseases should be used wherever the disease is severe. Rouging and destruction of affected plants help in checking the mosaic virus.

2.3.7 Harvesting and Post-Harvest Operations

Harvesting should be done at the right stage of maturity. Ripe fruits are to be harvested at frequent intervals. Retaining fruits for a long period on the plants causes wrinkles and colour fading. Soon after the harvest, the produce is to heaped or kept in clean gunnies for one day for uniform colour development for the pods. Sun-drying is the common practice in India. The preparation of drying floor differs from tract to tract. Levelled and compacted floor is to be made for drying. Fruits are spread on drying yards in layers. From the fifth day onwards, the produce is inverted on alternate days so that the pods in the lower layers are brought up to ensure quick and uniform drying. To avoid microbial activity and aflatoxin production, moisture in the dried pods should be brought down to 10%.

Since the produce is exposed to sun for 10-15 days on the open yards, it is likely to get contaminated with foreign matter. This also results in poor colour for the product due to bleaching effect of the sun rays. The produce can be dried within a period of 18 hours using air blown drier keeping temperature at 44-46°C. This method saves time, avoids the drying operations for 10-15 days but also imparts deep red colour and glossy texture to the fruits. Solar drier and tray drier can be used. While drying, the produce can be covered with polythene sheets during night time to avoid dew deposition and resultant colour fading.

Grading is to be done to remove defective and discoloured pods. Packing is done in gunny bags, or jute boras. Chillies should be properly stored to avoid infestation of pests. It is preferable to store dried chilli in refrigerated condition (cold storage) to retain colour.

2.4 BLACK PEPPER

Black pepper (*Piper nigrum*) is the most important item in export basket of spices from India. It is predominantly grown as an intercrop with coffee, arecanut and tea.

2.4.1 Climate and Soil

Pepper is a plant of humid tropics requiring adequate rainfall and humidity. It grows successfully between 20° north and south latitude and from sea level up to 1500 metres above mean sea level. The crop tolerates temperatures between 10°C and 40°C. A well distributed annual rainfall of 1250-2000 mm is considered ideal for pepper.

Pepper can be grown in a wide range of soils with pH 4.5 to 6.0. In its natural habitat, it thrives well in red laterite soils. The soil best suited is friable loam with high humus content and good drainage. In soil having pH less than 5.0, lime is to be applied to neutralize aluminium toxicity.

2.4.2 Choice of Varieties and Plant Material

Majority of the cultivated types of pepper are monocious (male and female flowers found in the same spike) though variation in sex expression ranging from complete male to complete female is found. Over 75 cultivars of pepper are known in India. Karimunda is the most popular of all the established cultivars of pepper in Kerala. The other important cultivars are Kottanadan, Narayakkodi, Aimpiriyan, Neelamundi, Kuthiravally, Balancotta and Kalluvally in Kerala. Cultivars Malligesara and Uddagare are popular in Karnataka. In terms of quality, Kottanadan has the highest oleoresin (17.8%) followed by Aimpiriyan (15.7%).

Varieties cultivated should be adapted to the soil and climatic conditions and be naturally resistant to pests and diseases of the region. When organic planting materials are available, they should be used. When certified organic planting materials are not available, conventional planting materials can be used initially.

Source of planting material: Pepper develops three types of aerial shoots, viz., (a) primary stem with long internodes, with adventitious roots which cling to standard; (b) runner shoots which originate from the base of the vines also have long internodes which strike roots at each node and (c) fruit bearing lateral branches with limited growth.

Cuttings are raised mainly from the runner shoots collected from mother vines grown organically. Terminal shoots can also be used. Cuttings from the lateral branches are seldom used since in addition to reduction in the number of fruiting shoots, the vines raised from them are generally short lived and bushy in habit. However rooted lateral branches are useful in raising pepper in pots (Bush pepper).

Initially, the cuttings can be collected from the conventional plantations in absence of purely organic source. Runners shoots are to be kept coiled on wooden pegs fixed at the base of the vines to prevent the shoots from coming into contact with soil and striking roots. The runner shoots are separated from the vine in Feb-March, and after trimming the leaves, cuttings of 2-3 nodes each are planted in nursery beds. The nursery techniques such as bed nursery or rapid multiplication using bamboo splits can be used for production of rooted planting materials. Adequate shade is to be provided and irrigated frequently. The cuttings will strike roots and become ready for planting in May-June.

Practices adopted for better results are that the soil should be solarised prior to use and inoculated with cultures of VAM and *Trichoderma* (250g mass multiplied media in 25 kg compost). The vines in rapid multiplications units may be sprayed with vermiwash (50 ml per unit) for enhancing growth. This will control the two important nursery disease i.e. leaf rot caused by *Rhizoctonai solani* and basal wilt caused by *Sclerotium rolfsi*. However, if incidence of these disease noticed, timely adoption of phytosanitary measures and spot application of Bordeaux mixture @ 1% may be sprayed. In area where, nematodes are likely to occur, addition of crushed neem seed is recommended.

2.4.3 Preparation of Land and Planting Shade Trees

Selection of site: North and north-eastern slopes are preferred for planting so the vines are not subjected to the scorching effect of the southern sun during summer.

Shade tree: Pepper requires support of sufficient heights to put forth lateral branches and to bear fruits. Hence live standards of fast growing nature, resistant to pest and diseases and having deep root system are selected. These live standards like *Garuga pinnata* (Indian Coral tree) or *Grevilea robusta* (Silver oak) are planted in April-May with spacing of 4x2m in sloppy land and 2.5x2.5m in flat lands. When *Garuga pinnata* is used, the primary stems/stem cuttings are cut in March-April and stacked in shade in groups. The stacked stems start sprouting in May. The stems are planted in the edge of the pits dug for pepper vines.

Planting: With the onset of monsoon, 2-3 rooted cuttings of pepper are planted individually in the pits on the northern side of each standard. (In the case of unrooted cuttings, about 4-5 cuttings per pit are to be planted and the number of nodes in this case may be 4-5). At least, one node of the cutting should go below the soil for proper rooting. At a spacing of 2.5m x 2.5m, there will be about 1600 standards per ha. Application of 2 kg compost mixed with 125 g rock phosphate at the time of planting is suggested.

2.4.4 Cultural Practices

As the cuttings grow, the shoots are tied to the standards as often as required. The young vines should be protected from hot sun during summer by providing them with artificial shade. Regulation of shade by lopping the branches of standards is necessary not only for providing optimum light to the vines but also for enabling the standards to grow straight. Adequate mulch with green leaf or organic matter should be given towards the end of north-east monsoon. The base of the vines should not be disturbed so as to avoid root damage.

During second year, practically the same cultural practices are repeated. However, lopping of standards should be done carefully from the fourth year onwards, not only to regulate height of the standards, but also to shade the pepper vines optimally. Excessive shading during flowering and fruiting encourage pest infestation.

From the fourth year, usually two diggings are given, one during May-June, and the other towards the end of south-west monsoon in October-November. Growing cover crops like *Calapogonium mucunoides, Mimosa invisa* are also recommended to provide an effective soil cover to prevent soil erosion during rainy season. Further they dry during summer, leaving thick organic mulch.

Conversion period: For existing plantation, a minimum of three years is required as conversion period for organic cultivation. For a newly-planted or replanted area raised through organic cultivation practices, the first yield itself can be considered as organic provided chemicals have not been used in the previous cropping. In case of cultivation on virgin land and for farms where records are available that no chemicals were used previously, the conversion period can be relaxed.

Buffer zone: An isolation distance of at least 25 m width is to be left from all around the conventional plantation in order to avoid contamination of organically cultivated plots from neighbouring farms/conventional farms. In sloppy lands, adequate precaution should be taken to avoid the entry of run-off water and drift from the neighbouring farms.

Diversity in crop production: Pepper is generally cultivated as intercrop in coconut/ arecanut and in coffee/tea plantation. Under this situation, all the crops are to be cultivated following organic farming principles. While selecting standards for new planting as many number of recommended species as possible may be used.

2.4.5 Manuring

Organic manure in the form of cattle manure or compost is applied at the rate of 4 kg per vine which can be gradually increased to 10 kg per vine during May-June. Compost made from green lopping, crop residues, grasses, cow dung, poultry droppings, fortified wood ash and or rock phosphate should be used regularly instead of farmyard manure alone. Based on soil test results, application of permitted inputs like lime at the rate of 600 g per vine during April-May in alternative years is also recommended.

2.4.6 Crop Protection

Organic farming systems should be carried out in such a way that it minimizes losses from pests, diseases and weeds. Various pests and diseases of pepper can be effectively managed by following timely surveillance of the pepper gardens, appropriate cultural practices, adoption of proper plant and field sanitation, measures and by application of Bordeaux mixture, soil conditioners (neem cake), bio agents (*Trichoderma* for disease control) and control of pest by mechanical means.

Pests: *Poll beetle* and leaf *gall thrips* managed by spraying neem at the rate of 400 ml/100 l of water or other neem preparations are suggested. Nematodes can be controlled by growing marigold as a trap crop and uprooting the trap crop at flowering stage and burning. Shade management during May-June and application of neem oil (0.6 per cent) during Oct-Nov control scale insects.

Diseases: For management of disease adoption of phytosanitary measures is very essential. This includes removal and destruction of dead vines along with root systems from the gardens as well as use of disease-free planting materials. Tillage operations are kept to the minimum to avoid soil disturbance and root damage. Adequate drainage is also required to reduce water stagnation. Application of *Trichoderma* at the rate of 5 kg per vine during May-June with suitable carrier medium such as coffee husk and well rotten cow dung followed by second round of application of the same quantity during August-September is suggested.

Nursery diseases: The disease is often serious in nurseries during April-May period when warm humid conditions prevail. It is caused by *Rhizoctonia solani*. The fungus infects both leaves and stems. Infection starts as small brownish lesions appear brittle and whitish to dark gray in colour. The adjacent infected leaves get adpressed to each other because of the fast spreading fungal threads (hyphae), which bind them together. On stems of rooted cuttings the infection occurs as dark brown lesions which spread both upwards and downwards. The new flushes subtending to the points of infection gradually drop and dry up.

Basal wilt (Sclerotium rolfsii): The disease is noticed mainly in nurseries during June-September period. The disease is caused by the fungus *Sclerotium rolfsii*. Water soaked spots are seen on the stem. Occasionally the round lesions show concentric zonation. On the stem, the infected portion shows gradually advancing felt-like whitish threads causing soft tarring of the tissues. Whitish fungal threads girdle the stem resulting in drooping of leaves beyond the point of infection and in advanced cases the rooted cuttings dry up. On the mature lesions appear small whitish to cream coloured grain like sclerotial bodies. Both these diseases in the nursery can be controlled, if noticed early, by observing strict phytosanitary measures. The affected cuttings along with defoliated leaves should be removed from the nursery and destroyed. Later all the cuttings should be given a spray with 1% Bordeaux mixture to check the disease incidence.

2.4.7 Contamination Control

All relevant measures should be taken to minimize the contamination from outside and within the farm. Accumulation of heavy metals and other pollutants should be avoided. For protected structure coverings, plastic mulches, fleeces, insect netting and silage wrapping, only products based on polythene and polypropylene or other polycarbonates are allowed. These should be removed from the soil after use and should not be burnt on the farmland. The use of polychloride based products is prohibited.

2.4.8 Soils and Water Conservation

In sloppy lands, soil disturbance should be minimum in all agricultural operations. Clearing of land through means of burning of organic matter e.g. slash and burn, straw burning should be restricted to the minimum and the clearing of primary forest is prohibited. Relevant measures may be followed to prevent salination of soil and water.

2.4.9 Harvests and Post-Harvest Operation

The crop takes about 6-8 months from flowering to harvest. The harvest season extends from November to January in the plains and January to March in hills. During harvesting the whole spike is hand picked when one or two berries in the spike turn bright orange or purple. The berries are separated from the spikes and dried in the sun for 7-10 days, on a concrete floor or bamboo mat under sun till it is crisp. During sun drying it is important to turn over the material periodically to facilitate uniform drying. Without turning, mould contamination may result in a poorly dried product with greyish unattractive appearance.

To improve the colour and appearance of dried berries and to reduce the drying time, harvested green pepper is to be soaked in hot water for a minute and then dried under sun. For this mature greenish yellow, pepper spikes are despiked after harvest. The pepper berries after cleaning are transferred to perforated aluminium vessel or bamboo basket and dipped in boiling water for a minute, drained and spread out on a clean cement floor or bamboo mat for sun drying. The water should be clean and uncontaminated. The blanched pepper requires only 3-4 days for drying in sun and this process also gives attractive and uniform colour. The moisture level to be maintained at 11 per cent. The process minimizes microbial contamination and gives hygienic product which can be preserved easily. Cleaning of pepper should be done by winnowing after drying. Only properly dried cleaned peppers should be stored after packing in suitable materials.

Making of white pepper

- Harvest only well mature spikes with at least 3-5 berries ripe. Separate immature ones from the lot. Then keep spikes in the corner of a room covered with clean gunny bags for one or two days to hasten ripening of remaining berries.
- Separate ripe berries from spikes. Fill berries loosely in clean gunny bags of 50 kg capacity.
- Keep bags immersed in a canal/stream where clean and uncontaminated water is flowing for 6-9 days till the outer skin of berries decays by fermentation.
- Take out the bags and empty the berries to a tank partially filled with water.
- Knead berries to completely separate out the skin and adhering loose tissue.

- Clean the corns by washing with fresh water and remove damaged ones.
- Dry immediately to avoid fungal infection and discolouration.
- For sun drying, spread thinly the corns on a bamboo mat or cement floor. Repeat heaping and spreading at two hours intervals for uniform drying. Once dried to a moisture level of 11 per cent may be stored in clean gunny polythene bags or bins.

2.4.10 Packaging

Care should be taken to see that organically-produced black and white pepper is protected from co-mingling with non-organic products and they should not be stored and transported together except when labelled or physically separated. All products should be adequately identified through the whole process. The dried black and white pepper is to be packed in eco-friendly packing materials such as clean gunny bags. Use of polythene bags should be minimized to the extent possible. Waste generating packaging materials is to be avoided. Recyclable/ reusable packing materials should be used wherever possible.

2.5 SMALL CARDAMOM

Cardamom (*Elettaria cardamomum* Maton) the "Queen of Spices", enjoys a unique position in the international spices market, as one of the most sought after spices. From time immemorial, India is known as the home of cardamom. Cardamom is indigenous to the evergreen forests in South India. Cardamom is used for flavouring various food preparations, confectionary, beverages and liquors. It is also used for medicinal purpose, both in Allopathy and Ayurveda systems.

2.5.1 Climate and Soil

The natural habitat of cardamom is the evergreen forests of Western Ghats. It is found to grow within an altitude ranging between 600 and 1200 metres above MSL. Considerable variations both in the total rainfall pattern and its distribution are noticed in the cardamom tracts. In most of the cardamom areas, the annual rainfall is between 1500 and 4000 mm. and the temperature ranges from 10 to 35°C.

Cardamom generally grows well in forest loamy soils. These soils are generally acidic in nature, with pH from 4.2 to 6.8. Analytical data of soil samples in cardamom-growing tracts indicate that they are high in organic matter and nitrogen, low to medium in available phosphorus and medium to high in available potassium.

2.5.2 Sources of Planting Material

Initially the seed can be collected from any elite plantation even if they are not grown organically. However, the methods followed for raising seedlings should conform to the organic standards. If rhizomes are to be used as planting material, the plantation should have been following organic methods of production at least one year prior to collection. Tissue culture plantlets should not be used as planting materials in order to keep integrity with natural methods of propagation. Acid treatment of seeds should be avoided. Treatment of seeds with *Trichoderma* culture (50 ml spore suspension for 100 g of seed) is desirable as prophylactic measures for managing nursery rot diseases. At the times of preparation of beds, incorporation of VAM multiplied in recommended organic medium may be done.

2.5.3 Nursery Management

In order to raise a cardamom plantation, seedlings or suckers of high yielding varieties are to be used. The different steps involved in raising the nursery are given below:

Primary nursery

Site selection: Select nursery sites on gentle sloppy area and preferably near to a perennial water source. Clean the area from all existing vegetation, stumps, roots, stones etc. In the cleared area, beds can be prepared with one metre width, 20 cm height and required length, generally six metres. Jungle top soil can be spread to a thickness of 2 to 3 cms on the beds. Solarisation of nursery beds help to eliminate the diseases, nematodes and other soil pests.

Seed collection: Fully ripened bold capsules from high yielding and disease-free mother clumps of known source can be collected from second and third harvests for seed extraction. Seeds after extraction should be washed using water to remove the mucilage. It is then mixed with wood ash and dried in shade. Storage of seed is not advisable for longer period, because it is experimentally proved that 15 days of storage decreased germination for about 20 % and 3 to 5 months storage decreased germination up to 94%. Therefore sowing of seeds after the extraction should not be delayed much. Sowing in September is the best for good germination. Sowing in winter and during south west monsoon should be avoided.

Sowing: Sowing can be done in rows at a distance of 10 cm. Seed rate is 30 to 50 gm per 6 x 1 m size bed. After sowing cover the seeds with a thin layer of fine soil. Then cover the bed with mulch material, either with pothagrass or paddy straw. Avoid the contact of mulch materials with the soil by supporting twigs laid across the bed. Water the beds to sufficient moisture conditions. Once sprouting is observed, remove the mulch and cover the bed with thinly sliced mulch material. To protect the seedlings from direct sunlight, provide overhead pandal, Germination commences 20 to 25 days after sowing and continues for further 30 to 40 days.

Secondary nursery: Prepare beds as in primary nursery. A layer of cattle manure and wood ash may be spread on the bed and mixed with soil. Seedlings of three to four leaf stages from the primary nursery beds can be transplanted in the secondary nursery at a distance of 20 to 25 cms. Mulching and watering of beds should be done immediately after transplanting. Coirmat pandal can be erected to protect seedlings from direct sunlight.

For raising polybag seedlings (preferably bio-degradable polybags), potting mixture may be prepared by using 3:1:1 soil rich in organic matter, well rotten cow dung or vermicompost and sand. To this VAM and *Trichoderma* can also be added (250g of mass multiple media mixed with 25 kg of well rotten cow dung). If growth of the seedlings is not adequate, spraying vermiwash once in a month is desirable (20 ml per plant). Regular surveillance and adopting phytosanitary measures may manage the diseases in the nurseries. Restricted application of Bordeaux mixture (1%) may be done to control rot diseases at the initial stage itself. Changing the nursery location is beneficial to ward off pests and diseases and for vigorous growth of seedlings.

Cardamom plants from secondary nursery or polybags can be transplanted to the main field during the last week of May after pre-monsoon showers or the first week of June soon after commencement of southwest monsoon.

2.5.4 Vegetative Propagation

This method is simple, reliable and facilitates easy multiplication of selected types. Plants raised from rhizome come to bearing earlier than the seedlings raised from seeds by about a year. Vegetative propagation is advantageous in areas where 'Katte' disease is not a problem. When rhizomes are planted in the midst of grownup plants for infilling of vacancies, the risk of the strangling effect of surrounding plants is minimised.

Propagation by suckers consists of splitting clumps into sections having one or two old suckers with two or three new shoots. The split suckers are planted in the small depression of the filled in pit and covered with soil and mulch. Sucker planting should be avoided during periods of heavy rain. Upright planting of suckers supported by stakes without root pruning is recommended for infilling of vacancies. In wind swept areas and in replanting, rhizomes may be planted flat on the ground.

2.5.5 Field Planting and Management

Before taking up the planting, field should be made ready. For planting in new area, ground should be cleared or if it is replanting area, old plants should be removed. Shade regulation, terracing and preparation of pits should be done during summer months.

Shade regulation: Shade regulation is one of the important practices. It should be attended during summer (March-April) in the new planting areas and during May-June after the receipt of summer showers in the existing plantations. If there is thick shade due to dense branches and bigger leaves, chopping off branches should be done to provide filtered light of 40 to 60 per cent of the open area.

South-Western slopes should be provided with more shade than North-Eastern slopes. Shade trees should have small leaves, tap root system and in summer it should not shed leaves. If area is open due to tree fall, planting of tree species viz. Karuna (*Vernonia arborea*) Corangati, Chandana Viambu, Njaval tree etc. should be taken up immediately to protect the plants from direct sunlight. Too much shade or too much openness of area is not advisable for cardamom cultivation as it affects growth and yield.

Field preparation: In areas having medium and steep slopes, soil preparation will be different from that of gentle slopes. In sloppy areas soil should be protected from soil loss (erosion) due to rains for which planting should be taken up in terraces. Terraces should be made across the slope at required distances depending on the spacing adopted. Almost 8 to 15 cm depth of top soil should be removed before making terraces and kept aside which can be used for pit filling. Width of terraces should be 1.5 to 1.8 m. Pits of 90 x 90 x 45 cm can be prepared before commencement of monsoon, about 1/3 of the pit should be filled with top soil and 1/3 should be filled with 1:3 mixture of organic manure and top soil. In low rainfall areas, trenches of size 75 cm width and 30 cm depth may be taken and plants may be planted at a spacing of 1 to 1.5m.

Planting: Organic planting material of high-yielding variety suitable for the areas may be selected for planting. They may be planted in the already prepared pits and plants should be protected from wind by staking. Plant to plant distance can be 2.4x2.4 m or 1.8 x 1.8 m when planted in high rainfall or irrigated areas. Immediately after planting, the plant base should be mulched well with available dried leaves to protect soil from erosion and conservation of moisture. Planting should be done diagonally to the slope which will be helpful as a self protector of soil.

Weed control: Weeds are potential competitors in the consumption of water and nutrients which will depress the cardamom growth. At the initial stage, if cardamom clump development is not enough, weed growth will be more. Two or three rounds of hand weeding at the plant base during May, September and December/January and slash weeding in other areas are advisable. Use of spade for weeding is not advisable because it will loosen the soil and cause soil erosion. The weeded materials may be used for mulching.

Irrigation: Judicious irrigation during summer months ensures increase in yield by at least 50%. Irrigation is required generally from February to April but at times from January to May depending upon rainfall. In case of dry weather, irrigation during March-August is advisable because this is the period in which development of young tillers and panicles takes place. If plant suffers during this stage, yield will be reduced. Water may be stored during rainy season wherever possible by constructing water harvesting structure without causing damage to the environment. This water can be used for irrigation. Irrigation can be done through different methods such as pot irrigation, sprinkler irrigation and drip irrigation depending on the facilities available in the plantation. Pot irrigation can be done at weekly intervals at the rate of 20-30 litre per clump depending, upon the clump size. In case of sprinkler, irrigation with amount of water equivalent to 35 to 45 mm rain at fortnightly intervals is recommended under average conditions. In case of drip irrigation, water at the rate of 4-6 litre per clump per day can be given.

Bee keeping: Mainly honeybees pollinate small cardamom and maintenance of honeybee colonies in plantation is found to enhance the yield. However, raising bee colonies should follow organic principles laid down for the purpose. The collection area should be organic and/or wild and should be as varied as possible to fulfil the nutritional needs of the colony and contribute to good health. If extra feed is to be supplied it should be fully organic. Hives are to be situated in organically-managed cardamom fields and/or wild natural areas. Hives should not be placed close to fields or other areas where chemical pesticides and herbicides are used. For pest and disease control and for hive disinfect ion, only products like caustic soda, lactic, oxalic or acetic acid, formic acid, sulphur, etheric oils and *Bacillus thuringiensis* may be used.

2.5.6 Soil and Water Conservation

Planting in trenches across the slope, mulching of soil, diagonal planting and rectangular silt pit opening (1.8 x 0.5 x 0.6m) in between four plants will help in soil and water conservation in gentle slopes. If slope is steep, construction of stone pitching walls at 10-20 m intervals across the slope and also making water collecting trenches along the wall will be helpful.

***Forking and mulching*:** Forking the plant base to a distance up to 90 cm and to a depth of 9-12 cm is found to enhance root proliferation and better growth of plants. As far as possible, the entire plantation and particularly the plant base are to be kept under mulch. It is very essential to keep the plant base mulched (5-10 cm thick) except during June to September to reduce the ill-effects of drought, for reducing evaporation loss and to maintain optimum temperature.

***Trashing*:** Trashing consists of removing old tillers and dry leaves and leaf sheaths. This operation may be carried out once in a year at any time one month after completion of the final harvest. These materials can also be used as mulch.

***Earthing up*:** This operation is not required in a normal plantation. However, due to erosion of soil or mismanagement, at times it is noticed that the topsoil covering the plant base is washed away and the rhizomes and roots are exposed and in such situations, earthing up of the plant base with top soil is recommended during Dec-Jan. While carrying out this operation, care should be exercised to ensure that only top soil is used, and it is evenly spread at the base covering only half the bulb portion of the rhizome. This operation helps to keep the top 10 to 15 cm soil loose and friable enabling easy root penetration and water percolation.

2.5.7 Manuring

Application of organic manures such as Neem Cake (@ 1-2 kg/plant) or Poultry manure/Farmyard Manure/Cowdung Compost (@ 2-5 kg/plant) may be made once in an year, during May/June along with Rock phosphate (180 g/plant). The manures should be thoroughly mixed with surface soil after application. For the subsequent application to be made in September, organic manures need not be applied.

2.5.8 Plant Protection

Pests: Removal of drooping dry leaves, dry sheath, old panicles and other dry plant parts is an important sanitation method recommended for reducing the pest inoculum in the plantation. Mechanical collection and destruction of egg masses of pests, larvae of hairy caterpillar (*Eupterote sp.*) and beetles of root grub (*Balepta fuliscorna*) are other effective methods in reducing the pest damage. As soon as bore holes of stem borer (*Conogethes punctiferalis*) are noticed, injection of *Bacillus thurigensis* preparation into the bore hole (0.5 ml in 10 ml water) will kill the larve so that subsequent resurgence can be reduced. Wherever organic methods of cultivation are adopted outbreak of white flies (*Dialeurodes cardamom*) is seldom observed. However in event of such outbreak, collection of adults using yellow sticky trap and control of nymphs by spraying neem oil with soft soap made up of minimum caustic soda (500 ml neem and 500 g soft soap in 100 litre water) is to be followed. In area prone to nematodes (*Meloidogine sp.*), application of crushed neem seed can take care of problems. Application of fish oil resin may be used for managing thrips (*Sciothrips cardamom*). Malabar varities are found to be tolerant to thrips to a certain extent. Regular surveillance is absolutely essential for timely detection and adoption of remedial measures against the pests affecting cardamom.

Disease: The Katte (Mosaic) disease is prevalent in all cardamom-growing areas in India. It is one of the major diseases of cardamom. 'Katte' disease affects plants of all ages. The first visible symptoms appear on the youngest leaf of affected tillers as spindle shaped chlorotic flecks. Later, these flecks develop into slender discontinuous stripes of pale green and dark green areas, running parallel to the veins from the midrib of leaf margin. As the disease advances, subsequent leaves show characteristic mosaic symptoms. The leaf sheaths and pseudostems also show mosaic pattern. Mature leaves formed before infections do not develop symptoms. The infection is systemic in nature and gradually spreads to all tillers of affected plant. Immediately after infection, there is no growth reduction but within one to two years after infection, there is a gradual reduction in clump size. In advanced stages, the affected plants produce shorter and slender tillers with a few shorter panicles. 'Katte' affected plants do not die but give only poor yield. The yield reduction has been found to be 70 per cent within 3 years after infection.

As long as katte inoculum is present in the field, any formulation of insecticides fails to prevent the disease spread. Diseased plants cannot be cured but the losses can be minimised by adopting the following measures:

- Keep a constant surveillance on the occurrence of katte disease.
- Use only healthy seedlings raised from 'katte' - free plants.
- Avoid rhizome planting using materials taken from disease-affected gardens.
- Practice regular rouging (uproot and destroy).
- Repeat tracing of affected plants and rouging at weekly intervals for at least 4 consecutive months.
- Replant the rouged areas with healthy disease-free materials.
- Destroy wild plants like Amomum, Alpinia, Curcuma, Colocasia etc. if they show symptoms of katte.
- Do not raise nursery near katte affected areas.

Biocontrol of rot diseases: Recent studies show that azhukal and rhizome rot can be controlled to some extent with the bio-agent *Trichoderma*. It is antagonistic soil fungus acting against the rot pathogens. The fungus is green in colour and grows abundantly on cow dung and organic crop residues such as coffee husk, tea waste, neem cake, coir compost etc. *Trichoderma viride* or *T. harzianum* specific to cardamom can be mass multiplied on carrier media for 30 to 45 days. These can be applied to plant basins at the rate of one kg per 5 kg of cow dung during May, August, September and October months after phytosanitation.

Contamination control: All relevant measures should be taken to minimize the contamination from out side and within the farm. For protected structure coverings, plastic mulches, fleeces, insect netting and silage wrapping, only products based on polythene and polypropylene or other polycarbonates are allowed. These should be removed from the soil after use and should not be burnt on the farmland. The use of polychloride based products is prohibited.

2.5.9 Harvest and Post-harvest Operations

After harvesting, the freshly harvested capsules need to be cleaned from dirt. Curing of cardamom capsules is done by reducing the moisture from 80 per cent to 10 per cent at an optimum temperature (50°C) for retaining green colour to the maximum extent. Cardamom can be cured by following method:

Conventional curing: This is the most commonly adopted method for curing cardamom. It requires a structure fitted with furnace, flue pipes and chimney ventilators etc. It is masonry structure consisting of two apartments, a curing room and a furnace room. Curing room is a tall one provided with ceiling at the roof and fitted with wire gauge on the beams at the middle of the room parallel to the ground floor, making the room into two compartments. Flue pipes having a radius of about 25 cms made of galvanized iron sheets are provided in the ground floor from one end to the other from the furnace to chimney pipe to expel the smoke through the roof. Racks holding rectangular trays are also fixed for accommodating larger quantities of cardamom.

Capsules are spread in a single layer on the racks and trays. After spreading, the curing room is closed and heating is done by burning firewood in the furnace and the heat produced is conducted. Only fallen trees and lopped branches should be used as fuel. The hot smoke passes through the pipes bringing the room temperature to 45-50°C. This temperature is maintained for three to four hours. At this stage capsules sweat and give off moisture. Ventilators are then opened for sudden cooling and sweeping out vapour from the drying capsules. Ventilators are closed after vapour escapes completely and temperature is maintained at 40° C for about 24-30 hours. In general quality of capsules cured by this method is very good.

Sun drying: Cardamom is directly dried under sunlight. Sun drying generally requires 5-6 days and rainy season should be avoided for drying. By this method, it is not possible to obtain good green colour of the capsules.

Processing methods: Processing methods be based on mechanized, physical and biological processes. Care should be taken to maintain vital quality of an organic ingredient through out each step of its processing. Processing method is to be selected in such a way that it will limit the number and quantity of additives and processing aids.

Packaging: Care should be taken to protect the organically produced cardamom from co-mingling with non-organic products and they are not stored and transported together except when labelled or physically separated. All products should be adequately identified through the whole process. The dried cardamom is to be packed in eco-friendly packing materials such as clean gunny bags. Use of polythene bags should be minimized to the extent possible. Recyclable packing materials should be used wherever possible.

SPICES BOARD INDIA

Spices Board India has prepared a document on production of organic spices. It features the organic concepts, principles, basic standards, production guidelines, documentation, inspection and certification. The document has been published after approval by the National Standards Committee constituted by the members of IFOAM in India. The Board provides assistance to overseas buyers in sourcing organic spices from India.

Chapter 13

Organic Fruits and Vine Production

Fruits produced organically are sold at a premium all over the world justifying the cause of organic farming. Most of the fruits crops cultivated in India are primarily export-oriented crops. They offer a good scope for switching the production of fruit over to organic farming. Secondly, as these crops are generally grown in ecologically fragile hilly tracts, adopting organic farming methods would protect environment and also prevent contamination thereby producing produce of high nutritional quality. Further, these plantation crops are more amenable for organic farming as they produce huge amount of waste biomass for recycling, which can meet major portion of nutrients required for such farming.

1. WHAT DO THE ORGANIC GUIDELINES SAY?

Growing systems: The shape of trees and the planting distance must ensure that sufficient light is available to the fruit during the whole growing period. The fruit species and varieties and the rootstock should be appropriate to the local soil and climatic conditions.

Pruning: To obtain good quality fruit, pruning promotes an open tree crown with slow-growing but strongly developed fruit-bearing wood. It must be adapted to the condition of the trees, and also their shape, vigour and age.

Soil management: Orchards must have ground cover present all the year round. It should be maintained so that it keeps a diverse flora and fauna. Cover by a monoculture should be avoided. The rows of trees may, particularly in young orchards, be kept cultivated mechanically or covered with material of organic origin (e.g. bark compost, rape straw) or by long-life plastic sheeting.

Fertilization and mulching: Amended organic material may be brought in and used as mulch, or it may also be worked under. Fertilization (organic) and mulching must be carried out sparingly and at appropriate times, so as not to distort the physiological balance of the trees and the quality of the fruit.

Pest and disease management: All tree cultivation measures, such as the choice of shape, distance between trees, variety and general care, also serve to increase the resistance of fruit trees. When new plantings are made, disease-resistant fruit varieties should be given preference.

Bearing and growth regulation: Accurate thinning by hand at the right time serves to improve fruit quality and prevent alternate bearing.

Berries and miscellaneous fruit: These standards are valid wherever relevant also for bush fruits and any other fruit crops.

2. PRODUCTION TECHNOLOGY

The conventional fruit production systems can be easily converted into organic production systems. Cost can be minimized by selecting appropriate lands, understanding the history of the land, natural extremes that affect the land, pests and diseases and limitations of the land. Thereafter, follow the relevant standards and basic guidelines of IFOAM. A conversion plan and a schedule are compulsory for effective management system to minimize the sudden yield drops, to rejuvenate and conserve the soil, to improve and conserve bio-diversity and to formulate weed and plant management schedule.

Conversion of the traditional orchards to organic management system should take certain important steps. All the records should be maintained to date and labels or tags for identification. The conversion should follow the certification requirements at the proper time and the inspection should be done according to the schedule.

2.1 Land Development, Land Preparation and Planting

Except for pineapples, land development for other fruit crops does not employ farm machinery. Land clearing, soil conservation, and field preparation (digging and filling pits) and drainage system can be developed manually. Soil conservation measures such as contour land preparation and planting, drainage and runoff control, and methods for rain water harvesting are implemented during land development stage of fruits (e.g. banana, sweet orange, mango, litchi, cashew, coconut, etc.) in higher elevations. Entire operations of land preparation, selection of suitable planting material and field planting are also done manually. The operations are labour intensive and costly but assures the basis for sustainable fruit production.

No artificial material is recommended of temporary shelter for the plants established in the field. Leaves of coconut (*Cocos nucifera*), *Borassus flabellifera*, *Aleurites triloba*, gliricidea (*Gliricidia maculata*), and wild sunflower (*Tithonia diversifolia*) can be used for shading the plants soon after planting. Live supports are recommended for passion fruit trellising.

2.2 Nutrient Management

Soil moisture management and plant nutrition are important aspects of fruit production. Use of organic manure is one of the main requirements of organic fruit production. Many potential raw materials have been suggested by research organisation as organic fertilizer for fruit crop production and plant propagation.

Table 13.1: Organic nutrition of some unique fruit plants

Crop	Time of application	Type of organic manure	Amount (kg/plant)
Banana	Basal dressing	Compost + kitchen ash, cattle manure or Farm Yard Manure Coconut husk layer beneath the pit	5
	Soon after planting	Coir dust layer as a mulch,	
	Once in 2-3 years	Ground dolomite,	0.50
	After planting	Mulching with straw, coir dust, saw dust, paddy husk, banana residue	ample amount
	Top dressing with organic manure at 3 month intervals	Poultry manure or	3-5
		cattle and other animal manure	5-10
		or green manure	7-10
Papaya	Basal dressing	Poultry manure or	5-10
		cattle manure and	4-5
		ground dolomite	5
	Top dressing with organic manure 3 month intervals plus mulching	compost or	5-10
		cattle manure +	4-5
		straw or/and crop residues or/and any type of animal manure or live mulch	ample amount
Pineapple	Basal dressing	any organic manure	
	Mulching as top dressing	Coir dust, saw dust, leaves, rice husk, banana trunk chops or straw	ample amount
		ground dolomite	2-5 t/ha
Passion fruit	Basal dressing	Cattle manure or poultry manure	5-10
	Top dressing	Cattle manure or poultry manure	3-4

A balanced humus supply is essential. The total amount of nitrogen fertilizers applied must not exceed 90 kg N/ha fruit cultivation area per year.

Nutrient should be applied as per age of various fruit plants. In general the following nutrient management is required to keep the soil fertile.

- Apply organic manures (10-20 kg/tree).
- Growing of legumes for green manuring or as inter/cover crops during

establishment of orchards (initial 5-6 years).

- Mulching after application of FYM/compost and release of earthworms in presence of proper moisture.
- Need-based foliar spraying of biodynamic liquid manures/vermiwash, etc.

2.3 Pest and Disease Management

In organic agriculture, one of the most important goal is the achieving of healthy plants by supporting a balanced ecological proportion of pest and beneficial species. Essential measures to prevent disease are suitable stocking densities as well as the selection of healthy and hardy plants, species and varieties. The hardiness of copses can also be strengthened and the risk of infection can be lowered by using appropriate soil management and cultivation measures (shape pruning, rootstock building, cut part, foliage work, line spacing, maintenance underneath of trees, etc.). Some of the important pest and disease control measures for fruit cultivation is given below:

- Spraying of biodynamic liquid pesticides prepared from cowurine, neem, karanj *(Pongamia pinnata)*, calotropis, dhatura, casto, *Thevtia nerrifolia*, *Vitex* species leaves.
- Nettle leaves extract sprays to control hard pest like mango hopper.
- Sprays of biodynamic solution at flowering and fruit development stages.
- Biodynamic tree paste/cowdung paste to control gummosis and dieback.
- Spraying of casuarina leaves extract to control fungal disease.

2.3.1 Pest control in some important fruit plants

Mango

- Mango stem borer can be managed by timely pruning and destruction of affected branches by inserting piece in the holes. In the holes, insert cotton soaked in petrol or kerosene oil and plug the hole with mud.
- Pruning of overcrowding branches to reduce hopper population. During January-February, 2 % alcoholic extract of *Alpiniagalanza* rhizome or 0.5% neem products or 0.125 % citronellas oil or lemon grass oil or spearmint oil is effective.
- Wrap 20-25 cm wide alkathene (400 gauges) around tree trunk and seal the edges with mud to avoid mealy bug crawling up. Spray of fungal suspension *Beuveria basiana* (1 x 10^7 spores/g) during January is found effective.
- Racking up and hoeing the soil around mango serve to destroy pupae of fruit fly. Collect and destroy fallen infested fruits by fruit fly. Do not retain birds damaged on tree as they attract fruit fly. Harvest fully mature fruits one week early to escape egg laying. Grow Tulsi *(Ocimum sanctum)* plant around mango orchard which attract adult fly from mango tree.

- Apply sticky band at upper end of trunk to prevent migration of weevils to branches for egg laying on fruits during Feb-March.
- Post harvest of dipping mature fruit in 5% NaCl in water for one hour kill 90 % fruit fly eggs. It also removes externally present pesticide residue. Immerse mango fruits in hot water at 48±1°C for 45 minutes also kills 100% eggs and helpful in controlling stone weevil too.
- For leaf webber, prune the infested shoot with webbed leaves and destroy.

Guava

- For fruit fly control, apply the same methods as described in case of mango.
- In case of Tea mosquito bug infestation, collect infested and fallen fruits and destroy it. Moderately resistant varieties like Lucknow-47, smooth green and Sharanpur seedless can be grown on endemic areas.
- For the control of fruit-sucking moths, destroy weeds like *Tinospora spp.* and *Coculus spp.*, which provide food for larva of fruit sucking moths. Harvest fruits little early. Generate smoke in orchard during dusk.
- In case of bark eating caterpillar prune death shoots, remove silken gallery and inject DDVP (0.1%) into bore holes. For the control of guava trunk borer (*Aristobiaa testudo Voet*), inject DDVP 0.05% @ 5ml/bore holes.

Banana

- Banana stem weevil infested plant should be uprooted and burnt. After harvesting remove pseudostem.
- For aphids which also act as vector of bunchy top disease in banana uproot and burn infested plants, avoid taking ratoon.

Citrus

- Pluck citrus leaf miner infested leaves in early stage and destroy.
- Install light trap for destruction of fruit-sucking moth. Use biat traps having Gur (1kg) + vinegar (60 g) in 10 litre of water.
- Collect and destroy nymph of citrus psylla.
- Hand-picking of egg masses and larva of lemon butterfly.

Papaya

- Collect and destroy all fallen plant to reduce fruit fly infestation. Do not retain damaged fruits on tree trunk because they attract fruit fly.
- Aphid also transit mosiac virus disease. Destroy virus infected plants.

Pineapple

- For the control of mealy bug (*Dymicuccus brevipus*) plant healthy suckers.
- Fruit eating beetles can be suppressed by not allowing fruit to ripen in the field.

Pomegranate

- Destroy the anar butterfly infested fruits. Bag the fruits with cloth or paper bag.

Jackfruit (Shoot and fruit borer)

- Cover the young developing fruits with alkathene bag (perforated). Remove and destroy the effected shoots, flower buds and fruits in initial stage of attack.

3. CULTIVATION OF EXPORT POTENTIAL FRUITS

Organic fruit production is a sustainable, economic and eco-friendly, since there is no risk of residual toxicity. It improves soil fertility and produces quality including self-life. Addition of compost prepared from farm wastes, appropriate crop rotation and inclusion of legume helps maintain organic matter in soil. Detailed knowledge of the history of the land use, soil properties and their characterization is important. Ready access to organic matter sources or provision for on-farm generation of biomass will be required to be able to build up and maintain soil fertility. Cultivation techniques for some export potential fruits crops are discussed here.

3.1 CITRUS

Citrus can grow well in wide range of soils. Citrus fruits flourish well on light soils with a good drainage. Deep soils with pH range of 5.5 to 7.5 are considered good. However, they can grow in pH range of 4 to 9. Presence of calcium carbonate concentration within feeding zone may adversely affect the growth.

The sub-tropical climate is the best suited for citrus growth and development. Temperature below 4^0C is harmful for the young plants. Soil temperature around 25^0C seems to be optimum for root growth. Dry and arid conditions coupled with well defined summer having low rainfall (ranging from 75 cm to 250 cm) are most favourable for the growth of the crop. High humidity favours spread of many diseases. Frost is highly injurious. Hot wind during summer results in desiccation and drop of flowers and young fruits.

3.1.1 Selection of Site and Land Preparation

The site selected should preferably be nearer to the market centres and roads, besides being agro-climatically suitable. Windbreaks should be provided on the sides from which high winds are expected. Plants suitable for providing wind brakes are jamun, mulberry, shisam, etc.

Land needs to be ploughed, cross-ploughed and levelled. In hilly areas, planting is done on terraces against the slopes. In such land, high density planting is possible as aerial space available is more than that in flat land.

3.1.2 Planting Material and Varieties

Planting material is produced by 'T' budding or seedlings may be used for orange plantation. Cutting and air layering are done in lemons. Nursery beds are prepared on light fertile soils. Selection is done by eliminating weaklings, off types and non-uniform seedlings in 2-3 stages in the nursery beds. If certified bud wood is not available for propagation, nuclear seedlings may be selected in the nursery beds, as they are more vigorous, uniform and virus-free. Seedlings may be grown in polythene bags also. They become ready for plantation in the main field after attaining the height of about 30-40 cm after one year.

Varieties: The important Indian varieties of different types of citrus and their respective suitable regions are as under:

Mandarin Orange : Kurg (Kurg & Wyned regions), Nagpur (Vidharba region), Darjeeling (Darjeeling region), Khasi (Meghalaya region), Sumthira (Assam), exotic variety, Kinnow (Nagpur, Akola regions, Punjab & adjoining States).

Sweet Orange : Blood Red (Haryana, Punjab & Rajasthan), Mosambi (Maharashtra), Satgudi (Andhra Pradesh), exotic varieties - Jaffa, Hamlin & Pineapple (Punjab, Haryana, Rajasthan), Valencia.

Acid Lime : Pramalini, Vikram, Chakradhari, PKM1, Selection 49, Seedless lime, Tahiti

Sweet Lime : Mithachikra, Mithotra

Lemon : Eureka, Lisbon, Villafranca, Lucknow Seedless, Assam Lemon, Nepali Round, Lemon 1

Plant Density: The best season of planting is June to August. Pits of the size of 60 cm x 60 cm x 60 cm may be dug for planting seedlings. 10 kg of FYM and 500 g of rock phosphate may be applied per pit while planting. With good irrigation system, planting may be done in other months also.

Table 13.2: Planting density for citrus

Spacing	Orange	Sweet Lime	Lime/Lemon
Normal spacing	6 m x 6 m	5 m x 5 m	4.5 m x 4.5 m
Plant population	275 per ha	400 per ha	494 per ha

In very light soils, spacing may be 4 m x 4 m. In fertile soils and in high rainfall areas spacing may be 5 m x 5m.

Intercrops: Leguminous vegetables like cow peas, french bean, peas, etc., may be grown in citrus orchards. Intercropping is advisable only during the initial three years.

3.1.3 Irrigation

Quantity of water and frequency of irrigation depends on the soil texture and growth stage. Citrus requires life-saving watering in the first year during winter and summer. Diseases like root rot and collar rot may occur under over irrigated condition and if the collar region is wetted. Light irrigation with high frequency is beneficial. Irrigation water containing more than 1000 ppm salts is injurious. Under unirrigated condition chances of damage to spring blossom is high and the next crop maturing in October-November may be heavier. Partial drying out of the soil in spring season is found good.

3.1.4 Manuring

Citrus plants should be manured in three equal doses three times in a year in February, June and September. Depending on the soil, age and growth of plants, the dose varies. The doses mentioned in the following tables may be considered judicious.

Table 13.3: Doses of manures and fertilizers (per tree per year)

FYM/Nutrients	One year old	Eighth year onwards
FYM or Cowdung	10 kg	50 kg
Nitrogen	50 g	400 g
Phosphate	50 g	400 g
Potash	50 g	500 g

The dose should increase every year proportionately to reach full quantity on the eighth year. Source of nitrogen, phosphorus and potash may be vermicompost, biocompost, rock phosphates, wood ash, etc. These organic fertilizers are spread on the ground up to leaf drip and mixed with soil by light spading. Irrigation should be applied if there is moisture stress after application of fertilizers. Two sprays of biodynamic solution may be given to fulfil micronutrient requirements.

3.1.5 Interculture and Pruning

Ploughing, spading of basins, weed control, etc., are important inter-culture operations for soil aeration and health.

Trimming: In order to allow the growth of a strong trunk, all shoots in the first 40-50 cm from ground level developed in the early stage should be removed. The centre of the plant should remain open. Branches should be well distributed to all sides. Cross twigs and water suckers are to be removed early. The bearing trees require little or no pruning. All diseased, injured and drooping branches and dead wood are to be removed periodically.

3.1.6 Pests and Diseases Management

Pests: Important pests of citrus are citrus psylla, leaf miner, scale insects, orange shoot borer, fruit fly, fruit sucking moth, mites, etc. Other pests attacking citrus particularly mandarin orange, specially in humid climate are mealy bug, nematode, etc.

Table 13.4: Control measures of major pests of Citrus

Crop/ pest	Biotic agents	Dosage/ ha	Frequency of application	Application method
Icerya purchasi	*Rodolia cardinalis*	10 beetles/ infested plant	Once on noticing the infestation	Releasing adults
Planococcus citri	*Cryptolaemus montrouzieri*	10 beetles /infested plant	After the blossom	Releasing adults
Planococcus citri	*Leptomastix dactylopii*	3,000 adults	Need-based under expert supervision	Releasing adults
Papilio demoleus	*Bacillus thuringiensis var. kurstaki*	0.5 %	Single application for each generation	Spray with knapsack sprayer
Phytophthora spp.	*Trichoderma ciride/ T. harzianum*	100 kg/ha	Incorporate around the tree base covering up to the area covered by canopy	Soil application along with organic matter

Note: Ant suppression operations should be adopted in the orchard.

***Diseases*:** The important diseases of citrus are tristeza, citrus canker, gummosis, powdery mildew, anthracnose, etc.

Control measures of these diseases are stated briefly below:

1. *Tristeza*: Control of aphids and use of cross-protected seedlings are recommended.
2. *Citrus canker*: Cutting of effected twigs followed by spraying of 1% Bordeaux mixture.
3. *Gummosis:* Scraping of the effected area and application of Bordeaux mixture.
4. *Powdery mildew*: Dead twigs are to be pruned first.
5. *Anthracnose:* Dried twigs are pruned off first.

Note: Bordeaux mixture should be used in consultation with Certifying Agency

3.1.7 Harvesting

Economic life of orange is 25-30 years and lime is 15-20 years. Mature fruits are picked up in 2-3 cycles. There may be 2 or 3 crops in a year in summer, rainy season and autumn. Orange is picked when colour starts developing.

Citrus may yield fruits as follows depending up on the soil, climatic and organic management.

Mandarin Orange : Commences from the 4th/5th year with 40/45 fruits per tree. Yield stabilises in the 10th year. Average production is about 400-500 fruits per tree after stabilisation.

Sweet Orange : Commences from 3rd/4th year with 15 to 20 fruits per tree. Yield stabilises around the 8th year. Average production is about 175-250 fruits per tree after stabilisation.

Lime/ Lemon : Commences from the 2nd/3rd year with 50-60 fruits per tree. Yield stabilises in the 8th year. Average production is about 700 fruits per tree after stabilisation

3.1.8 Cost of Development of Mandarin Orange

The details of the cost and yield estimate are furnished below:

Table 13.5: Projected income from citrus cultivation

Item/Year	6 yr	7 yr	8 yr	9 yr	10 yr onwards
No. of fruits/plant	80	150	250	350	450
No.of fruits/ha	22,000	41,250	68,750	96,250	123,750
Gross Income (Rs/ha)	16,500	30,938	51,563	72,188	92,813
Expenses (Rs/ha)	12,050	13,410	14,900	14,900	14,900
Net income (Rs/ha)	4,450	17,528	36,663	57,288	77,913

Unit Cost of Orange cultivation is about Rs.60,200 per hectare. The cost presented here is indicative only.

Table 13.6: Estimated cost (Rs/ha) for citrus fruits cultivation

Item	1 yr	2 to 3 yr	4 yr	5 yr	6 yr	7 yr	8 yr onwards
MATERIALS							
Planting materials	2,121	-	-	-	-	-	-
FYM	825	2,888	2,063	2,475	2,888	3,300	4,125
Nutrients supply	510	2,546	2,036	2,536	3,054	3,667	4,278
Plant Protection measures	400	975	575	775	775	900	900
Irrigation	700	1400	700	700	700	700	700
Biodynamic solution	-	-	400	400	550	700	700
Workable fencing	2,000	-	-	-	-	-	-
Sub Total - I	6,556	7809	5,774	6,886	7,969	9,267	10,703

Item	1 yr	2 to 3 yr	4 yr	5 yr	6 yr	7 yr	8 yr onwards
LABOUR							
Land preparation	1,050	-	-	-	-	-	-
Digging of pits	1,200	-	-	-	-	-	-
Plantation	1,500	-	-	-	-	-	-
Application of fertilizers, pesticides, etc.	750	2,100	1,320	1,620	1,620	1,620	1,620
Interculture	900	1,440	600	600	600	600	600
Pruning/Training	120	300	-	300	300	300	300
Irrigation	1,800	2820	1,200	1,200	1,200	1,200	1,200
Harvesting & packing	-	-	-	300	360	420	480
Sub Total (II)	7,320	6,660	3,120	4,020	4,080	4,140	4,200
Total (I + II)	13,876	14,496	8,894	10,906	12,047	13,407	14,903

3.2 PINEAPPLE

Pineapple can grow in sandy, alluvial or laterite soil. Slightly acidic soil with pH range of 5.5 to 6.0 is considered optimum for pineapple cultivation. The soil should be well drained and light in texture. Heavy clay soil is not suitable.

Areas with a heavy rainfall are best for pineapple growth. Optimum rainfall is 1500 mm per year although it can grow in areas having 500 mm to 5550 mm of rainfall. The fruit grows well near the seacoast as well as in inland, so long as temperature ranges from 15.5 to 32.5^{0} C. Low temperature, bright sunshine and total shade are harmful.

3.2.1 Land Preparation

The lands should be thoroughly ploughed and pulverised to a good tilth. It should be properly levelled to avoid water stagnation. Adequate drainage channels should be provided.

3.2.2 Manuring

Pineapple requires abundant supply of Nitrogen and Potash. Manuring should be done in 2-3 equal doses every year, once at the onset of monsoon (May-June) and again at the end of the rainy season (September-October) after the fruit are harvested and slips and suckers are removed. The fertilizers doses recommended for obtaining higher yield are 10g nitrogen, 5g phosphorus and 10g potash per plant per year in addition to 500g FYM. Source of nitrogen, phosphorus and potash may be vermicompost, biocompost, poultry manure, rock phosphates, wood ash, neem cake, rock potash etc. These organic fertilizers are spread on the ground up to leaf drip and mixed with soil by light spading. Fertilization is followed by earthing up around the stem.

3.2.3 Planting Material

Pineapple can be propagated by sucker, slip and crown. Growth is best with suckers and slips. Plants from crown bear flowers after 3 to 20 months later than suckers and slips depending on the climatic conditions. Therefore, crowns are not normally used. Suckers should be planted within 2 weeks after removing from the mother plant. The planting material should be selected from healthy diseased free plant.

Varieties: The most popular commercial pineapple variety in India is Giant Kew. Other important verities are Queen, Kew, Mauritius, Charlotte, Rothchild, Jaldhup, Desi, Lakhat, etc. Qualitatively, Queen is the outstanding table variety. Kew and Giant Kew variety are better suited for canning, while Queen and Mauritius cultivars are good for juice production.

3.2.4 Planting and Plant Density

The best time of planting pineapple is in the early rains or early winter. With irrigation it may be any time of the year. Suckers should be planted at 10-15cm depths in 15-20cm deep holes. About 500g FYM or cow dung is to be added to the soil of each hole. Planting may be done in single or double row following triangular or rectangular system.

In high rainfall areas the density may be around 40000 to 44000 plants per hectare. In low rainfall areas with cool weather, even higher density of 63000 to 64000/ha is recommended.

Spacing: Pineapple is planted in double hedge system for convenient intercultural operations. For a density of 44000 plants per ha, the spacing should be 90 cm x 30 cm x 60 cm i.e. 90 cm between two rows of adjacent beds, 30 cm between plants in a row and 60 cm between two single rows.

Intercrop: To protect the fruit from sunburn, partial shade may be provided by planting arhar (*Cajanas cajan*) in between the beds. The spacing has to be adjusted accordingly. Covering the maturing fruits with rice straw or pineapple leaves will reduce both sunburn and bird damage.

Crop cycle: One main crop followed by two ratoons is the usual crop cycle in pineapple. After the fourth year the plot needs to be uprooted and replanted.

3.2.5 Irrigation

Pineapple is mostly grown as a rainfed crop, but supplementary irrigation gives higher production. Irrigation in summer and winter keeps the plants healthier. About 8 to 12 light irrigations during winter and summer months are beneficial. Irrigation is essential after planting or after manuring if there is deficiency in soil moisture.

3.2.6 Plant Protection

No serious pest or disease of pineapple is prevalent in India. However, Mealy bug and Heart rot are important pest and disease respectively. Use of EM Solution can also be used for pests and disease control as described in the "Bio-control of Pests and Diseases"

Mealy bug: For the control of mealy bug (*Dymicuccus brevipus*) plant healthy suckers

Heart rot: Apply of Bordeaux mixture

3.2.7 Interculture and After Care

Controlling weed is the major interculture operation in pineapple field. Weeding should be done at least 3 to 4 times a year.

After care of the ratoon crop: Desuckering immediately after fruit harvest is important. Keeping one or two suckers on the mother plant near the ground level, all others are removed. Slips also should be removed. After desuckering, plants should be fertilized and earthed up.

Staggering of harvesting: It is almost throughout the year is possible by the following means:

- using different planting materials.
- planting suckers and slips at regular intervals (July-Dec.)

3.2.8 Harvesting

The fruit is ready for harvest when the dark green colour becomes lighter and the deep-seated eyes become shallow. Harvesting for local markets should be done at full maturity stage and for distance markets at 75-80 % maturity stage.

Fruits are mainly harvested during July-August. However, a small crop is harvested during December to March also. By regulating the crop, harvesting is possible almost 8 month a year. Harvesting should be done with a sharp knife severing the fruit stalk with a clean cut and retaining 5-7 cm of the stalk. Slips and a part of the crown are removed. Any mechanical injury on the fruit skin may cause the fruit to rot quickly. Hence, great care is necessary in handling the fruit. For canning industry, only the cylindrical fruits of about 1.5kg size are considered good. Such fruits are obtained when peduncles are upright and plant density is around 40000 per ha.

3.2.9 Post Harvest Management

The yield from a plant population of 35000-40000 per ha is about 40-50 tonnes and that from a plant population of 43000-50000 per ha normally varies from 50 to 60 tonnes. Economic life of a pineapple plantation is expected to be around 4 years. After this the plot should be uprooted and replanted.

Pineapples are sensitive to injuries due to compression or impact. Rotting starts and spread quickly from bruised or injured parts. The produce should be cooled as soon as possible after harvest. Cooling temperature is 7-10^{0}C for ripe and half-ripe ones. The relative humidity should be 85-95 per cent. The same temperature and RH are kept in storage and transport. Cooling is done by forced air method. Storage period of pineapple is 2-4 weeks depending on the cultivars and ripening stage.

3.2.10 Cost of Development of Pineapple

The unit cost for the development of pineapple in 1 ha of land works out to be Rs.98600/-. The cost of organic fruits will be at least 2-fold of the present market price.

Table 13.7: Projected income from Pineapple cultivation

Item	1 yr	2yr	3yr	4yr
Yield (No. of fruits)	--	30,800	28,600	24,200
Gross income (Rs.)	--	123,200	114,400	96,800
Cost (Rs.)	72,900	36,700	35,160	34,500
Net Income (Rs.)	--	86,500	79,240	62,300

Table 13.8: Estimated Cost (Rs./ha) for pineapple cultivation

Item	1yr	2yr	3yr	4yr
MATERIALS				
Planting materials	24,200	--	--	--
FYM	6,600	6,600	6,600	6,600
Nutrient supply	12,100	12,100	12,100	12,100
Irrigation	1,600	1,600	1,800	1,800
Plant protection	600	700	800	800
Sprayer	2,000	--	--	--
Workable fencing	2,000	--	--	--
Sub - total I	49,100	21,000	21,300	21,300
LABOUR				
Land preparation, fixing slabs etc. (125 md*)	7,500	--	--	--
Preparing suckers and planting (115 md)	6,900	--	--	--
Application of manures (60 md)	3,600	3,600	3,600	3,600
Plant protection. (12 md)	720	900	900	900
Interculture (Weeding- 40 md, Earthing up- 25 md)	3,900	3,900	3,900	3,900
Irrigation (20md)	1,200	1,200	1,200	1,200
Harvesting @ 400 fruits/md	--	4,920	4,260	3,600
Sub - total II	23,820	14,520	13,860	13,200
GRAND TOTAL	72,920	36,720	35,160	34,500

**md- Mandays*

3.3 MANGO

Mango can be grown under both tropical and sub-tropical climate from sea level to 1400 m altitude, provided there is no high humidity, rain or frost during the flowering period. Places with good rainfall and dry summer are ideal for mango cultivation. It is better to avoid areas with winds and cyclones which may cause flower and fruit shedding and breaking of branches.

Mango comes up on a wide range of soils from alluvial to laterite provided they are deep (minimum 6 ft) and well drained. It prefers slightly acidic to alkaline soils (pH 5.5 to 7.5).

3.3.1 Plant Material and Varieties

Farmers should always get one year old healthy, true to type straight growing grafts from reliable sources/recognised nurseries. Inarching, veneer grafting, side grafting and epicotyl grafting are the popular methods of propagation in mango.

The important varieties of mango grown in India are, Alphonso, Bangalora, Banganpalli, Bombai, Bombay Green, Dashehari, Fazli, Fernandin, Himsagar, Kesar, Kishen Bhog, Langra, Mankhurd, Mulgoa, Neelam, Samarbehist, Chausa, Suvarnarekha and Vanaraj.

3.3.2 Land Preparation and Planting

Land should be prepared by deep ploughing followed by harrowing and levelling with a gentle slope for good drainage.

Spacing: Spacing varies from 10 m x 10 m, in the dry zones where growth is less, to 12 m x 12 m, in heavy rainfall areas and rich soils where abundant vegetative growth occurs. New dwarf hybrids like Amrapali can be planted at closer spacing. Pits are filled with original soil mixed with 20-25 kg well rotten FYM, 2.5 kg rock phosphate and 1 kg rock potash.

Healthy and straight growing plant material can be planted at the centre of pits along with the ball of the earth intact during rainy season in such a way that the roots are not expanded and the graft union is above the ground level. In the initial one or two years, it is advisable to provide some shade to the young plants and also stake to make them grow straight.

***Inter cropping*:** Inter crops such as vegetables, legumes, short duration and dwarf fruit crops like papaya, guava, peach, plum, etc. depending on the agro-climatic factors of the region can be grown. The water and nutrient requirements of the inter crops must be met separately.

***Training and pruning*:** About one meter from the base on the main trunk should be kept free from branching and the main stem can be allowed thereafter spaced at 20-25 cm apart in such a way that they grow in different directions. Branches which cross over/rub each other may be removed at pencil thickness.

3.3.3 Manuring

In general, 75 gm nitrogen, 25 gm phosphorus and 70 gm of potash per plant per year of the age from first to tenth year and thereafter 750 gm, 250 gm, and 700 gm respectively of NPK per plant per year can be applied (June-July and October). To supply nitrogen, the natural and organic materials like farm yard manure (FYM), vermin-compost, oil cakes (neem, castor and karanj) and green manures. Bone meal (crushed) can be used as organic phosphorus nutrition. These organic fertilizers are spread on the ground up to leaf drip and mixed with soil by light spading. Foliar spray of biodynamic solution is effective before flowering.

3.3.4 Irrigation

Young plants are watered frequently for proper establishment. In case of grown up trees, irrigation at 10 to 15 days interval from fruit set to maturity is beneficial for improving yield. However, irrigation is not recommended for 2-3 months prior to flowering, as it is likely to promote vegetative growth at the expense of flowering.

3.3.5 Plant Protection

Mango is prone to damages by a large number of pests, diseases and disorders. The suggested control measures for other most important pests already described above *(para 2.0.3)*. Controls of malformation and fruit drops in mango are as follows:

Malformation: One spray of 200 ppm NAA in October followed by deblossoming at bud burst stage in December - January.

Fruit drop: Regular irrigation during fruit development, timely and effective control of pests and diseases and spraying 20 ppm NAA at pea size of fruits. *Use of NAA must be done after approval from Certifying body.*

3.3.6 Post Harvest Management

Graft plants start bearing at the age of 3-4 years (10-20 fruits) to give optimum crop from 10-15th year which continues to increase up to the age of 40 years under good management.

Storage: Shelf life of mangoes being short (2 to 3 weeks) they are cooled as soon as possible to storage temperature of 13 degree Celsius. A few varieties can withstand storage temperature of 10 degree Celsius. Steps involved in post harvest handling include preparation, grading, washing, drying, waxing, packing, pre-cooling, palletisation and transportation.

Packaging: Mangoes are generally packed in corrugated fibreboard boxes 40 cm x 30 cm x 20cm in size. Fruits are packed in single layer 8 to 20 fruits per carton. The boxes should have sufficient number of air holes (about 8 % of the surface area) to allow good ventilation.

3.3.7 Cost of Development of Mango

The cost presented here is indicative only. The unit cost estimated for this model scheme is Rs.34400/- per ha.

Table 13.9: Projected income from mango cultivation

Item	6 yr	7 yr	8 yr	9 yr	10 yr	11 yr	12 yr	15th yr onwards
Yield (q/ha)	25	35	50	60	70	80	90	120
Gross value @ Rs. 4.50/kg	11250	15750	22500	27000	31500	3600	40500	54000
Maintenance (Rs/ha)	4600	6000	6000	6200	6200	6200	6200	6200
Net Income (Rs/ha)	6650	9750	16500	20800	25300	29800	34300	47800

Table 13.10: Estimated cost (Rs/ha) for mango cultivation

Item	1 yr	2 yr	3 yr	4yr	5 yr	Total
Planting material	2200	--	--	--	--	2200
Nutrient supply	3000	1100	1100	1400	1400	8000
Plant protection	1100	600	600	700	700	3700
Implements	1500	--	--	--	--	1500
Fencing	2500	--	--	--	--	2500
Irrigation	1800	500	500	500	500	3800
Labour	3200	1200	1200	1500	1500	8600
Intercropping	1500	--	--	--	--	1500
Miscellaneous	600	500	500	500	500	2600
Total	17400	3900	3900	4600	4600	34400

3.4 LITCHI

Litchi requires a moist sub-tropical climate without heavy frost and dry winds. The four essentials for litchi cultivation are (a) freedom from frost, (b) High humidity, (c) rich deep soil, and (d) abundant moisture. In litchi growing tracts, the maximum temperature varies from 20°C to 32°C between the flowering time and the fruiting season. Dry atmosphere and cold snaps between 1°C and 4°C in winter are beneficial. Litchi grows successfully on the hills up to an elevation of 1050 m.

Alluvial soil with good drainage is suitable for litchi orchards. It can grow in wide range of soils from light sandy to heavy clay. It cannot withstand water logging for long. Soils taken from the root region of old trees may be added to pits while planting litchi. This practice introduces mycorrhiza that helps growth of litchi.

Table 13.11: Potential States for Litchi expansion

State	District
Bihar	Muzaffarpur, East Champaran Samastipur, Vaisali, Bhagalpur
Uttar Pradesh	Saharanpur
Uttaranchal	Dehradun, Pithaurgarh, Nainital, Hardwar
West Bengal	Murshidabad, 24-Paraganas
Assam	Sonitpur, Kamrup, Bongaigaon
Punjab	Gurdaspur, Ropar, Hoshiarpur

Wind breaks should be provided on the sides from which high winds are expected. Plants suitable for providing wind breaks are seedlings of mango, jamun, eucalyptus, mulberry, shisham etc. Live thorny hedge preferably of 'Karonda' (*Caricas carandas*) or other locally available plant should be provided around the field in order to protect the young trees from animals etc.

3.4.1 Planting Material and Varieties

Good Quality air layered nursery plants purchased from organic plant nurseries or organically approved nurseries should be planted. Normally 10 m x 10 m spacing will be provided with a density of 100 plants per ha. Bombai and Muzaffarpur varieties which grow in to huge trees may also be planted at this density but good pruning will be needed. China and Elachi can be comfortably planted at 124 plants per hectare. High density planting and pruning or thinning of plants is the most modern technology.

Table 13.12: State wise (India) commercial varieties

State	District
Bihar	Shahi, Rose scented, China, Kasba, Purbi, Early Bedana, Late Bedana.
Uttar Pradesh	Rose scented, Dehradun, Calcutta
West Bengal	Bombay Green, Kalyani Sellection
Assam	Muzaffarpur, China, Shahi
Punjab	Muzaffarpur, Dehradun, Seedless, Late Bedana

3.4.2 Manuring

Table 13.13: Nutrient requirement for Litchi

FYM / Nutrients	One year old tree	Full bearing tree (10th year onwards)
FYM	20 kg	80 kg
Nitrogen (N)	100 g	1000 g
Phosporous (P_2O_5)	50 g	500 g
Potash (K_2O)	50 g	1000 g

To supply nitrogen, the natural and organic materials like farm yard manure (FYM), vermicompost, oil cakes (neem, castor and karanj) and green manures. Bone-meal (crushed) can be used as organic phosphorus nutrition. Biofertilizers can also be used for better growth and fruit yield. The above quantities of NPK manures and organic fertilizers (after laboratory test) should be applied in two splits one in January and another in September.

Roots of litchi plants have good association with vesicular arbuscular mycorrhiza (VAM) which improves water and nutrient uptake efficiency of litchi plants. Extent of benefits of mycorrhizal association depends upon the phosphorus requirement and root system of plants. In case of phosphorus deficient soils the VAM has the capacity to mobilise the native phosphorus in soil, which otherwise unavailable to plants. Combined inoculation of azotobactor and VAM proves beneficial for nitrogen and phosphorus nutrition of litchi plants.

3.4.3 Pruning and Training

Litchi requires light pruning. As the fruits are borne on the growth of the previous year, the common practice is breaking off a metre or so of a branch along with the bunch of the fruits at the time of harvesting.

3.4.4 Plant Protection

Litchi is free from very serious pests and diseases. However, mites and trunk borers may cause some damage for which contact bio-insecticide spray will be the effective control mechanism.

3.4.5 Irrigation

Life-saving irrigation is needed for young plants at least in the first 2 years during winter and summer. Irrigation of the young trees should be done by basin system. During dry season irrigation schedule will be 30 times during 1st year, 20 times during 2nd to 4th year and 10 times from 5th year onwards.

3.4.6 Harvesting

Litchi fruits are ready for harvest as soon as they start changing colours towards reddish in the 1st to 2nd week of May. Litchi is very short lived and remains in the market for about 4 to 6 weeks. During harvesting a small portion of twigs are cut off along with bunches.

Trees raised from layers start bearing from the 5th year of age. Yield continues to increase up to the 10th year and thereafter it stabilizes. Economic life of litchi is 50 to 60 years.

3.4.7 Post-harvest Management

Litchi fruits are graded and packed in baskets or boxes with green leaves as cushioning materials in the orchard itself. These are sent to the wholesale and to retail markets very fast. Shelf life of litchi is poor under normal conditions. The fruits keep for 3 days only and hence fruit baskets cannot be held in store for long. Fruits are sensitive to ethylene exposure.

Post harvest management for export market

Step 1: Spray 0.6% calcium chloride

Step 2: Harvest bunches where red blush is appearing

Step 3: Grade according to size and colour. Reject smalls and defectives. 2.4 to 4 cm long and above 20 g fruits may be exported. Cherry red and dark brown fruits are chosen for export.

Step 4: Dip in copper sulphate (0.5%) for 3 min

Step 5: Packaging in perforated CFB boxes. Thinner boxes may be used. Litchi leaves may be inserted.

Step 6: Sulphitation. Expose for a shortwhile in fume by burning 1.3 kg sulphur in 20 m^3 space for 5 tonnes of Litchi

Step 7: Pre-cooling to 1.5% to 2^{0}C at 90 - 95% RH by forced air method. Transporting by refrigerated containers at 1-2^{0}C with 90-95% RH.s

Note: Consult Certification body before using copper sulphate, sulphur and calcium chloride.

Storage is done in poly bags at 1.5-2^{0}C and 90-95 %. RH. The fruits may be kept in store up to 6 weeks. Storing is done at 7^{0}C when the storage life is less than 3 weeks.

3.4.8 Cost of Development of Litchi

In the present model, the unit cost (indicative only) for development of Litchi in one hectare is works out to be Rs.44600/-. Maintenance cost from 6th year onwards is Rs.8000/- per ha. The cost of organic fruits will be at least 2-fold of the present market price.

Table 13.14: Projected income from litchi cultivation

Item	1yr	2-4 yr	5 yr	6 yr	7 yr	8 yr	9 yr	10-30 yr
Capital cost	19700	5600	8000	-	-	-	-	-
Maintenance cost	0	0	0	8000	8000	8000	8000	8000
Total cost	19700	5600	8000	8000	8000	8000	8000	8000
Income	0	0	10000	20000	40000	60000	80000	100000
Net Benefits	--	--	2000	12000	32000	52000	72000	92000

Table 13.15: Estimate Cost (Rs/ha) for litchi cultivation

Item of expenditure	1yr	2 yr	3yr	4 yr	5yr	Total
Preparation of land	3000	-	-	-	-	3000
Digging and filling pits and planting	1000	-	-	-	-	1000
Cost of planting material	2500	-	-	-	-	2500
Manures & fertilizers	2500	2800	3300	3600	4000	16200
Live fencing	1500	-	-	-	-	1500
Irrigation	600	700	800	900	1100	4100
Plant Protection	1200	400	500	600	800	3500
Intercropping (income recycled)	5000	-	-	-	-	5000
Interculture	600	600	600	600	600	3000
Watch & ward	-	-	-	-	1000	1000
Misc	1800	500	500	500	500	3800
Total	19700	5000	5700	6200	8000	44600

3.5 CASHEW

Cashew is a tropical plant and thrives at high temperatures. Young plants are sensitive to frost. Areas where the temperatures range from 20 to 30 degree Celsius with an annual precipitation of 1000 - 2000 mm are ideal for cashew growing. Heavy rainfall, evenly distributed throughout the year, is not favourable though the trees may grow and some times set fruit. It needs a climate with a well defined dry season of at least four months to produce the best yields. Coincidence of excessive rainfall and high relative humidity with flowering may result in flower/fruit drop and heavy incidence of fungal disease. Cashew is regarded as "essentially coastal tree" but that is not true. It also grows well at considerable distance from the coast.

Cashew is very modest in its soil requirements and can adapt itself to varying soil conditions without impairing productivity. But cashew performs much better on good soils than on poor soils. The best soils for cashew are deep, friable well-drained sandy loams without a hard pan. Cashew also thrives on pure sandy soils although mineral deficiencies are more likely to occur. Water stagnation and flooding are not congenial for cashew.

3.5.1 Planting Material and Varieties

Cashew is a cross pollinated crop and exhibits wide variations in respect of nut, apple and yield of seedling progenies. Therefore, vegetative propagation has been advocated to mitigate this problem. Air-layering has been quite successful but survival percentage seems to be low and it has been reported that the plantations raised from air-layers are more susceptible to drought and the life of such plantation is shorter as compared to that of grafted or seeding ones. The anchorage has also been observed to be poor, especially in cyclone-prone areas. Number of other methods of propagation such as budding and grafting has been found successful. Epicotyl grafting and softwood grafting are found to be successful because it is easy to produce large number of grafts in a short time. The percentage of field establishment is also reported to be high with these grafts. Some important varieties with yield potential ranging between 20-25 kg per tree are:

BPP-1	High percentage of perfect flower, 13.2%, fruit set high, yield 17 kg (25 year old plant) per plant, shelling, 27.5 % nuts of 5 g average weight
Vengurla-3	Nut weight 9 g
BPP-2	Yield 19 kg/tree (25 years), shelling 26%, nuts 4g average
Vengurla-1	Average yield 23 kg/plant at 28 year's age, nut weight 6g, shelling 31%.
Vengurla-2	Yield 24 kg/plant at 20 year's age, nut weight 4 g, shelling 32 %.
Other varieties	BPP-3,4, 5, Vengurla-4,5, VRI-1,2, Ullal-1,2, Anakkayam-1, BLA 39-4, K-22-1, NDR 2-1
Export variety	K-22-1 and NDR 2-1

3.5.2 Preparation of Land, Layout and Planting

It is better to select the land having good drainage and devoid of sub-surface hard rock or hardpan, for successful cultivation of cashew. The land should be ploughed thoroughly and levelled in case of agricultural lands. In case of sloppy land, land is to be terraced or bunds constructed in. Pits of 1 m^3 are to be dug and allowed to wither.

Layout: Cashew trees are generally planted with a spacing of 7 to 9 meters usually by the square system. It is, however, preferable to plant by the equilateral triangular system, especially on the slopes, as this accommodates 15 % more plants without affecting the growth and development of the trees. In undulating areas they are preferably planted along the contours, with cradle pits or trenches provided at requisite spacing in a staggered manner to arrest soil erosion and help moisture conservation.

Planting: The grafted plants obtained from the superior mother plant are usually planted after filling the pits (1 m^3 size) with topsoil and farm yard manure (FYM) at the onset of monsoon. It is desirable to dig the pits well in advance and allow sufficient time for withering. Burning of the debris and forest wastes inside the pits before planting is advantageous.

It is essential to provide stakes and temporary shade with the locally available materials wherever necessary (especially in the south west aspects in case of forest plantation) to reduce the mortality rate and achieve quicker establishment. If the monsoon rains are inadequate, one or two pot watering is done during the initial stages to ensure establishment. Mulching with black polythene is beneficial to increase the growth and yield of cashew.

3.5.3 Manuring

Table 13.16: The nutritional requirements of Cashew

Nutrient	Grams per tree per year		
	1 year	2 year	3 year & after
Nitrogen	170	340	500
Phosphorus	45	85	125
Potassium	45	85	125

To supply nitrogen, the natural and organic materials like farm yard manure (FYM), vermin-compost, oil cakes (neem, castor and karanj) and green manures. Bone meal (crushed) can be used as organic phosphorus nutrition. These organic fertilizers are spread on the ground up to leaf drip and mixed with soil by light spading. The manuring should be done in pre and post-monsoon period. Foliar spray of biodynamic solution is effective before flowering.

3.5.4 Inter-cropping

Considering the long pre-bearing period and low income in the early period of bearing and fluctuations in the yield and price from year to year it is recommended to take up inter-cropping in cashew plantations. Leguminous crops such as groundnuts and beans are very suitable for inter cropping. Tall growing inter crops like certain varieties of sorghum and millet should not be encouraged between young cashew, as they provide too much shade. Besides the annual crops, the arid zone fruit crops having less canopy especially anona, phalsa, etc., can be thought of, depending on the suitability.

Top working: The rejuvenation of unthrift cashew plantations through top working involves beheading of trees, allowing juvenile shoots to start-out and taking up of *in-situ* grafting using procured scions of high yielding varieties. Periods from 'November to March' and 'February to June' have been found to be ideal for beheading and *in-situ* grafting respectively. It has been observed that the top worked trees within a period of two years have not only put forth a canopy of 3-4 m in diameter and 5-6 m in height (as that of 8-10 year old trees) but also have given an yield of 3 to 5 kg nuts per tree in their first bearing itself.

3.5.5 Weeding

Weeding with a light digging should preferably be done before the end of the rainy reason. Hoeing, cutting the weeds off underground, is more effective than slashing.

3.5.6 Pests and Diseases

Pests: It is observed that there are about 30 species of insects infesting cashew. Out of these tea mosquito, flower thrips and stem and root borer and fruit and nut borer are the major pests, which are reported to cause around 30 % loss to the yield. Tea mosquito population builds up during the beginning of the rainy season, when the cashew tree is full of new flush. In case of thrips, both nymphs and adults suck and scrape at the underside of the leaves, mainly along main veins, causing yellowish patches, latter turning grey. These thrips are more active during the dry season. In Stem and Root Borers, the young white grubs bore into the fresh tissues of the bark of the trunk and roots and feed on the subsequent sub-epidermal tissues and make tunnels in irregular directions. Due to severe damage to the vascular tissue the sap flow is arrested and the stem is weakened. The characteristics symptoms of damage include the presence of small holes, in the collar region, gummosis, yellowing and shedding of the leaves and drying of the twigs. The young caterpillar borers (Fruit and Nut Borers) through the apple and nut causing deformity and /or loss of kernel weight.

Risk of infection can be lowered by using appropriate soil management and cultivation measures. Spraying of biodynamic liquid pesticides prepared from cowurine, neem, karanj *(Pongamia pinnata)*, calotropis, dhatura, casto, *Thevtia nerrifolia, Vitex species* leaves, etc. are helpful.

Diseases: Fortunately cashew crop does not have any serious disease problem, except the powdery mildew caused by a fungus, which affects the young twigs and inflorescences and make it wither. This disease generally appears when the weather becomes cloudy. Control can be obtained by dusting with 2 % sulphur W.P. in consultation with certifying body.

3.5.7 Harvesting and Yield

Normally harvesting consists of reaping the nuts that have dropped to the ground after maturing. However, if the nuts are also used for making jam, juice, syrup, fenni, etc., the fruit has to be harvested before it falls naturally. Plantations of unknown origin or seeding progenies with conventional methods of cultivation yield less than one kg of raw nuts per tree. However, there is a chance to increase the yield up to 4 to 5 kg per tree with the adoption of improved production techniques, over a period of 4 to 5 years. In new plantations, with the use of elite planting material coupled with a package of improved agronomic practices, an yield of 8-10 kg per tree could be achieved.

3.5.8 Processing, Grading and Packing

The processing of cashew involves- preliminary cleaning, roasting, shelling and separation, drying, peeling, etc. A preliminary cleaning of cashew nuts is done by manual picking of large objects and by sieving. The cleaned cashew nut is roasted in open pan or earthen ware or rotary cylinder or hot oil bath. The rotary cylinder method is more hygienic and efficient, despite a major portion of the CNSL (Cashew Nut Shell Liquid) would be lost. The hot oil bath process combines good roasting and recovery of shell liquid. The cleaned cashew nuts are placed in wire baskets and immersed in a tank containing CNSL, boiling at a constant temperature of about 180-200 degree Celsius for about 60 to 90 seconds. The CNSL in the tank should be stirred continuously to avoid local overheating and excessive polymerisation and clogging. However, the hot oil bath processing is costlier, and is resorted to only by a few processors. The most common method adopted is roasting by rotary cylinder method. After roasting, the shells are removed and the nuts extracted manually. In manual shelling, recovery of whole kernels is more compared to the mechanical shelling. The kernels are dried in hot air chambers which facilitates peeling of the outer coating. To prevent breakage, the kernels are to be handled very carefully, as they are brittle at this stage. The shelling percentage of cashew varies from 20-25.

Grading and Packing: Grading is done for export purposes based on 'counts' or number of kernels. Sound kernels are named as "wholes" and broken ones as 'splits'. The wholes are again classified as whole white kernels, whole scorched kernels, whole dessert kernels (a) and whole dessert kernels (b). The splits are also further graded into white pieces, scorched pieces, dessert pieces (a) and dessert pieces (b) based on certain physical characters. The wholes are packed in several grades viz., 210, 240, 280, 320, 400, 459 and 500; the popular grade is 320. The specifications for graded kernels are that they should be fully developed, Ivory white in colour and should be free from insect damage and black and brown spots. Packing is done in time by Vita pack method (exhausting the air inside the packing tin, pumping in carbon dioxide and sealing).

3.5.9 Cost of Development of Cashew

The unit cost varies from location to location. The cost presented here is indicative only. The unit cost estimated is Rs.29600/- per ha.

On an average the raw cashew nuts fetch a price of Rs.10-12 per kg in the internal market. The cost of organic fruits will be at least 2-fold of the present market price.

Table 13.17: Projected income from cashew cultivation

Item	4 yr	5 yr	6 yr	7 yr	8 yr	9 yr	10 yr onwards
Yield (kg/ha.)	150	300	500	800	1,200	1,600	2,000
Gross Income (Rs/ha)	5,250	10,500	17,500	28,000	42,000	56,000	70,000

Table 13.18: Estimated Cost (Rs/ha) for cashew cultivation

Item of Expenditure	Cost per year					
	1 yr	2 yr	3 yr	4 yr	5 yr	Total
Land Preparation	800	0	0	0	0	800
Digging of pits and filling	1,600	0	0	0	0	1,600
Planting Material	2,400	240	0	0	0	2,640
Planting - Staking	200	40	0	0	0	240
Live Hedge	1,000	0	0	0	0	1,000
Manures and Organic Fertilizers	701	1,357	1,938	2,514	2,514	9,022
Plant Protection	556	713	869	1,181	1,181	4,500
Irrigation	900	1,000	1,100	1,200	1,200	5,400
Interculture/training	600	660	726	799	799	3,582
Harvesting & Packing	0	0	0	200	400	600
Miscellaneous	43	90	67	7	7	216
Total	8,800	4,100	4,700	5,900	6,100	29,600

3.6 PASSION FRUIT

Passion fruit *(Passiflora spp)* is found to be growing wild in many parts of Indian western ghat such as Nilgiris, Wynad, Kodaikanal, Shevroys, Coorg and Malabar as well as in Himachal Pradesh and North-Eastern states like Nagaland, Manipur and Mizoram.

Passion fruit prefers tropical and sub-tropical climate and grows well up to 2000 m altitude with an annual rainfall of 1000-2500 mm. The crop requires an optimum temperature of 20 to 30°C and below 15°C it restricts vegetative growth and flowering. It can grow in lightly sandy soils with pH of 6-7 provided there is good drainage.

3.6.1 Planting Material and Varieties

Passion fruit is propagated by seeds, cuttings and grafting on resistant root stocks. Seedlings or grafted plants are more vigorous than cuttings. Propagation by seed should be avoided as it includes variability. For cuttings, select mature vines and prepare 30-35 cm long cuttings with minimum three nodes. Raise the cuttings in nursery bed/polythene bags with suitable potting media. It takes one month for rooting and can be transplanted to main field after three months. Provide a spacing of 3 m between plants and 2 m between rows. The best time for planting is during monsoon. Seasonal vegetables could be grown as intercrops during the first year of planting passion fruit.

There are two varieties. One is purple *(Passiflora edulis* Sims) which is more productive and suited for higher elevation (hills). More prolific varieties though, bear large fruits and poor in juice content and flavour, higher in TSS, acidity and carotenoid pigments. The other one is yellow type *(Passiflora edulis* var *flavicarpa)* which performs well in plains and has field tolerance to collar rot, wilt, brown leaf spot, thrips and nematodes where as the purple one is susceptible. Kaveri, a hybrid between purple and yellow is high yielding, tolerates collar rot, wilt, brown leaf spot and nematodes. Another useful species is *P. quadrangularis.*

3.6.2 Manuring and Irrigation

The plant requires 35 kg Nitrogen, 32 kg Phosphorus and 27 kg Potash per hectare per year. To supply nitrogen, the natural and organic materials like farmyard manure (FYM), vermin-compost, oil cakes (neem, castor and karanj) and green manures may be used. Bone meal (crushed) can be used as organic phosphorus nutrition.

If there is prolonged dry spell it needs to be irrigated at fortnightly intervals to get optimum yield.

3.6.3 Plant Protection

Major insect pests are fruit fly, mealy bugs, and aphids. Among the disease, brown spot, root rot and wilt are serious. Appropriate soil management and cultivation measures, spraying of biodynamic liquid pesticides prepared from cowurine, neem, karanj *(Pongamia pinnata),* calotropis, dhatura, casto, *Thevtia nerrifolia,,* etc. are helpful to control the pest and diseases. Proper drainage of excess water is necessary to keep root rot under control.

3.6.4 Training and Pruning

Training is quite important in regulating yield, as it has to support considerable weight for five years. Weak and faulty construction of trellis may result in sagging and loss of vines. Two-arm *kniffin* system is ideal method of training in crop. The trellis should always run across the slope or in north-south direction to facilitate uniform exposure to sunlight. Once the vines reach the wire, the tips should be pinched to facilitate leader formation (branching). Two leaders are to be directed on either side of the wire in opposite direction which in turn develops laterals. These laterals need to be trained downwards hanging from the wire to form the fruiting area of the vine.

Systematic pruning is also important as the crop bears fruits only on current season's growth. Pruning should be done after harvesting the crop in April and November. It can be done by cutting back of the laterals to the nearest active bud as otherwise with increase in age of lateral, the basal buds become dormant or sterile.

3.6.5 Harvest and Post-harvest Management

It starts yielding fruits from 10 months after planting and bearing reaches optimum by 16-18 months. There are two main periods of fruiting from August to December and March to May. Fruits take 80-85 days to reach maturity. Slightly purple coloured fruits along with a small portion of the stem/pedicel should be picked up. Average yield of purple variety is 8-10 t/ha and that of the hybrid Kaveri is 16-20 tonnes/ha. The fruits should be disposed of quickly to prevent loss in weight and their appearance. Storing in polythene helps to prevent deterioration.

3.6.6 Commercial use of Passion Fruit

Various products prepared from passion fruit include, jam, squash, carbonated drink and juice concentrate. The juice of passion fruit with an excellent flavour, is quite delicious, nutritious and liked for its blending quality. The juice is extensively used in confectionery and preparation of cakes, pies and ice cream. To enhance the flavour of the final product passion fruit juice is often mixed with juices of pineapple, mango or ginger. It is rich source of vitamin A and contains fair amounts of sodium, magnesium, sulphur and chlorides. The plant with attractive flowers and dark green dense canopy is also grown as ornamental plant.

These products got a great demand in both domestic and export market. There is a scope for export of fresh fruits also. One can earn an income of up to Rs 29,000/- per hectare of passion fruit through sale of fresh fruits. When processed in to squash, the returns could be upto Rs 2,50,000 (considering the average yield as 12-15 tonnes/ha and 3 kg fruits make 750 ml squash worth of Rs 60/- in the market).

3.7 VANILLA

Vanilla planifolia is a fleshy perennial orchid, climbing up trees or other supports by means of aerial adventitious roots. Vanilla comes up well in loose and friable soil with very high organic matter content and of loamy texture. It prefers land with gentle slope and well-drained soils. Vanilla requires w- m and moist conditions with well distributed rainfall of 150 to 300 cm with a temperature range of 25 to 32°C. It comes up well from sea level to around 1500 m above MSL. The crop requires more than 50 % shade and thrives best under filtered light. Considering these conditions, Vanilla can be cultivated in shade houses, fitted with micro sprinklers/foggers.

3.7.1 Planting Material

The crop is usually established by planting *in situ* shoot cuttings each, preferably having 8 to 10 inter nodes as these flowers earlier than the shorter cuttings. However, the cuttings with less than five to six inter nodes and 60 cm in length should not be used directly for planting. Such shorter cuttings properly rooted in the nursery, establish well in the field, compared to the stem-cuttings. Micro-propagated plants can also be used for planting. Vanilla is propagated mainly by shoot cuttings or rooted cuttings. Strong, healthy and actively growing vines are selected, cut into pieces of one meter long with three or four leaves removed from the bottom. The cuttings are kept in a shady place for one week. Alternatively, 3-4 noded rooted cuttings are also used for planting.

Though more than 50 species of vanilla exist, only three are important as sources of vanillin. *Vanilla planifolia* Andrews is the most preferred commercially and therefore, cultivated widely.

Planting: Vanilla is planted in a medium rich in organic matter. Decomposed organic manure/vermi-compost is filled in the trenches made at a spacing of 8 ft. In these trenches, support pillars of 7 ft long will be placed at a spacing of 6 ft, and two cuttings each, will be planted around one support pole. The plant density per acre thus works out to 2400. The wines are trained on GI wires tied between the support pillars at a height of 5 ft.

3.7.2 Creation of Micro-climate

For optimum growth of these plants, a controlled environment is created by establishing suitable green-house/ shade net house which provides the appropriate amount of light, temperature and humidity which are essential for commercial production of vanilla. HDPE net providing 50-60% shade can be supported with stone pillars of 12 ft height to provide the required shade. Micro-sprinklers with both irrigation and misting/ fogging facility need to be installed in the shade house, which will ensure the irrigation as well as humidity requirements.

The cultivation of the crop can also be done as mixed crop in the existing coconut/ arecanut/ coffee gardens. The main crop provides the requisite shade conditions and also the support for the vines.

3.7.3 Aftercare

The main source of nutrients for the crop is from organic sources viz., decomposed leaf mould or dry/decomposed FYM/vermi-compost. A thick layer of organic debris also helps to retain enough moisture and gives a loose soil structure for the roots to spread. Hence, it is important that easily decomposable organic matter is applied around the plant base at least 3-4 times in a year. Besides, manuring, spray biodynamic solution to give a full coverage of the foliage and stem to enhance the growth of the vines. Presently, farmers are getting very good response in growth and yield by spraying vermi-wash to the foliage.

3.7.4 Flowering and Pollination

The flowering commences from the 3rd year after planting, during January-February months. Flowers are borne in inflorescence. Up to 20-30 flowers come in a bunch.

Moisture stress for about one month i.e., irrigation is stopped during the month of December and the tips of the vines are pruned. These operations induce flowering in the plant and once 10 % of the flowers appear, copious irrigation is given to induce profuse flowering.

Self-pollination by natural means is not possible in India due to the absence of specific pollinating agents. Artificial pollination is carried out by hand with the help of a pointed bamboo splinter, a stiff grass or a sharpened toothpick to get fruit set. The ideal time for pollination is morning to midnight (7 am to 12 p.m.). On an average, a skilled worker can pollinate 1200-2000 flowers a day. It is ideal to pollinate only the first formed 8-10 flowers on the lower side of the inflorescence. It is also recommended to maintain only 10-12 inflorescences per vine in order to get beans with maximum length and girth and of high quality standards. Generally, one flower in an inflorescence opens in a day. The flowering is spread over a period of 3 weeks.

3.7.5 Cost of Development of Vanilla

The indicative cost of cultivation of Vanilla as a pure crop under shade house with irrigation facilities like micro-sprinklers has been worked out for a unit size of 1 acre. While working out the cost, it is presumed that irrigation facilities like source of water and lifting devices are available with the promoter.

Table 13.19: Cost of cultivation of vanilla in shade house

Unit Area	1 acre
Spacing between supporting pillars	8' x 6'
No. of plants trailing on one pole	2
Total plant population	2400

The price of fresh vanilla beans in the market is in the range of Rs 3000-3500/kg. However, for working out the economics, a price of Rs 500/kg of green pods is assumed.

Table 13.20: Item wise cost of vanilla cultivation

Item-wise cost	Qty	Rate	Amount (in Rs)			
			1yr	2 yr	3 yr	Total
Land Development						
Land clearing/ levelling	LS		10000	-	-	10000
Construction of shade house						
Shade net (in sq m)	4200	19	79800	-	-	79800
Stone pillars (12', 8", 4")	200	220	44000	-	-	44000
Fixing of pillars	200	20	4000	-	-	4000
Erection of net (sq m)	4000	3	12000	-	-	12000
Irrigation structures						
Irrigation equipments	LS		30000	-	-	30000
Misc. equipments	LS		5000	-	-	5000
Erection of support system						
Stone pillars (7',4",4")	1200	65	78000	-	-	78000
GI wire for trailing of vines	LS		10000	-	-	
Erection charges	1200	10	12000	-	-	12000
Cost of planting materials	2400	60	144000	-	-	144000

Item-wise cost	Qty	Rate	Amount (in Rs)			
			1yr	2 yr	3 yr	Total
Cost of cultivation till bearing stage						
Cost of FYM (tones)	15	1000	15000	-	-	15000
Vermi-compost	10	4000	40000	-	-	40000
Top dressing				5000	5000	10000
Foliar spray	LS		5000	10000	10000	25000
Cost of irrigation (Rs 1500 per month)			16500	16500	16500	49500
Labour charges			20400	22200	28200	70800
Total capital cost			515700	53700	59700	629100

Table 13.21: The estimated cost of vanilla cultivation

Details	1 yr	2 yr	3 yr	4 yr	5 yr	6 to 15 yr
Yield						
Yield of fresh beans per vine (kg)			0.250	0.500	0.750	1.000
Total yield (kg/acre)			600	1200	1800	2400
Income						
Income (Rs 500/kg)			300000	600000	900000	1200000
Maintenance (in Rs)			-	65000	65000	65000
Net Income (Rs)			300000	535000	835000	1135000
Cost						
Capital cost (Rs)	515700	53700	59700	-	-	-
Recurring (Rs)				65000	65000	65000
Total expenditure (Rs)	515700	53700	59700	65000	65000	65000
Benefit (Rs)	0	0	300000	600000	900000	1200000
Net benefit	--	--	240300	535000	835000	1135000

3.7.6 Harvesting and Processing

After pollination and fertilization, the beans develop very quickly and obtain full size in about 5-6 weeks but it takes 9-11 months for the same to mature. Around 75-90 mature beans make one kilogram. The beans are harvested when the distal end turns pale yellow in colour. The aroma and flavour develops only after the curing process. The different stages of curing include Killing (by dipping the beans in hot water at 63-65°C for 3 minutes, Sweating (through exposure to sunlight for 1-1 1/2 hours by spreading them on a raised platform every day for 5-7 days); Drying (by keeping the beans spread on racks in an airy room for up to 30 days) and Conditioning (keeping the dried beans bundled and covered in butter paper, in wooden boxes for about 2-3 months).

4.0 Fruit Selection and Quality Assurance

The fruit crops selected to start with should be those for which world markets are biggest and which are universally important in all tropical regions: pineapple, mango, and citrus. Other fruit crops of more regional importance could be jackfruit, litchi, guava, passion fruit, etc.

Crop and agri-processing residues are a potential source of organic matter and some nutrients but their characteristics need to be known, and in many cases pre-treatment is required (e.g. half-charring of rice husks) so that they can be used with crops.

Empirical experience also alerted to the risk of contamination in some organic waste materials sourced off-farm that might otherwise be used for compost, e.g. heavy metals in sawdust from treated timber, coir dust with tanning residues, oily substances in oil extraction residues.

The nitrate content of the products has to be minimised by an adequate cultivation (location, variety, manure). The quality obtained by ways of cultivation has to be preserved by the choice of careful harvesting, preparation and storage methods.

Hygiene and sanitation are very important in all stages of fruit crop management, starting from attention to the health status of planting stock, and carrying through to the control of any pest-promoting effects of crop residues in the field.

An analysis of critical points in the production chain proved to give a useful framework for developing organic cultivation methods through an integrated approach.

Chapter 14

Permanent Tropical Plantations

Organic cultivation of permanent tropical plantations are fulfilled by creating agro-forestry systems of wide agricultural bio-diversity and appropriate to the crops' habitat. The important contribution made by trees in tropical eco-systems to the fertility of the soil, the supply of nutrients, the water balance and bio-diversity are to be enhanced by the integration of shade trees in the cultivation system. Suitable species are those adapted to the habitat and, for preference, indigenous strains. The aim is to introduce diversity of shade tree species and to plant legume too.

Measures are to be taken to provide protection from erosion (e.g. planting vegetation round the borders, or dense undergrowth). Organic substances, especially the leaves which fall from the shade trees, are particularly important. Management of this secondary vegetation must ensure good soil coverage and the preservation of the mulch layer.

1. PRODUCTION TECHNOLOGY

Humus Balance and Fertilization

A fundamental means of maintaining and increasing soil fertility is to plant trees and grass. They provide diverse habitats and encourage the establishment of beneficial insects. An ideal means of breaking down the soil coverage is to sew such undergrowth is leguminous plants and herbs. No area should be entirely free of vegetation or other coverage the whole year round.

To further improve the supply of humus, organic fertilizers may be applied. Where permanent crops are intensively cultivated and thus, where necessary, entailing a greater turnover of substances in the soil, a higher fertilization rate (over 170 kg N/ha/year) is possible, upon consultation with organic certificate body. To ensure that the soil suffers neither from under- or over-supply, the soil, leaves and substrata should be analysed at least every three years to determine their nutritional and humus content.

Pests, Diseases and Weeds Control

The primary aim of organic agriculture is to achieve healthy vegetation by striving for ecological equilibrium between pests and beneficial insects. Important means of preventing diseases are suitable plant density and the choice of healthy and resistant plants.

- The farming intensity has to be matched to local ecological conditions. Excessive plant density, which prevents shade trees from growing (especially in the cultivation of coffee) and favours the spread of diseases, has to be avoided.
- The ability of the undergrowth to resist disease can be increased and the incidence of infection further reduced by suitable attention to the soil and specific cultivation measures (pruning, planting shade trees to increase the height of the vegetation).
- Suitable conditions to achieve a healthy micro-climate for permanent tropical crops are to be established.

Contamination Control

All relevant measures should be taken to minimise contamination from outside and within the farm. Suitable buffer zone should be maintained between conventional and organic block to prevent possible contamination with chemicals. The by-products like leaves, flowers, spadices, husk etc. should be recycled in the field after composting. Products based on polyethylene/polypropylene or other polycarbonate should be burnt or left out in the field. Accumulation of polychloride-based products is prohibited.

Cultivation techniques of some important organic plantation crops are discussed below:

1.1 COCONUT

The coconut palm thrives well under an evenly-distributed annual rainfall ranging from 1000 mm to 3000 mm. The palm requires an equitable warm and humid climate neither very hot, nor very cold. The mean annual temperature for optimum growth and maximum yield is stated to be 27 degree Celsius with a diurnal variation of 6 ^{0}C to 7 ^{0}C. The coconut palm thrives well up to an altitude of 600 m MSL.

The coconut palm can tolerate a wide range of soil conditions. But the palm does show certain growth preferences. A variety of factors such as drainage, soil depth, soil fertility and layout of the land has great influence on the growth of the palm. The major soil types that support coconut in India are laterite, alluvial, red sandy loam, coastal sandy and reclaimed soils with a pH ranging from 5.2 to 8.0.

Shallow soils with underlying hard rock, low lying areas subjected to water stagnation and clay soils should be avoided. Proper supply of moisture either through well distributed rainfall or irrigation and sufficient drainage are essential for coconut.

1.1.1 Planting Material and Plantation

Coconut seed nuts and seedlings should be certified organic. Availability of seed nut from certified plots would take at least another five years. Till such time seed nuts from conventional mother palms which are not treated with any chemical could be used. Mother palms in organically-grown plantations or in plantations not treated with chemicals, should have consistent yield record of 80 to 120 nuts palm per year in rainfed and irrigated conditions respectively.

***Nursery raising*:** Fully mature nuts (12 months old) are collected and sown in nursery. Seed nuts are spaced at 30 to 40 cms either vertically or horizontally in 20 to 25 cms deep trench. Coir pith and soils are used as substrates. Inoculation of Azospirrilum and Phospho-bacteria results in vigorous seedling growth.

***Varieties*:** The tall varieties are extensively grown throughout India while dwarf is grown mainly for parent material in hybrid seed production and for tender coconuts.

Table 14.1: Prominent variety of coconut

Variety/Hybrid	Yield Nut/Palm/ Year	Copra Content (g/nut)	Oil Content %	Oil Yield (t/ha.)
Indigenous				
West Coast Tall	81	176	68	1.69
East Coast Tall	86	100	63	0.96
Banavali Green Round	151	151	68	2.74
Kappadam	90	283	67	2.99
Exotic				
Fiji Tall	106	199	65	2.41
Philippines Ordinary	108	196	66	2.65
Chandrasankara (COD X WCT)	98	208	68	2.47
Lakshganga (LO X GB)	108	194	73	2.47
Anandganga (AO X GB)	95	216	68	2.47

The tall varieties generally grown along the west coast is called West Coast Tall and along the east coast is called East Coast Tall. Benaulim, Laccadive Ordinary, Laccadive Micro, Tiptur Tall, Kappadam, Komadan and Andaman Ordinary are some of the tall varieties. Gangabondam is a semi-tall type.

Chowghat Dwarf Orange, Chowghat Dwarf Yellow, Chowghat Dwarf Green, Malayan Yellow Dwarf and Malayan Orange Dwarf are some of the dwarf varieties grown in India. Many hybrid combinations of tall and dwarf are also grown in the country.

***Land preparation and planting*:** The land should be cleared of wild growth. If the land is sloppy and uneven, contour bunds and contour terraces are to be made. The land should be divided into blocks of convenient size by laying out footpaths and roads in-between. The land preparation should be completed well ahead of the commencement of the monsoon rains.

Size of the pit depends on the soil type and water table. In laterite soils large pits of the size 1.2 m x 1.2 m x 1.2 m may be dug which are filled with coconut husk for moisture conservation. The husk is to be buried in layers with concave surface facing upwards. In laterite soils, common salt @ 2 kg per pit may be applied, six months prior, on the floor of the pit to soften the hard pans in consultation with Adviser of Certifying body.

Spacing and planting: In general square system of planting with a spacing of 7.5m to 9 m is practised. This will accommodate 177 to 124 palms per hectare. Seedlings are planted after filling the pits up to 60 cm with topsoil amended with compost or biofertilizer. One-year-old seedlings with five leaves with early split of leaves with a girth at collar about 10 cm are used for planting. Planting the seedlings during May with the onset of pre-monsoon rain is ideal.

Intercropping: Cropping systems in coconut plantations promote agricultural bio-diversity which encompasses the variety and variability of animals, plants and micro-organisms. Cultivation of short duration vegetables and fruit crops like ginger, yam, pineapple, banana, spices could be adopted during early years coconut growth to augment income during the pre-bearing stage of coconut. In established plantations, planting of compatible crops like black pepper, vanilla, cocoa, banana, tree species like nutmeg, cinnamon and clove etc., as per local situation. Other crops which have been found suitable, are medicinal and aromatic plants.

1.1.2 Manuring and Irrigation

The nutritional requirement of coconut and its associate crops should be met by proper recycling of all the by-products available within the farm. Regular manuring from the first year of planting is essential to ensure good vegetative growth, early flowering and bearing and high yield. Organic manure at the rate of 25-50 kg per palm per year may be applied with the onset of monsoon when soil moisture content is high. Different forms of organic manures like compost, farmyard manure, bonemeal, fishmeal, neem cake, groundnut cake, gingerly cake, etc. could be used for this purpose. Green manure crops like sunhemp, gliricidia, dhaincha, etc. could also be grown as intercrops to incorporate in the coconut basins later.

Table 14.2: Manure and fertilizer requirement of coconut plant

Manure and Fertilizer (g/plant)	1 yr	2 yr	3 yr	4 yr	5 yr	6 yr	7 yr	8 yr onwards
Nitrogen	150	200	250	300	350	400	450	500
Phosphorus	200	120	170	220	270	320	370	400
Potassium	300	400	500	600	700	800	900	1000
FYM	4000	2000	2500	3000	3500	4500	5000	5000

Irrigation: The newly-planted seedlings should be shaded and irrigated properly during the summer months. Under basin irrigation, 45 litres per palm once in four days will be beneficial in sandy soil. In areas where water is scarce, drip irrigation system can be adopted. Provision of drainage is equally important in areas subject to waterlogging.

1.1.3 Pests and Diseases

Pests: The major insect pests of the coconut palm are the rhinoceros beetle, the leaf-eating caterpillar, red palm weevil, the root eating white grub, etc. These pests can be controlled by adopting the following measures:

Rhinoceros beetle: The beetle attacks fronds and cuts the leaves before opening. Killing the beetles by hooks mechanically is the most effective. The breeding places such as decaying organic matter, FYM, dead palms, etc. should be treated with insecticides. Biological control by release of exotic predator and bacteria infected beetles are effective measures. Releasing 10-15 *Baculovirus* infected beetles per hectare has been found to be sucessful in checking the population of the rhinoceros beetle. *Matarrhizium anisopliae*, or the green muscardine fungus is another bio-control agent, which can infect all stages of rhinoceros beetle.

Red palm weevil: The larva of the weevil bores into the trunk and feeds on the inner tissue making large holes. Externally exudation of reddish gum is only visible. The palm may die if the attack is severe. Pyrecon-E (plant derivative Pyrethrum) and Entomopathogenic Nematodes (EPN) has been found effective for the control of red palm weevil (*Rhyncophorus ferrugineus*).

Other pests: The parasitoids *Bracon brevicornis* and *Apantilis taragamae* are used for the effective control of coconut leaf-eating catrpiller. Neem oil-garlic-soap emulsion has emerged as one of the potent measures for the control of the eriophyid mite (*Aceria guerreronis*). *Hisutella thompsonii*, a fungus has been found to be an effective microbial agent for annihilating this pest. Increased use of organic amendments like neem and marotti (*Hydnocarpus*) oil cakes and farmyard manure help in the integrated management burrowing nematode, *Radopholus similes* in coconut.

Diseases: Coconut palm is affected by a number of diseases, some of which are lethal while others gradually reduce the vigour of the palm causing severe loss in the yield. Important diseases are bud rot, root wilt, leaf rot, leaf blight, mahali or fruit rot and nut fall, stem bleeding, ganoderma wilt, crown choking disease, etc. Control measures of some of these diseases are stated briefly below:

Bud rot: Young plants are damaged most. Benign entophytic bacteria viz. *Bacillus subtilis, B. macerans* and *B. amyloliquifactens* have all been found to be checking the growth of *Phytophthora palmivor*, the causal organism of bud rot of coconut. Bordeaux mixture may also be used in the leaf area to control the disease.

Stem bleeding: Exudation of reddish brown liquid through cracks on trunk which turn brown later is observed. Cavity may develop beneath the affected area. Scraping the affected area and then application of Bordeux mixture is recommended.

Gliocladium virens, Trichoderma hamatum and *Tharzianum* have been found to be very effective in reducing the population of *Tehilaviopsis paradox* (causal agent of stem bleeding) in the soil. These antagonistic fungi thrive very well in neem cake supplemented with a small quantity of rice/wheat bran thus affecting their multiplication an easy task.

Trichoderma harzianum has been found to control the multiplication of *Ganoderma lucidum,* the causal organism of basal stem rot in sick soils. When combined with Phosphobacterium or plant growth promoting Rhizobacteria (PGPR), synergistic effects have been noticed.

1.1.4 Harvesting

Coconuts are harvested at varying intervals in a year. The frequency differs in different areas depending upon the yield of the trees. In well maintained and high-yielding gardens, bunches are produced regularly and harvesting is done once a month.

Coconuts become mature in about 12 months after the opening of the spathe. It is the ripe coconut which is the source of major coconut products. Nuts which are eleven months old give fibre of good quality and can be harvested in the tracts where green husks are required for the manufacture of coir fibre. Economic life of the coconut palm can be considered as 60 years.

Utilisation of Coconut: Coconut industry in the country is mainly confined to traditional activities such as copra-making, oil extraction, coir manufacture and toddy tapping. Products such as desiccated coconut, coconut-based handicrafts, shell powder, shell charcoal and shell-based activated carbon are also manufactured in the country on a limited scale.

1.1.5 Cost of Development of Coconut

The unit cost estimated is Rs.64,470 per ha. The cost presented here is indicative only. The cost of organic fruits will be at least 2-fold of the present market price.

Table 14.3: Estimated Cost (Rs/ha) of coconut cultivation

Particulars	1 yr	2 yr	3 yr	4 yr	5 yr	6 yr	7 yr	8 yr onwards
Planting Material	2880	--	--	--	--	--	--	--
FYM	1900	963	1200	1450	1685	1925	2200	2300
Fertilizer	1300	1278	1660	2050	2435	2823	3200	3450
Irrigation	914	868	1100	1100	1265	1265	1400	1400
Plant protection	275	275	370	370	483	483	600	600
Fencing	750	--	--	--	--	--	--	--
Sub Total	8020	3386	4333	4974	5873	6502	7400	7750
Operation	4500	1750	1750	2000	2000	2500	2500	3100
Intercrop	7000	6300	5600	4900	4075	--	--	--
Grand Total	19520	5136	6083	6974	7873	9002	9900	10850

Table 14.4: Projected income from coconut cultivation

Particulars	7 yr	8 yr	9 yr	10 yr	11 yr	12 yr onwards
Nuts No/ha.	1,750	3,500	7,000	8,750	10,500	12,250
Gross sale value @ Rs.5.50/nut	9,625	19,250	38,500	48,125	57,750	67,375
Maintenance (Rs/ha)	10,850	10,850	10,850	10,850	10,850	10,850

1.2 OIL PALM

Oil palm is the highest oil producer among perennial oil yielding crops. It produces two distinct oils viz., palm oil (extracted from mesocarp of fresh fruits) and palm kernel oil (from kernel). Palm oil are used in formulation of margarine and cooking fat such as vanaspati. Palm kernel oil has variety of industrial uses. It can be used in manufacture of biscuits, ice creams, soaps, detergents, shampoos and also as frying fat.

Deep well-drained medium loam soil, rich in humus is the most suitable for oil palm cultivation. Oil palm requires a well distributed rainfall of 2500 to 4000 mm per annum and a temperature range of 19-33° C. It is a water-loving crop and it requires adequate irrigation. The crop responds well to drip irrigation and yields are reported to increase by at least 20%.

1.2.1 Planting and Varieties

Oil palm is planted in triangular system at spacing of 9 m x 9 m x 9 m accommodating 143 plants in a hectare. Planting can be done in any season. However, the best period is June to December. Seedlings of 10-14 months age are best suited for planting.

***Varieties*:** Broadly, there are three varieties viz., Dura, Piscifera and Tenera. Tenera, a hybrid of Dura and Piscifera is characterized by a thin shell and medium to high mesocarp (65-90%) and high oil content (16-20%). It is a commercially cultivated variety.

1.2.2 Manuring and Intercrop

Manuring includes all the varied operations to be attended in a plantation including application of farmyard manure, compost, leaf manure. Cultivation of green manure *in situ* and application of more leaf manure should be taken up systematically in order to get increased yield. Fertilizers are preferably applied in 2 split doses. Application of green leaf manure or compost is advantageous.

Table 14.5: Fertilizer requirement (kg/palm/year) of oil palm

Year	FYM	Nitrogen	Phosphorus	Potassium	Magnesium
1	50	0.40	0.20	0.40	0.012
2	50	0.80	0.40	0.80	0.024
3	50	1.20	0.60	1.20	0.05
4 onwards	50	1.40	0.80	2.40	0.05

***Inter crop*:** During the initial stages of plantation in oil palm i.e., up to 3rd year, some of the light feeder inter crops such as pulses, cereals, vegetables, grasses etc., can be grown. Inter crop should be grown 1 m away from the basin in 1st year of oil palm plantation. In two-year-old plantations, it should be grown 2 m away, followed by 3 m in the third year plantations.

Oil palm starts bearing from 4th year onwards and its economic life varies from 30 to 35 years. The yield of oil palm varies according to age and management. Under average management conditions in a mature plantation (8 to 9 years old), yield of 15-18 tonnes of fresh fruit bunches (FFBs) per hectare is expected. Under good maintenance and management, yield up to 25-30 tonnes of FFBs per hectare is possible.

1.2.3 Irrigation

Oil palm requires adequate irrigation, as it is a fast-growing crop with high productivity and biomass production. A minimum of 150 litres per day is required from 3 year onwards. In older plantations the requirement goes up to 20 litres per day.

The crop responds well to drip or micro sprinkler irrigation particularly when water is limited. If drip is installed four drippers have to be placed for each palm. If each dripper discharges 8 litre per hour, 4-5 hour of irrigation is sufficient to discharge 160 litre per day. Drip irrigation increases the productivity by 15-20 per cent, reduces wastage of water, and requires less power/fuel per irrigation compared to conventional irrigation methods. It is important to note that any physiological stress shifts sex ratio in favour of male flowers and consequently the productivity is reduced.

1.2.4 Cost of Development of Oil Palm

The cost of cultivation (indicative) of one hectare of oil palm works out to Rs. 74800 including installation of drip system of irrigation.

Table 14.6: Projected income from oil palm cultivation

Particulars	1-3 yr	4 yr	5 yr	6 yr	7 yr	8 yr	9 yr	10 yr
Yield (tonne/ha)	--	4	6	8	12	14	16	18
Income (Rs)	0	14000	21000	28000	42000	49000	56000	63000
Expenditure (Rs)	57200	13300	13700	13700	14300	14600	14900	15200
Net income (Rs)	--	700	7300	14300	27700	34400	41100	47800

Table 14.7: Estimated cost (Rs/ha) from oil palm cultivation

Item of expenditure	Units	1 yr	2 yr	3 yr	4 yr	Total
Material						
Plant material	Seedlings	7865	770	--	--	8635
Manure and fertilizer		3152	4874	6685	8852	23563
Drip irrigation system		21675	--	--	--	21675
Plant protection		600	600	1200	1200	3600
Subtotal		*33292*	*6244*	*7885*	*10052*	*57473*
Labour						
Land preparation	LS	1750	--	--	--	1750
Planting @ Rs. 15/plant		2145	--	--	--	2145
Manure and fertilizer	4	200	200	200	200	800
Bio Control	2	100	100	100	100	400
Irrigation	5	250	250	250	250	--
Interculture operations	10	500	500	500	500	2000
Harvesting	150	--	--	--	600	600
Watch and ward		--	--	--	1000	1000
Subtotal		*4945*	*1050*	*1050*	*2650*	*9695*
Misc.	5%	1912	365	447	635	3359
Total (rounded off)		*40100*	*7700*	*9400*	*13300*	*70500*

1.3 TEA

Tea is grown in over 30 countries, India continues to hold the number one position in production, consumption and exports. Today more than 10,000 hectares has been certified under organic tea cultivation.

Well-distributed rainfall ranging around 2000 mm to 5000 mm is considered suitable for successful tea plantation. The monthly average maximum temperature ranging between 28^0 C and 32^0C during April to September, with occasional rise up to 36-37^0 C is good for the plantation. Tea is planted in flat and slightly undulating land at elevation ranging from 20 to 250 m above sea level in major part of the plains of NE India. On hill slopes of Darjeeling and South India, it is planted up to a height of 2000 m above sea level.

1.3.1 Field Management

Tea requires strongly acidic soil with pH around 5.0 and soil depth of 1.5 meters to 2.0 meters and good drainage. Sandy loam to silty loam type of soil with pH range of 4.5 to 5.5 is ideal for growing tea. Soil should possess a minimum 1% of organic matter, 70 kg/ha of P_2O_5 and 160 kg/ha of K_2O for successful establishment of tea. The area needs to be sufficiently isolated by maintaining a buffer zone of 100 meters width on all sides of the field, depending on the topography to ensure that there is no possibility of pollutants and contaminants drifting into from any known or unknown source.

Land planning: Topographical planning for laying out drains and bunds, culvert and plucking access would generally help in soil water management, easier supervision, cheaper maintenance and better utilisation of mandays for plucking, spraying, cultivation, etc. Planting in hilly areas should be done along the contour.

Pit digging: After sub soiling, ploughing and harrowing, pits of 45 m width and 45 cm depth are to be dug. Further loosening of heavy soil by 15 cm helps in better establishment due to improved subsoil drainage. Old roots, stone, etc. are to be removed from the pits. At the time of planting application of compost @ 0.5 to 1.0 kg per pit is required. No oil cake should be used. The young plants should be provided with cross staking to prevent wind damage.

1.3.2 Planting Materials

Good number of clones has been released from Tocklai Experimental Station (TRA) and United Planters' Association of South India (UPASI) for using as the planting materials by the industry. About 200 clones selected jointly by the TRA, UPASI and the industry are also available at present. Besides, about 100 industry clones have been released. Clones proven to be successful in a particular locality and recommended by TRA/UPASI only should be chosen by the growers.

Only well grown plants of 50 to 60 cm height with basal stem diameter of not less than 0.5 cm should be planted in the main field. Results from the use of planting materials raised in sleeves are always better than those raised in beds.

Source of planting materials: The sources of planting material must be very reliable. The seeds and vegetative propagated (VP) cuttings for raising nursery should be preferably collected from the organic estates. If not available, seeds from conventional estates not treated with any chemicals can be used. The organic tea nursery should be clearly separated from the conventional nursery, if both are located in same estate.

The quality of the planting material will primarily decide the health of the plant and to a great extent the style of bush formation. The beneficiary may develop his own nursery or may obtain the materials from any neighbouring reputed organic nursery. The nearby established tea gardens may be contacted well ahead of the planting season to ensure procurement of planting materials in time.

Planting seasons: Generally 12-15 months old seedlings/cuttings are planted either in spring (May-June) after the first few showers of rain or in autumn (October-November) while the soil is still moist. However, spring planting may be preferred for better establishment of the plants.

Planting may be done either in single hedge or in double hedge system. The latter may be preferred for the convenience in field works combined with higher plant population. The soil around the sapling should be pressed hard. It needs to be ensured that no depressions are left around the plants. Mulching should be done immediately after planting to avoid mortalities.

Spacing and plant population: Various spacing is followed by the industry. The spacing normally advisable is 105 cm x 65 cm, 105 cm x 75 cm, 105 cm x 60 cm x 60 cm, 105 cm x 75 cm x 60 cm, etc. For small growers, TRA advises the first (14650 plants/ha single hedge) and the third (20,200 plants/ha double hedge) ones. However, in view of the immense popularity of the spacing of 105 cm x 65 cm (the first one) amongst the small growers, a population of 15000 plants/ha is considered to be the best.

1.3.3 Bush Development

Frame forming in the initial stages is very important to build up the bush to give higher productivity. In general, the leader stem of young plants is cut 4-6 months after planting, leaving 8-10 leaves below the cut. Later these plants are tipped at 35 cm height and second tipping will be at 50 cm. After this initial training, the bushes are brought under plucking. Young plants are subjected to formative pruning in the 5th or 6th year from planting.

Mulching: Following planting, the area is thoroughly mulched preferably with Guatemala grass, water hyacinth, citronella grass, etc. Mulching materials may be collected from nearby forests also. Mulching helps conservation of soil moisture, prevents erosion, reduces weed growth and adds organic matter to the soil.

Practices for proper formation of tea bush: Some practices stated below can be followed for encouraging the proper formation of the tea bush:

1. Breaking the stem using thumb and forefinger 25 cm above ground in such a way that at least ¼th of the stem is remained in tact. The dried broken branches can be removed just below the broken portion by clean pruning using 7.5 cm (3 inch) pruning knife.
2. Decentre selectively between 15-20 cm.
3. Debudding by nipping/removing all the growing buds from the main stem/ growing shoots above 25 cm from the ground. Once the branches below this height are developed, the top debudded portion is removed by clean pruning using a 7.5 cm (3 inch) pruning knife.

Pruning: Generally in plains 3 to 4 year pruning cycle is followed. A 3-year cycle for vigorous teas and 4-year cycle for average tea may be followed. While forming a sound pruning programme, the factors need to be analysed are- crop and quality, crop distribution, annual fluctuations, incidence of pests and diseases, frame height, availability of labour, type of tea to be manufactured and marketed, availability of water for irrigation, and shade status.

The suitable time for light prune (LP) and deep skiff (DS) is between December and mid-January and that of medium skiff between January and early February. Normally, a 3-year pruning cycle of LP-UP-UP (unpruned) may be adopted if irrigation arrangement is there. Without irrigation, however, LP-UP-DS is advisable.

Thereafter, 3 to 4 year pruning cycle may be followed depending on the elevation and factors stated under 'Pruning'. Considering the factors such as the growth and region of growing tea and climatic conditions of the region, the sequence and type of different pruning operations stated above can be modified.

The following operations are recommended for bringing up of young tea, depending upon the season of planting:

(a) Spring Planting

Year	Operation
0 year (May-Dec)	May-June - Plant tea. September - Finger prune/Decentre at 25 cm from the ground and tip at 60 cm.
1st year (Jan - Dec)	Keep unpruned (UP) and continue tipping at the previous level.
2nd year	Keep UP Pluck over one leaf in the beginning.
3rd year	Cut across between 35-40 cm and open up the centre selectively tip over 25 cm. (It is called the Frame Forming Pruning).
4th year	Keep UP (1st UP year)
5th year	Deep skiff / Medium skiff
6th year	Keep UP
7th year	Light Prune 5-6 cm above last cut across mark and tip over 25 cm.

(b) Autumn planting

Year	Operation
0 year	October - November - plant tea.
1st year (Jan- Dec)	May - June - Finger prune / Decentre at 25 cm. Tip 60 cm from ground.
2nd year	Keep UP. Continue tipping at previous level.
3rd year	Keep UP. Pluck over one leaf in the beginning.
4th year	Cut across between 35-40 cm and tip over 25 cm (Frame Forming Pruning)
5th year	Keep UP (1st year UP)
6th year	Deep skiff / Medium skiff
7th year	Keep UP
8th year	Light Prune 5-6 cm above last cut across mark and tip over 25 cm.

Table 14.8: Tipping and plucking of tea bush

Type of prune	Tipping heights
Medium Pruned (MP)	25-35 cm from MP mark which generally comes between 50-65 cm from ground.
Light Pruned (LP)	On the average height of 5 leaves or 25 cm whichever is lower.
Deep Skiff (DS)	Average height of 2 leaves between 7-10 cm.
Medium Skiff (MS)	Over 1 leaf, between 3.5-5.0 cm
Light Skiff (LS), Level Off Skiff (LOS) and Unpruned (UP)	Over janam*, raising the plucking table by a leaf on certain occasions

* *janam* = When shoots on a bush start growing after a period of dormancy, small leaflets (cataphylls or bud scales) are first unfolded before the true leaves appear. These bud scales are called *janam* (birth scales).

The plucking interval should be maintained between 7-8 days for quality.

1.3.4 Planting of Shade Tree

Alongwith planting of tea, a combination of temporary (*Indigofera teysmanii*) and permanent shade trees (*Grevillea robusta, Albizzia lebbeck, A. odoratizsima, A. chinensis, Deris robusta, Acacia lenticularis*, etc.) should also be planted. Green crops like *Crotolaria anagyroids* and *Priotropis cytisoides* are useful in young tea areas and their seeds should be sown simultaneously with tea in lines. When the permanent shade trees establish well, the temporary shade trees need to be removed.

Initially shade trees should be planted at 6 m x 6 m spacing and later thinned to 12m x 6m after 8-10 years of planting and finally to 12 m x 12 m by about 12-15 years. Application of 100 g rock phosphate and 400 g dolomite per pit at the time of planting is suggested for successful establishment of shade trees. A spacing of 3.6 m x 3.6 m is suggested for the temporary shade trees. As a routine, the poorly shaded areas should be renovated by planting new shade trees.

1.3.5 Manuring

In order to meet the nitrogen demand of young fields it is suggested to apply 10 MT of compost, 2.5 MT of neem cake and 2.5 MT of castor cake per hectare every year. Application of 25-35 g of rock Mussories phosphate/plant would take care of 'P' requirement of young tea bushes. For mature fields, 5MT each of neem cake and castor cake/ha/year may be applied as a source of nitrogen and 60-80 kg/ha of rock phosphate during the first and third years as source of phosphorous based on soil test values. Requirement of potassium is met through application of wood ash @ 500 kg/ha in matured fields, during dry season.

Foliar application: Weed extract prepared in a barrel (200 litres) by filling one-third with plant materials (*Erythrina, Crotolaria* and other weeds) and remaining with water and allowing to ferment for 3-4 months is sprayed at a dilution of 20 litres in 200 litres water.

Cow manure solution can be prepared by filling one-third of the drum with fresh cowdung and topping the rest of the volume with water. After 10 days the solution is turned everyday by maintaining the volume by adding water. After three months the solution is filtered and sprayed @ 20 litres in 200 litres of water per hectare.

Liming: Liming is not recommended for new clearings. However, in mature tea fields, the soil pH should be maintained at about 5.0 by application of lime or dolomite.

Biofertilizers**:** The bioinoculants are mixed with soil at 1:5 ratio (one part *Azospirillum* with five parts of soil) and applied in the root zone of the bushes. Phosphate solublisers viz., *Pseudomonas* and *Bacillus* sp. Can be used @ 25kg/ha.

In tea fields, the average annual biomass contribution through shade trees is estimated at 2.5-5.0 MT/ha and through the pruning of tea bushes at 14 MT/ha. Effective recycling of the above biomass would add a substantial amount of nutrients of the soil.

1.3.6 Pests, Diseases and Weed Control

Intensive management of pests and diseases is required particularly during the formative years for early establishment. The common pests like thrips, greenfly and mites, caterpillar pests (like looper and red slug), crickets, cockchafer grubs and termites normally create problem during the early years. In mature tea also pests like red spider, pink, purple and scarlet mites, jassids, thrips, helopeltis, caterpillars, termites, etc., cause a great deal of crop loss.

Among the diseases, red rust is a common one in young tea areas. In mature tea, besides this red rust, black rot, poria, brown root rot, charcoal stump rot, blister blight, etc. are very common.

Some of the biocontrol of important pests and diseases are described below:

Soil pests: Application of neem cake @ 250 g/pit in new clearings can be effectively control the attack of cockchafer.

Stem borers: Phassus borer and red coffee borer is important pest in new clearings. These can be controlled by application of neem products and regulations of shade.

Shot hole borer**:** Major pests in young and mature tea fields. The incidence could be minimized by removing affected branches, avoiding pruning during dry weather followed by spraying of the fungal pathogen *Beauvaria bassiana* @ 1.5 kg/ha during May-October when humidity levels are high and also by placing cut stems of *Montonneoa bipinnatifida* in infested plantation during 3rd and 4th year.

Leaf feeders**:** Manual removal of the caterpillars and pupae while harvesting reduces the incidence. In case of severe infestation, spray neem formulations @

1000ml/ha in 200 litres of water. Light traps can be also be used to attract the moths of these caterpillars.

Blister blight: Incidence of the disease can be reduced by using resistant clones, by thinning shade, timely weed control, pruning at right time.

Root disease: Four types of root disease viz., Root splitting, Black, Red and Brown root disease may occur in our patches. The following measures are suggested to prevent the spread of root disease.

Black root disease can be controlled by avoiding burial of pruning in infested field and incorporation of *Trichoderma virens* @ 200g/pi at the time of planting. For the control of root diseases, isolate the infected area, take trenches of 1.3 m deep and 45 cm width, put the soil for the trenches inside the infected patch, uproot and burn the bushes. Rehabilitating the soil with Guatemala grass and use bio control agents @ 200 per planting pit as follows.

Trichoderma harzium	: Red root and root splitting
T. Viride	: Black root disease
T. harzium, T. hamatum, *T. Viride, T. resei and T., koningii*	: Brown rot disease
Gliocladium vires	: Red, Brown and Root splitting

Weed control: For high productivity, effective control of weeds is very essential. Till the plants are about 1 year old, the young tea should be kept under manual weeding by hand pulling, slashing, cultivation of green manure crops/cover crops/ grain legumes and mulching with weed slashing and shade litter. Once the tea bushes cover the field the weeds are suppressed naturally.

1.3.7 Fencing

Depending on the perimeter of the plot of land, the length of fencing required would vary. A fencing with barbed wire and locally available wooden posts, preferably live, may be erected. A map of the plot measured to the scale indicating the line of proposed fencing would help assessing the actual length of the fencing required. The following formula may be applied for working out the approximate length of fencing in meter.

Area (ha)	Approximate perimeter (m)
0.5	310
1.0	440
1.5	540
2.0	620
2.5	700
3.0	760
4.0	880
5.0	980

Fencing length (m) = (Square root of the plot size in sqm X 4) + 10%

1.3.8 Drainage and Irrigation

For laying out drains, bunds, culverts, etc., topographical planning is helpful. The drain design, i.e. depth, width, batter, spacing, etc. would be governed by the topography, soil type, soil structure, etc. Waterlogging reduces the crop besides causing root diseases. Existing drains should be renovated as a routine practice to keep them functional. Regular cleaning and desilting of drains to 90 cm depth is desirable. Protection of drain-edges with vegetation is more important than batter or depth.

Irrigation: Sprinkler irrigation is recommended where necessary twice monthly from December to April providing depth of irrigation between 5 and 10 cm depending upon the degree of moisture stress.

1.3.9 Peak Plucking and Post Harvesting Practices

As regards the plucking operation, about 18% of the annual crop in a single month may be considered as peak plucking. The quantum of leaf during peak plucking in one hectare mature plantation (6 years and above) would be as under:

High Yielding areas = 1980 kg of green leaf

Low Yielding areas = 1620 kg of green leaf

This 1980/1620 kg can be plucked by 80/65 mandays.

Yield: The following yield estimate may be considered depending upon the agro-climatic conditions.

Table 14.9: Yield estimates of high and low yielding variety of tea

Variety	1 yr	2 yr	3 yr	4 yr	5 yr	6 yr onwards
High yielding	Nil	1,800	4,200	6,500	9,000	11,000
Low yielding	Nil	700	1,800	4,200	6,500	9,000

(Figs. in kg green leaf/ha)

Table 14.10: Estimated cost for development of Tea plantation by small growers

YEAR	1 yr	2 yr	3 yr	4 yr	5 yr	6-15 yr
Development Cost	136900	25860	21930	16230	16440	164400
Plucking Cost	NIL	4320	10080	15600	21600	264000
Total Cost	136900	30180	32010	31830	38040	428400
Benefit	0	21600	50400	78000	108000	1320000
Net Benefit	(136900)	(8580)	18390	46170	69960	891600

Unit Size: 1.0 ha Plant Population: 15000/ha

Manufacturing practices: For manufacturing only mechanical and physical processes are allowed with natural fermentation. It can be manufactured as orthodox, CTC Oolong and green tea.

Isolation of manufacturing facility: Manufacturing of organic tea should be carried out in separated factory to eliminate all chances and possibilities of coming into contact with the conventional tea.

Storage, packing transport and shipment: There should be separate store for organic tea where no fumigants, insecticides or fungicides are used. Vacuum steam or high-pressure water cleaning is permitted. Biodegradable packaging materials are to be used as packaging material. The organic teas should be transported separately and there should not be any chance of it coming into contact with conventional tea and the organic tea should be stored in a separate place away from conventional tea.

1.3.10 Organic Conversion Period

For an existing plantation, the minimum conversion period should be three years from the last usage of synthetic chemicals. For a newly planted or replanted area raised through organic cultivation practice, the first harvest year itself can be considered as organic produce, provided the record is available for NO use of chemicals in the previous cropping.

1.4 COFFEE

Coffee, is the most traded agriculture commodities in the world, cultivated in about 80 countries. Coffee throughout the world including the Asia-Pacific region supports millions of small farmers and their families. With world coffee prices at their lowest levels for many years, the very existence of many small farmers is a stake representing a real threat to the industry. There is an opportunity to improve farm incomes through sustainable production of high value, high quality, specialty coffee and organic coffee which is in strong demand in the world market.

Arabica coffee grows well at an elevation of 1000-1500 m MSL, while robusta coffee comes up well at lower altitudes of 500-1000 m MSL.

The package of practices for organic coffee production prescribed by Central Coffee Research Institute and Coffee Board are described hereunder:

1.4.1 Selection of Site and Buffer Zone

The site of coffee plantation should be provided with appropriate isolation distance of 100 meters from the conventional estates/blocks, to prevent contamination with chemicals.

In choosing a site for a new plantation due consideration should be given to elevation and weather parameters. Locations with gentle to moderate slopes covered with a good canopy of evergreen trees are to be preferred. Southern and western aspects should be avoided especially at lower elevations. The area should be provided with more shade to protect coffee from afternoon sun. In wind prone areas, wind belts consisting of tall trees like silver oak, tree, etc. should be raised.

Land preparation should be completed well ahead of commencement of monsoon (June) by retaining selective evergreen shade trees at a spacing of 30-40 ft for providing filter shade.

Soil conservation measures: Opening of staggered pits (cradle pits) (20 inch wide, 10-12 inches deep) across the slope in between the coffee rows would not only help in prevention of soil erosion but also in better harvesting of rain water and conservation of moisture for dry months. Additionally these pits also act as *in situ* compost pits which are to be renovated every year just before the onset of south-west monsoon.

1.4.2 Planting Materials and Varieties

Raising a nursery: Seeds for raising nursery should be collected from organic estates/blocks only. However, if not available, seeds from conventional estates/ blocks not treated with any chemical can be used. The organic nursery should be clearly separated from conventional nursery, if both the activities are carried out in the same estate.

Varieties: The varieties selected must be well adapted to local conditions tolerant/ resistant pest/disease. In case of Arabica, varieties such as S.795, Sln.5-B, Sln.6 and Sln.9, while, in case of robustas improved varieties like S.274 and C x R may be selected.

Preparations for planting: The spacing suggested for Arabica (tall varieties) 6 ft x 6 ft and for robusta 10 ft x 10 ft. Before planting, pits of size 45 cu.cm. (L x B x D) are to be opened during the months of April-May and exposed to sun for about a fortnight to kill soil pests like cockchafers (root grubs), nematodes etc. Later, pits are filled with top fertile soil and well-decomposed farmyard manure or compost (1-2 kg/pit) prepared on the estate.

Planting of coffee: Planting of coffee seedlings should be taken up during August-September. It is advisable to add about 50g of rock phosphate to each pit, for encouraging root growth and better establishment of plants at the time of planting. While in cockchafer and nematodes infested fields, application of neem cake @ 250 g per pit is advocated.

Inter cropping: Cultivation of short duration vegetables and fruit crops like ginger, elephant foot yam, pineapple, banana, papaya etc. can be practical to augment income during the pre-bearing stage of coffee.

Mixed cropping and diversified farming: Planting of compatible crops like pepper and orange in the coffee blocks and associate crops like cardamom, arecanut etc., in valleys is a common feature in coffee estates in India. Of late, cultivation of a high value crop vanilla, medicinal and aromatic crops is gaining popularity in coffee areas. It is also advisable to go far diversification, by maintaining subsidiary activities like apiculture, etc.

Planting of shade trees: Shade plays a vital role in maintaining the ecosystem a required microclimate in coffee plantations. A two-tier mixed shade canopy consisting of lower canopy of temporary shade trees and a top canopy of permanent shade trees recommended for coffee. The optimum shade requirement would be 50 % for arabica coffee and 30 % for robusta coffee. The canopy of shade trees should be regulated once in two years to provide optimum shade.

It is advisable to plant temporary shade trees like dadap (*Erythrina lithosperma*) at closer spacing (15 ft - 20 ft apart) initially, for providing optimum shade to young coffee plants. In large open spaces, evergreen permanent shade trees such as *Ficus sp., Albizzia sp., Artocarpus* etc., should be planted at suitable intervals (30ft - 40ft apart).

1.4.3 After Care of Young Plantations

After planting, the coffee seedlings should be provided with staking and mulching to protect against wind damage and to conserve soil moisture. During the commencement of dry period, temporary shade butting is to be provided using jungle tree twigs to the young plants in open areas.

Pruning and training: Pruning of all unproductive branches, criss-cross branches and certain types of undesirable branches should be done immediately after harvesting of the crop once in a year especially in arabica. In case of robusta, regular pruning is not required as it has a self-pruning habit in shedding of laterals after 3 to 4 crops. But, in irrigated fields, light pruning every year is suggested to regulate the cropping wood.

Generally, single system of training is recommended for coffee in India where topping (capping) restricts to the plant height. The tall arabica varieties are topped at two stages (two-tier system) where 1st topping at 2:5 ft height and 2nd topping at 4.5 ft to 5 ft height. Second topping is done after harvesting 4-5 crops, when the spread of lower canopy is complete. While the dwarf arabicas as well as the robustas are capped at single level (single tier system). The prescribed topping height for dwarf arabicas is 3 ft to 5 ft depending on soil fertility and wind proness and for robustas is at 4.5 ft to 5 ft.

Handling, centering and desuckering: Handling is thinning out of young flush that arises after main pruning. It should be carried out at least once or twice in a year after main pruning. First handling should be carried out during June-July and in case required a second round during September.

1.4.4 Nutrient Management

It is estimated that shades trees contribute a biomass of about 10 MT/ha/year in terms of leaf litter that would contribute about 40-60kg N, 30-33kg P_2O_5 and 40-60kg K_2O to the soil on decomposition. Additionally, coffee processing wastes like pulp and cherry husk are available on the farm for composting. For every tonne of clear coffee produced on the farm, about 1 MT of pulp or cherry husk are obtained as byproduct which when recycled in the form of compost would contribute about 14-15 Kg N, 3-37 kg P_2O_5 and 29-37 kg K_2O. Further, recycling of cherry husk contributes 1.66-2% N, 0.4–0.5% P_2O_5 and 2.4–2.6% kg K_2O (0.18 MT actually). On the whole it is estimated that nearly 84-95 kg N, 40-42 kg P_2O_5 and 108–123 kg K_2O are available for recycling in the coffee fields in an hectare.

***Application of manure and fertilizer*:** Application of farmyard manure or compost prepared on the farm @ 1500 kg/ha per year in two splits is suggested, while the deficiency in nutrient supply can be met by application of rock phosphate, bone meal, wood ash and other permitted products. Use of bi-fertilizers may also be resorted to improve the availability of nutrients. Correction of soil pH using agricultural lime or dolomite based on soil test values should be taken up at least once in two years to enhance proper availability of applied nutrients.

***Green manuring for soil enrichment*:** In newly planted fields, green manure crops (*Crotolaria striata; C. anagyroides, Tephrosia vogelii*) and legumes (cow pea and horse gram) may be cultivated for 2-3 years during *Kharif* season (June-September) and incorporated into soil before flowering to build up the soil fertility. Green manuring also helps to suppress weed growth in the early years.

1.4.5 Pest, Disease and Weed Management

International regulations stipulate that pests and diseases should be primarily tackled by use of tolerant/resistant cultivars; manipulation of microclimate through shade regulation and pruning etc.; use of eco-friendly approaches like biological control. Only in exceptional cases should other permitted compounds be used. Some of the approaches for pest and disease control in organic coffee estates would be as follows:

Pest management: No serious pest attack has been reported in young coffee plantation except for sporadic incidence of soil-borne foliar and suckling pest. While the sucking pests (mealy bug, green scale) and foliar pests (leaf minor and grasshoppers) can be effectively controlled by spraying neem kernel extract, other plant based extracts and other permitted products. Application of neem cake @ 250g/plant can be very effective against soil borne pest like cockchafers (root grubs) and nematodes.

The control measures of some of the common pest of old coffee plantation are given below:

<u>White stem borer</u> *(Xylotrechus quadripse)*: White stem borer, is the most serious pest of arabica coffee in India. The pest can be controlled by adopting integrated measures:

- Maintenance of optimum shade
- Regular tracing and destruction of infested plants.

- Scrubbing of main stem and thick primaries with coir gloves to remove loose scaly bark.
- Repeated application of neem oil at higher concentrations (2-5 %) at short intervals of 1-2 weeks.
- Use of Pheromone traps for monitoring incidence.
- Application of 10 % lime on stem
- Mass release of parasitoid *Apensia sp.*, etc.

Coffee berry borer *(Hypothenemus hampeii):* This pest attacks the berries of both arabica and robusta coffee. Thus pest could be effectively tackled by adopting the following measures:

- Maintain optimum shade and good drainage
- Thorough and clean harvest of the crop in time
- Use gunny bags/picking mats/polythene sheets while harvesting
- Collect the gleanings and dip the infested in boiling water for a period of two minutes or in cold water for a period of 48 hrs to kill all the stages of the borer inside the berry.
- Spread the polythene sheet smeared with castor oil over the harvested coffee heap with the oiled surface facing the fruits during the day to trap all the beetles.
- Maintenance of trap plants around drying yard followed by clean harvest from the traps plants and treating the berries in boiling/cold water.

Biological measures:

- Timely spray of entomapathogenic fungus *Beauveria bassiana*
- Mass trapping of beetles using lures
- Fumigate the storage material and transport structures.

Shot hole borer *(Xylosandrus compactus):* This is a major pest on robusta coffee. The incidence may be minimised by providing optimum thin shade, good drainage, regular pruning and burning of affected twigs and suckers, especially during April-May and September-December.

Disease management: Young coffee plants are usually free from major diseases. However, in exposed areas, brown eye spot disease may cause defoliation. Providing adequate shade in exposed areas, mulching to conserve soil moisture and spraying with 1% Bordeaux mixture can take care of this minor disease.

Control measures of some common diseases generally observed in coffee plantation is described below:

Leaf rust: This disease caused by *Hemeleia vastatrix* generally attacks arabica coffee. Maintaining optimum shade, judicious pruning and spraying of 0.5% Bordeaux mixture as pre-monsoon, mid-monsoon and post-monsoon applications can effectively control the disease.

Black rot *(Koleroga noxia):* Black rot disease occurs only in endemic patches in the valleys under thickly shaded conditions on both arabaica and robusta during the rainy season. Preventive measures like thinning out the shade, regular pruning, handling, centering (opening up the center of the bush) followed by prophylactic sprays with 1 % Bordeaux mixture in the black rot prone endemic patches would effectively check the disease.

Root diseases: Four types of disease viz., Brown, Red, Black and Santavery root disease occur in endemic patches only. Of these the first three diseases attack both arabica and robusta, while the latter is specific to arabica coffee only. The following measures are suggested to prevent the spread of root diseases to neighbouring areas and to recover the infected areas.

- Uprooting of shade tree stumps after timber extraction followed by uprooting and burning of affected coffee bushes.
- Making an isolation trench around the affected plants by enclosing few healthy plants and application of 1-2 kg of agriculture lime to the uprooted pit and exposing the spot for at least six months before replanting.
- Application of 5 to 10 kg/plant of well-decomposed farmyard manure/compost fortified with bio-control agent *Trichoderma* to the surrounding healthy plants.

***Weed control*:** Weeds poses a serious problem in new coffee plantation. They can be controlled by following cultural practices such as cover digging (12 inches deep) during second year of planting and scuffling (4 to 6 inches deep) for the next two or three years during post-monsoon season. Weeds are to be controlled only by manual (slashing) weeding. Mulching with weed slashing and shade tree leaf litter etc. would also suppress the weed growth. Cultivation of green manure crops/cover crops/grain legume crops and mulching with weed slashing and shade tree leaf litter etc. would also help in smothering of weeds. Once the coffee bushes, cover up the weed growth would naturally get suppressed and manual slash weeding alone would be sufficient. Use of all kinds of herbicides is strictly prohibited.

1.4.6 Harvesting and Processing

Coffee fruits can be processed by wet method and dry/natural method. In case of holdings having both conventional and organic farming activities, the processing, drying and storage facilities should be distinctly separate for each kind of coffee. Only mechanical and physical process with natural fermentation should be adopted for processing.

1.4.7 Conversion of Established Plantation

When an existing coffee plantation is proposed for conversion to organic, it is essential to know about the consequences and requirements. Traditionally cultivated plantations with a low to medium yields levels (i.e., 250-400 kg/acre) can be easily brought under organic farming without any significant yield reduction. However, when the intensively managed estates with higher yield levels (i.e., above 500 kg/acre) are converted, there will be substantial reduction in yield up to 30 % in the initial 1 or 2 years. However, if managed systematically, these plantations would reach original high productivity level within 3-4 years.

There are certain requirements for organic conversion of existing plantations.

Conversion period: For existing plantation, a minimum of three years is required as conversion period for organic cultivation. For newly planted or replanted area, the first yield itself can be considered as organic, as the coffee has a pre-bearing period of 3-4 years.

Choice of varieties for replantation: It is advisable to replant blocks with recommended varieties in a phased manner during the conversion period. In case of robusta, the need for replanting will arise in case of very old, senile plantations with low productivity. However, the off-type robusta plants may be rejuvenated by top work with S.274 or CxR varieties.

2.0 SUSTAINABILITY OF THE CULTIVATION SYSTEM

The sustainability of the cultivation system of permanent tropical plantation is ensured by the following measures:

i) The requirements made of the organic cultivation of permanent tropic plantation are fulfilled by creating agro-forestry systems of wide agricultural bio-diversity and appropriate to the crops' habitat.

ii) The important contribution made by trees in tropical eco-systems to the fertility of the soil, the supply of nutrients, the water balance and bio-diversity are to be enhanced by the integration of shade trees in the cultivation system. Suitable species are those adapted to the habitat and, for preference, indigenous strains.

iii) Measures are to be taken to provide protection from erosion (e.g. planting vegetation round the borders, or dense undergrowth). Organic substances, especially the leaves which fall from the shade trees, are particularly important. Management of this secondary vegetation must ensure good soil coverage and the preservation of the mulch layer.

Chapter 15

Commercial Crop Production

The acreage and the market for organically-produced food are growing in many countries commanding nearly a double price than the conventional food. There is good demand for organic commercial crops and their products. Organic cotton and sugarcane are in good demand. Big textile corporate houses are using cotton for blending with conventional cotton in various proportions to improve their brand image and to meet social obligations.

1. PRODUCTION SYSTEM

The guideline for organic production of commercial crops like cotton and sugarcane is described in this chapter.

1.1 COTTON

Cotton is the most important commercial crop and plays a vital role in the national economy. Cotton is a tropical and sub-tropical and is commonly grown as rainfed crop. It needs an average temperature between 25°C and 33°C. Low temperature during fruiting results in poor fibre development, poor boll opening and inferior quality and lower fibre yields. An annual rainfall of 50 cm to 75 cm is most ideal for the crop but rains during boll opening results into deterioration in fibre quality.

1.1.1 Soil and Land Preparation

Cotton may grow on a variety of soil types possessing good drainage facilities, as the crop cannot tolerate waterlogging. The rainfed crop does well on heavy soils like black cotton soils or medium black soils but the irrigated crop can be grown on light alluvial soils having pH of 6.5 to 7.8.

Land reparation: Plough and harrow the field to obtain fine tilth. Use the field, which is levelled and well drained.

Seed bed: Bed and furrow or ridge and furrow

1.1.2 Varieties and Sowing

The American and Desi cotton varieties observed to be more stable and economical under organic conditions than hybrids. The research experience shows that hybrid cotton requires minimum 6 years in conversion of organic farms. A rotation of various crops is required to minimize the losses in first three years. The hybrid varieties that are found suitable with minimum losses are LRA 5166, AKA 8401,

NHH-44, etc. Resistant varieties like MCU-5VT (wilt), Anjali (Jassid), Supriya (White fly), etc. may also be selected for organic cotton production.

Seed dressing and seed treatment: Soak the seed in water for 3-6 hours, dry it by spreading in shade. After drying, the seed should be treated using the following methods:

Azospirillum and Phosphobacterium	:	10 kg of seeds should be treated with 500 g each of Azospirillum and Phosphobacterium.
Pseudomonas fluorescens	:	10 kg of seeds should be treated with 40 g of *Pseudomonas fluorescens.*
Trichoderma viride	:	10 kg of seeds should be treated with 40 g of *Trichoderma viride*

Spacing: The spacing for high yielding varieties 90-150 x 30-60 cm under irrigated condition, and 45-75x15-30 cm under rainfed condition. If required, thinning of crop may be done after 25-30 days to maintain the proper spacing. More spacing is required in heavy black soils and assured rainfall areas, and lesser spacing in light soils and low rainfall areas.

Seed sowing and de-topping: Seed should be sown by dibbling 2-3 seeds at each place. Seed drill may also be used for dibbling. Break the apical terminals when plants start growing beyond 4.5-5 ft tall. This helps in more branching in plants.

1.1.3 Manuring

Cotton crops are very rarely fertilized and mostly cotton growing fields are found to be deficient in nitrogen and organic matter content, therefore, application of organic manure every year is essential to get better yields.

Well decomposed FYM/compost @ 5-10 tonnes/ha should be applied at the time of land preparation. Application of organic fertilizers may be done as below:

- Bio-manure/oilcake: 500-1000 kg/ha
- Rock phosphate: 60-120 kg/ha
- Neem cake @200 kg/ha and FYM @12 tonnes/ha should be applied at the time of first weeding i.e. three weeks after sowing.

For improving the soil, grow a green manure crop such as *Sesbania sp*. In between rows sole crop and mixed it with soil when start flowering. Follow crop rotation with legumes. Some ideal crop rotations are-

Cotton + Cowpea/Blackgram (Kharif)- Groundnut

Cotton+Pigeanpea (Kharif)-Legume or fallow (Summer)-Groundnut

1.1.4 Pest and Disease Management

Always grow pest/disease resistant or tolerant hybrids/varieties. Monitor pests at regular interval and use recommended bio-pesticides at recommended dosages.

Some of the control measures for pest and disease are described below:

Mechanical control: Hand-picking of larvae of *H. armigera* and *S.litura, S. litura* egg masses, clipping of terminals at 65-75 days to remove *H. armigera* eggs.

Behavioural control: YST 50/ha for Whiteflies, one light trap/ha for Jassids and aphids, 12 pheromone traps/ha for bollworms and *S. litura*, blue cloth for *S. litura* (late stage larvae) may be used.

Botanical control: Spraying of Neem, Pongamia and Madhuca oil 3 %; NSKE, PSKE and MSKE 5 %; Vitex, Ipomea, Jatropha leaf extracts 10 %; may be effective to control the pests of cotton. Application of Neemcake @ 250kg/ha is effective for soil-borne pests and diseases.

Biological control: *Trichoderma chilonis* egg parasitoid 10 cc/ha in 40bits/cc, *Chrysoperla* and ladybird (*Cheilomenes*) predators as inoculative release, *in-situ* culturing of *Telenomus rileyi* using bamboo cage, dragon/damsel flies, red ants, conservation of spiders and Anthocorids. Larval parasitoids (*Bracon, Brachymeria, Apanteles, Coesia,* etc.) for bollwarm are some of the biological measures for controlling pests.

Microbial control: NPV for *H. armigera, S. litura, Amsacta albistriga, Spilosoma oblique* (500LE/ha; LE = 6×10^9 POB). B.t. (Tacturin 1000g/ha), *Beauveria bassiana* for sucking pests and bollworms (7.5 lit. broth culture in 800 lit water for 1 ha) are some of the important microbial control measures.

Inter crops: Cowpea, sunflower, marigold, bhendi, maize to enhance natural enemies and to serve as trap crops. Intercrop cotton with cowpea or blackgram may be done as 1:2-4 rows or strip crop with cotton with sorghum. Grow 6.0-7.5 meter strip of each crop alternatively. Cowpea and sorghum encourages natural enemies of pests.

Border crop: Maize 2 rows at 30 cm between rows and within the row. Plant Castor at every 6 m distance to trap *S. litura.*

Weed management: The crops needs two weedings and hoeing at 20 and 40 days after sowing which could be done by running cultivator, harrow or indigenous plough. In case of broadcast sown crop, hand weeding is the only alternative.

1.1.5 Irrigation

The crop requires light irrigations. The water requirement of cotton is 700-1200 mm depending upon the soil. Normally 4-5 irrigations are required depending on the rainfall. Heavy irrigation as well as drought condition leads to boll shedding. The crop should be given one pre-sowing heavy irrigation followed by first light irrigation after 3rd week and subsequent irrigation at the time of flowering or as and when needed by the crop depending upon atmospheric humidity and soil moisture conditions.

1.1.6 Cost of Production of Organic Cotton

Table 15.1: Cost of production of organic cotton

Activity	**Cost (Rs./hectare)**
Land preparation	1500
Seed	900
Seed inoculants	120
Sowing	1100
Intercultural operation	800
Weeding	1100
Manures	3000
Biological control agents	4100
Harvesting	1900
Total	**14520**
Yield (1100 kg/ha)	29700
Net Profit	**15180**

1.1.7 Harvest and Storage

Cotton harvesting should be done on sunny days when bolls are matured and fully opened. About 3-4 pickings are normally required at 15-20 days. Harvested cotton should not be contaminated with hairs, plant material and soil or stained with colour.

After picking, the farmer should store cotton at clean and dry place. Avoid storage of cotton on the farm. Harvested cotton should be sent to ginning mill for making bales.

1.2 SUGARCANE

Sugarcane is the main source of sugar. It requires warm and humid climate for its growth. It may grow successfully in the areas which receive an annual rainfall of between 75 cm and 120 cm. The crop needs high temperature, bright sunny days and high humidity for grand growth but cool, dry (least humidity) and frostless nights for maturity. It is observed that temperature above 50°C and below 20°C arrests the plant growth whereas frost has been found to be fatal for the crop. Drought results into production of more fibrous canes which possess lesser juice and give lower yields.

The crop needs heavy soil, rich in organic matter and plant nutrients but it should be free from salts and should have soil pH between 6.5-7.5. However with assured water supply it may be grown on light sandy loam soils and with better drainage facilities it can be taken in heavy clays. The soil should be fairly deep, preferably 120-150 cm.

1.2.1 Land Preparation

Sugarcane needs thorough seed-bed preparation so that a well pulverized and clean tilth may be obtained for planting the canes. Mixing of organic manures like compost, FYM, or green manuring is done much in advance so that by the time the crop is planted, the material is completely decomposed and well mixed in soil and it does not invite termite.

Sugarcane can be planted in furrows or in deep trenches. The furrows are made using plough, harrow and trenches are made with double mould board ridger. The depth of furrows should be about 10 cm and that of trenches comes to 20 cm.

1.2.2 Plant Material and Planting

The sugarcane is vegetative propagated therefore, one has to take enough care in selecting healthy organic planting material of top portion of the canes. The cane should be taken to stripped off and cut into 3 budded or 2 budded sets carefully without any damage to the eyes. 25-35 thousands of 3 budded sets and about 40-50 thousands of 2 budded sets are required for planting one hectare area.

Seed canes treatment: Seed canes should be dipped in 0.5% *Trichoderma* solution for 15 minutes.

Spacing: The spacing between two furrows is normally 75 cm and between trenches is about 90-100 cm. In case of trenches the edge of trenches are made with the help of manual labour by using spades.

Sowing: The canes containing 2-3 buds are placed horizontally 8-10 cm deep in the furrows or trenches either in end-to-end system (if the top portion is to be used) and eye-to-eye system (if middle portion of the cane is to be used). The bottom portion and ratoon canes should not be used as seedling material because germination of such canes is found to be very poor. The planting should be done during spring (Feb-Nov) and autumn (Sept-Nov).

Earthen up: The crop should be earthen up at the time of maximum growth phase to protect the growing plants from lodging followed by tying of stems together by gathering them from adjacent clumps and rows. The tying is done by using the trash and dried leaves of the plants. Some times bamboo pegs may also be used to tie the stems.

Crop rotation: Pigeonpea, chickpea, soybean, and groundnut rotation

1.2.3 Manuring

Sugarcane is a heavy feeder and needs constant supply of nutrients almost throughout the growing period. The crop requires about 5-6 tonnes of FYM or compost per hectare. Use of green manure (e.g. *sesbania*) prior to cultivation of sugarcane helps in minimizing the requirement of FYM or compost.

It is advisable to apply about 500 kg of well-powdered neem cake in furrows or trenches as it provides nutrition to the plants as well as it acts as an insecticide to protect the seedlings. The bottom of these furrows/trenches should be well pulverized up to next 10-15 cm and the cake powder should be mixed therein.

Basal dose	After 18 weeks
Well decomposed FYM: 5-6 tonnes	Bio-manure: 500 kg
Bio-manure: 500 kg	Rock Phosphate: 50 kg *(Restricted use)*
Rock Phosphate: 100 kg	Neem cake: 100 kg
(Restricted use) Acetobactor biofertilizer	
(Seed sets dip treatment)	
Neem cake: 100 kg	

1.2.4 Irrigation

The seed-canes should be irrigated from about 30 days before planting. There should be sufficient moisture in the soil at the time of sowing and thereafter it should be given first irrigation after 12-15 days of sowing and almost same interval should be maintained thereafter in subsequent irrigations. Under restricted water supply the crop could be given trash mulching which also controls weeds in addition to economising water use.

1.2.5 Weeding

The crop is found to be infested with nut grass and Johnson grass which are most competitive and cause a drastic reduction in growth and yield of the crop. Therefore, the crop should be given one blind hoeing after about one or two weeks of planting followed by 3-5 hoeing and weeding depending upon interval and frequency of irrigations and recurrence of the weed growth. Mechanical weeding and hoeing are most suitable method of weeding in sugarcane.

1.2.6 Plant Protection

Biological control of the pests and diseases should be followed. Neem for pyrilla, verticellium for woolly aphid, etc. may be used.

1.2.7 Harvesting

The crop needs fairly low temperature for maturity and accordingly the crop becomes ready for harvesting during winter (Dec-Jan). It is observed that crop should be harvested up to end of January as rise in temperature during February increases the Glucose content in the cane juice and extractable sugar percentage drops drastically and recovery becomes poor.

The stalks are cut from near ground level and leaves stripped off, the stalks are detopped and tied into bundles of the convenient sizes but after harvesting they should be processed as soon as it is possible. Delay in processing results in reducing the sugar contents of the juice.

1.2.8 Ratooning

First rationing is recommended because of very high cost required in seed material and planting operation. Second and subsequent rationings are not advisable due to the fact there is gradual population build up of insects/pest, disease pathogens and perennial weeds and the yield becomes very poor along with drastic deterioration in the soil fertility.

The following steps should be taken in successful production of ratoon:

i) The sugarcane stem should be cut from close to the ground.

ii) Trashes should be burnt out soon after harvesting to destroy all the weeds, eggs and young ones of insect/pests and disease pathogens. Rise in temperature helps in conversion of sucrose of stubbles into glucose which boost sprouting of stubbles into new shoots.

iii) The ridges and bunds should be destroyed for root pruning and proper distribution of irrigation water.

iv) The field should be irrigated soon after burning of trashes and destruction of bunds etc.

v) The field should be inter-cultivated by running plough or cultivator for improving aeration etc.

vi) The gap should be filled uprooting, splitting and transplanting stubbles (suckers) from crowded places.

vii) Ratoon should be fertilized with same doses of nutrients as that for main crop.

viii) Irrigation and plant protection measures should be properly adopted, as the crop is more prone to such infestation.

Normally ratoon crop gives somewhat lower yield than that of plant cane but the cost of seed, land preparation and planting is not required and the crop matures earlier to give more return than the plant canes, therefore, it becomes more remunerative.

1.2.9 Cost of Production

Table 15.1: Cost of production of organic sugarcane

Activity	**Cost (Rs./hectare)**
Land preparation	3400
Green manuring	1100
Seed	6000
Manures	18000
Sowing	2700
Weeding	1800
Biological control agents	2500
Irrigation	2800
Harvesting	2700
Total	41000
Yield (90 tonnes/ha)	85500
Net Profit	44500

Chapter **16**

Organic Livestock, Apiculture and Aquaculture

Animal agriculture was considered as ecological farming ever since practice of agriculture on this mother earth. Agriculture and animal husbandry always goes in hand in both mutual and beneficial co-existing as well as symbiotic ways. Organic livestock production intends to use native breeds of cattle, buffalo, sheep, goats and poultry; organic aquaculture for fisheries; and organic apiculture for beekeeping. These enterprises are time-tested individuals of excellence in their production in view of their adoption to the local terrain and vagaries of environmental conditions.

The number of animals kept on a particular farm or region is dependent upon the amount of fodder available, or the size of crop area used to grow fodder on the site itself. Planting fodder crops (especially legumes) is of particular help in improving the fertility of the soil, and in diversifying the crop rotation. Hedges can be useful not only as windbreaks and as protection against erosion, they can also act as constant source of forage for cattle.

Strategies

Strategies will always have to be adapted according to the prevailing site and regional conditions, and these also include the way that animals are kept. For example, it is important to take care that over-grazing or damage through bites do not occur, as these could cause problems with erosion. If animals are to be introduced into a functioning farm unit, then care must be taken not only to choose the correct race of animal suited to local conditions, but also to make sure that the species itself can be adapted to the existing eco-system without causing any damage. Cattle rearing is a special case as they have no direct natural competition for food because they eat plants that are not consumed by humans; they produce protein in form of milk and meat, and also leather and wool; they can be used for transport purposes; they produce dung.

Cattle rearing have a long tradition in many regions worldwide, and it should be remembered that large areas are required (stables, outdoor runs and forage area e.g. pasture or arable forage cultivation). Pigs can be easily adapted to a small farm structure, as they will feed on crop residues and organic waste products.

Nevertheless, stalls must be provided with plenty of space for them to roam outside, to protect the environment - especially delicate eco-systems- against the animal's natural tendency to root around.

Animal husbandry in organic farming rejects in principle the concept of maximising a short-term performance of the animals. Instead, an attempt should be made to achieve an optimum life performance of the animal. It is also important only to keep animals suited to the region that will be able to cope with the local climatic conditions, and are capable of withstanding the potential presence of diseases. Medication used in a preventive manner is forbidden in organic farming (with the exception of vaccinations).

1. LIVESTOCK: GENERAL REQUIREMENTS

1.1 Breeding

Landless animal husbandry systems are not permitted. The health and performance of animals is to be promoted by appropriate rearing, the choice of suitable races and the breeding methods adopted. Animals should be bred which fit in well with the ecological limits of the various requirements and conditions to be found on an organic farm. The target is a high longevity performance by the animals. Genetic manipulation, embryo transfer and synchronisation of oestrus cycles are forbidden. Artificial insemination is allowed, but all other forms of artificial or embryo transfers, spermasexing, etc. are forbidden.

Provenance of animals: The livestock, not used for food production and except for male breeding animals, must principally come from approved organic farm operations.

1.2 Embryo Transfer

Animals deriving from embryo transfer must not be kept on the farm. Bulls resulting directly from embryo transfer (ET) and their sperm are not to be used. As an interim measure, a maximum of 50 % of the male breeding stock or their semen may derive directly from ET. On organic farms, natural reproduction techniques should be preferred whenever possible. In selecting animals, particular attention should be paid to the longevity performance of their forbears.

1.3 Housing and Welfare

The specific ethological needs of all species of domestic animals must be taken into account with the provision of appropriate housing and the opportunity for movement and activity. The National Animal Protection Law must be observed in entirety. Calves being reared and fattened are to be kept in free groups, on litter, with access to the outdoor area, from their third week on to four months of age. Keeping the animals in herds is to be preferred to keeping them separated. A journal must be kept on the grazing and free-range periods of all animals.

Minimum surface areas indoor and outdoors

(A) Cattle, Sheep, Goats and Pig

Table 16.1: Indoor and outdoor area for cattle, sheep, goat and pig

Species or strain of animal	Indoor area		Outdoor area in m^2 (exercise area, excluding pasturage)
	Live weight (in kg)	Minimum size (m^2/animal)	
Breeding and fattening cattle	up to 100	1.5	1.1
	up to 200	2.5	1.9
	up to 350	4.0	3.0
	over 350	5.0 minimum of 1 m^2/ 100 kg	3.7 minimum of 0.75m^2/ 100kg
Dairy cattle	--	6	4.5
Bulls for breeding	--	10	30
Sheep and Goat	--	1.5 per sheep/goat 0.35 per lamb/kid	2.5 2.5 with 0.5 per lamb/kid
Furrowing Sows with piglets up to 40 days	--	7.5 per sow	2.5
Fattening pigs	up to 50	0.8	0.6
	up to 85	1.1	0.8
	up to 110	1.3	1.0
Piglets	30 kg (over 40 days)	0.6	0.4
Brood pigs	--	2.5 female 6.0 male	1.9 8.0

Source: EEC 2092/91 and Naturland Standards, 01/2004

(B) Poultry

Each poultry house must not contain more than:

Chickens	4800
Laying hens	3000
Turkeys	2500
Ducks	4000 (female)/3200 (male)

Total usable area of poultry houses for meat production on any single production unit must not exceed 1600 sq. meters. Poultry housing must have entry/exit pop-holes of a size adequate for the birds, and these pop-holes must have a combined length of at least 4 meters per 100 square meters area of the house available to the birds. Surface areas for indoor and outdoor are as follows:

Table 16.2: Indoor and outdoor area for poultry

Poultry Species	Indoor area			Outdoor area in m^2
	Number of poultry per square meter	Centimetre of perch per animal	Nest	
Laying hens	6	18	8 laying hens per nest	4
Fattening poultry (in fixed housing)	10 with a max. of 21 kg live weight per m^2	20	--	4 broilers and Guinea fowl 4.5 duck 10 turkey

Source: EEC 2092/91 and Naturland Standards, 01/2004

1.4 Stables

Stable floors with fissures or holes are forbidden. The bedding areas for all animals must be covered with straw or be dry and well insulated. Daylight must be provided. Materials and paints in stables must be innocuous (non-toxic). As far as possible, the cleansing agents and disinfectants being used must be innocuous and biodegradable.

1.5 Feed

Animals are to be fed correctly according to species (according to their natural feeding behaviour). The feeding of animals should not be in direct competition with the growing of food for humans. The basis of animal nutrition is the feed produced on the farm itself. At least 50 per cent of the feed must be produced on the individual organic farm. Conversion-fodder produced on the farm may be used in a proportion of maximum 50 % of the fodder surface (conversion-farms: up to 100 %). Purchased fodder serves only as supplement to this home-produced feed and fodder and should be organically produced as far as possible. The proportion of conversion-fodder should not exceed 30 % of the ration of each specific livestock category.

Non-organic fodder: The proportion of non-organic fodder contained in the total fodder ration must not exceed the following percentages:

- Ruminants: 10 % of the total dry matter consumption
- Non-ruminants: 15 % of the total dry matter consumption per animal category

In case of losses of fodder yield, in particular because of extraordinary weather conditions, the certification body can authorize a higher percentage of non-organic fodder for directly affected livestock keepers, in a specific area and for a limited time.

Specific feeding regulations: Young mammals must be fed on the basis of natural milk, preferably mother's milk. All mammals must be fed with natural milk during a minimum period that is defined according to each specific animal species. For cattle it amounts to three months, for goats and sheep to 35 days and for pigs to 40 days. Ruminants must receive at least 60 % of dry fodder matter as fresh, dried or ensilaged roughage. As for poultry, the fodder administered during the fattening period must consist of at least 65 % grains and grain legumes (their products and by-products) and oil seeds (their products and by-products).

Forbidden fodder: The fodder components must be of natural condition, and the processing methods must be as natural and as energy saving as possible. Fodder must not contain any traces of genetically modified organisms. In addition are forbidden:

- chemical-synthetic additives (urea, antimicrobial fattening promoters, enzymes, synthetic amino acids etc.)
- therapeutic chemical medication used prophylactically (sulphonamides and others), antibiotics, hormones, coccidiostatic preparations etc. Excepted are worming cures in areas of high risk (alpine meadows, permanent meadows)
- Fattening methods with forced feeding and the keeping and housing of animals in conditions that could lead to anaemia.

The addition of animal proteins, animal fats, patented preparations of fats and proteins, propylene glycol, propionic acid and any similar substances and additives that do not suit the digestion of ruminants to their fodder is forbidden. Mixtures of mineral substances or trace elements as well as vitamin preparations are allowed to meet the demand only. Natural products are recommended.

1.6 Conversion Periods

In order to be regarded as organic animals, livestock not coming from organic farm operations that are purchased after the beginning of the conversion to organic farming must be kept in accordance with the rules of these standards for at least:

- 12 months as for cattle used for meat production; in any case at least during three quarters of their life
- 6 months as for small ruminants and pigs
- 6 months as for milk producing animals
- 10 weeks as for poultry for meat production, which have been put in the chicken-coop before reaching the age of 3 days
- 6 weeks as for poultry for egg production

1.7 Purchase of Non-organic Animals (brought-in Animals)

About 80 % of the animals must derive from certified organic farms, which means that they must have spent at least the previous twelve months on an organic farm. Purchase of animals should be from certified organic farms whenever possible. However, breeding stock may be brought in from conventional farms with a yearly maximum of 10 % of the adult animals of the same species on the farm.

For the development of a new livestock, if there are not enough animals available from organic farms, they may be brought as per following age:

- laying hens up to an age of 18 weeks
- broilers must be put into the chicken-coop on the 3rd day at the latest
- buffalo from the moment of their weaning, but at the latest at the age of 6 months
- calves up to 4 weeks old which have received colostrums and are fed a diet consisting mainly of full milk
- piglets up to 6 weeks from the moment of their weaning; they must not be heavier than 25 kg
- 2 week old for any other poultry

Newly-purchased animals must be kept on the farm in order to be regarded as animals from organic production, is as follows:

- for pigs: 4 months
- for milk producing animals: 3 months

Certification programmes should set time limits (not exceeding 5 years) for implementation of certified organic animals from conception for each type of animal.

In order to develop a stock, calves and small ruminants for the production of meat may be marketed as animals from organic farming, if they are kept in accordance with the rules of the standards during the following periods before they are slaughtered:

- calves for at least 6 months
- small ruminants for at least 2 months.

If the quantity of animals available to supplement the natural increase or renewal of the stock, the livestock owner may, in agreement with the certification body, put in stable yearly female young cattle (animals that have not yet thrown), in a quantity coming up to 10 % of the grown up cattle, as well as up to 20 % of the stock of grown up pigs, sheep or goats.

1.8 Animal Stocking Density

The number of animals must be appropriate for the farmed area, the site and the climatic conditions. In lowland (plains), the stocking rate may not exceed 2.5 large animal equivalents per hectare of agricultural area. In mountain areas under poor conditions, the stocking rate is to be reduced. Animal stocking density is related to livestock units. To determine the appropriate density of livestock, the livestock units (dung plus urine) equivalent to 170 kg/ha/yr of agricultural area used for the various categories of animal should be set out. A list of suitable animal stocking density per hectare is given as below:

Table 16.3: Suitable animal stocking density per hectare

	Species or strain of animal/poultry	Suitable number of animals per hectare
Cattle	Female (1-2 years)	
	Male (1-2 years)	3
	Male (over 2 years)	3
	Breeding heifers	2
	Fattening heifers	2
	Dairy cattle	2
	Cows not suitable for breeding	2
	Nursing cows	2
	Calf	5
Goats	Mother goats	13
Pigs	Breeding sows (without farrows)	6
	Piglets	74
	Fattening pigs	14
	Other pigs	14
Poultry	Fattening hens	280
	Laying hens	140
	Fattening ducks	210
	Fattening turkeys	140

Source: EEC 2092/91 and Naturland Standards, 01/2004

Adjustments should be made for animals which produce different amounts of dung depending on their strain. If the animals are not kept the year round or if they are to be allocated differently because of their age or the purpose to which they are put, then the above figures will be calculated on the average of the animals kept annually.

1.9 Animal Health

Injured or sick animals must be treated. Natural remedies and healing methods have first priority, if experience shows that they have a therapeutical effect on the respective animal species or the disease that is to be treated. Chemical-synthetic allopathic treatments may be carried out on the veterinarian's prescription, if the disease or injury cannot be efficiently treated with alternative methods.

Withdrawal period: The withdrawal period between the last administering of a chemical-synthetic allopathic veterinary medicine and the marketing of foodstuff from such an animal amounts principally to the double of the legally stipulated time indicated on the package. Excepted are drugs for the draining of cows with udder problems. Before the use of draining agents, a bacteriological analysis of the milk must be carried out.

Prophylatic treatment: Prophylactic treatments with chemical synthetic allopathic medicines, antibiotics and hormones are forbidden. The uses of coczidiostatics, prophylactic iron-injections on pigs as well as the use of hormones or similar substances to control the reproduction (e.g. induction or synchronization of the heat) or for other aims are not permitted. Hormones may yet be administered to an individual animal in case of a therapeutic treatment by a veterinarian. Synthetic chemicals as worming agents and vaccinations are allowed on prescription by a veterinarian. Treated animals must be identifiable any time and unmistakably as such.

Number of treatments: If an animal is treated more than three times within a calendar year with chemical-synthetic allopathic veterinarian medicines or antibiotics (or more than once, if the productive life-cycles of less than a year), the respective animals or products of these animals must not be sold as organic. Zootechnical measures must be kept to a minimum. De-horning of young animals should only be carried out as a precaution for the safety of people and other animals and under local stunning. Permitted zootechnical measures must be carried out by qualified personnel. Castration is permitted. It may only be performed under anaesthetic and by a vet. Castration without anaesthetic of up to 14 days old piglets, lambs, goat kids and calves is tolerated.

Forbidden measures: Particularly forbidden without a vet's prescription are:

- tail-docking, except in the case of sheep when diarrhoea caused by their feed leaves no alternative and shearing does not help
- blunting of teeth
- clipping of beaks and claws in poultry
- trimming of wing-bones
- castration by means of elastic rings later than the third week of life for sheep and goats and the fifth week for calves
- caponising of roosters

2. LIVESTOCK: SPECIES SPECIFIC REQUIREMENTS

Cattle: Cattle are primarily to be fed with roughage. The appropriate structural balance in the daily feed ration must be observed (hay, straw, grain-whole-plant silage). Concentrates may only be used as a supplement. In summer, green feed must be provided. Exclusive year-round silage feeding is not permitted. Calves should not be taken off milk before they are 3 months old. Powdered milk substitute or powdered milk must not be given. Pure milk fattening without feeding roughage is not permitted.

Sheep: Sheep must be kept in flocks on pastureland, or in a free stable where they can move around freely and have access to the open air. Keeping them individually in boxes is only allowed for a maximum of seven days at lambing time and if they fall ill. Rams can be kept individually. During the season of vegetative growth, sheep must be put into the pastureland every day. In bad weather, it is sufficient to let them out in an outdoor run every day. In winter, all animals must have exercise area outside at least 13 times a month. Castration of males is allowed up to the age of 2 months, but the use of rubber rings only up to 14 days. Sheep should be primarily fed with roughage. The feeding Lambs are basically to be reared and fattened on mother's milk for at least 45 days. Pure milk fattening without feeding roughage is not permitted. The use of milk powder is forbidden.

Veterinary medicine: The keeping of sheep should be optimised so that worming with synthetic chemical medicaments is needed as little as possible. These medicaments are permitted by order of a vet. Preference should be given to individual treatment for claw diseases (cutting, disinfections). In bathing the feet, copper salts and formalin are to be used with discretion.

Goats: Goats must be taken to graze every day. During the birth period, mother animals must be able to move about freely for at least a day. Keeping goats individually in boxes is only permitted for a maximum of seven days after the lambs are born and in cases of illness. Male goats may be kept individually. Hormonal synchronisation of oestrus is forbidden. Goats are to be fed primarily with roughage produced on the farm. The feeding lambs are basically to be reared and fattened on mother's milk for at least 45 days. Pure milk fattening without feeding roughage is not permitted. The proportion of concentrates (non-organically or organically grown) may only comprise 10 % of the total consumption. The rearing of goats should be optimised so that worming with synthetic chemical medicaments is needed as little as possible. These medicaments are permitted by order of a vet. The use of persistent antibiotics to dry of lactation is forbidden.

Pigs: All pigs must have daily access to an outdoor-run from the 24th day of life. Farrowing sows are accepted from this in the first 24 days after farrowing. Non-lactating sows may only be shut in during feeding; otherwise they must be kept in groups, as a rule in pastures or rooting areas. Mother sows may be held individually in farrowing pens for the week before farrowing and during the suckling period. Tying up the mother sows is not allowed. The piglets may not be weaned before the age of six weeks. The pigs must have tall-growing roughage or straw available to keep them occupied. In the rearing of pigs on alpine meadows, the whey from cheese making may be fed to pigs without restriction. Pigs have to be fed roughage and succulent feed to fulfil the pig's specific needs. The copper content may not exceed 10 mg per kg of the feed (on dry matter basis). Roughage must be offered every day to breeding sows and pigs being fattened.

Poultry: The poultry must be provided with a sufficient number of places to drink and feed. On hot days, water should be offered in the runs as well.

***Laying hens*:** If the outdoor-run is restricted, each bird must have at least 5 sqm available. The run must be covered with vegetation and have structures providing shade and protection from enemies. In addition, a completely or partially roofed area (poor weather area) with a sand bath must be available, during the whole activity period of the hens. The maximum flock size is 250 birds. If the system incorporates special buildings, it is recommended that about 3 cocks per 100 hens are kept. In the stables, there should be no more than five birds per square metre of accessible floor space. Of the total ground area, 33% must be straw-covered scratching area. With systems that have an enclosed area like a conservatory (winter garden), the conservatory area must comprise 50% of the straw-covered scratching area if it is reckoned in as part of the total ground space. The hens must have sufficient nest boxes available, containing straw or soft pile matting. The birds must have sufficient raised perches. The manure heap must be separated off and the henhouse well and naturally lit. The light period must not exceed 16 hours (other than by natural daylight in the summer). Fluorescent lighting is not allowed other than high frequency lighting. Moulting should not be artificially induced.

Laying hens in the straw-covered areas must be given a suitable grain mixture. If possible, part of the grain should be offered as whole grains in the ground bedding. Care must be taken to ensure that all birds can feed at the same time (as a precaution against cannibalism). Drinking water systems must be constructed so that birds can drink from an open water surface.

Fattening chickens: Only birds from extensively to semi-intensively reared flocks should be kept. These must be allowed to conduct their basic behavioural needs, i.e. seeking the resting area and using the raised perches provided. Birds should have no physical defects caused by breeding and no skeletal deformities. Every bird should have at least 2 sqm of outdoor-run covered with vegetation. This run must contain structures providing the hens with shade and protection from enemies. After each batch of pullets, the position of the outdoor-run must be changed to avoid a build-up of parasite populations. The outdoor-run for poultry may be used twice a year. A minimum break of 12 weeks must be observed between the usages of the same outdoor-run. A completely or partially roofed court with a sand bath (bad weather free range) must be available during the whole activity period of the hens. The maximum flock number is 500 chickens in the end phase of the fattening; and several separate flocks per farm are allowed. Fattening chicks up to 21 days may be kept in separate pens with a maximum of 1000 chicks per flock. Four flocks per pen unit are allowed. For the 4 week-old chicks, access to the outdoor-run and a pasture must be permanently available. In the stables, the population density should be kept to a maximum of 20 kg live weight/sqm. The scratching area and lighting is analogous to that required by laying hens.

Turkeys: For the rearing and feeding of turkeys, the same rules apply as for laying hens and chickens. The following deviations should be noted:

- From the 7th week onward, every bird must have 10 sqm of run available. This comes down to 9.6 sqm for flocks larger than 125 birds, and to 9 sqm when more than 150 birds are being reared. In the run, birds must be able to peck around. They must have a shaded area of at least 1 sqm for every three birds.
- When penned for fattening, the population density must be no more than three birds (half males, half females) per sqm of accessible floor space, corresponding to a maximum of 36.5 kg live weight/sqm. In the rearing phase (1-6 weeks old) this should be reduced to 30 kg live weight/sqm. The ground and flooring must have enough straw to keep it dry.
- The turkey must be able to peck around in the pens as well. The buildings must be adequately lit by natural daylight. During the night, one or more dim lights must be kept on (maximum 25 Watt, red light).

If there are more than 50 birds, an inventory control card must be kept. The birds must have been on an organic farm at least since their eighth week of life.

Table 16.4: Minimum ages for slaughtering of poultry

Poultry species	Minimum Age in days
Chickens	81
Ducks	49/70/84 (Peking/Female/Male)
Turkeys	140

Source: Naturland Standards, 01/2004

3. ORGANIC MILK PRODUCTION

Conversion of conventional dairy farm into organic dairy farm takes two years. Healthy animals are vital for good milk production. It is prohibited to keep the dairy cattle tied up permanently. They must be given opportunity to graze (throughout grazing season) or to go outdoors the whole year. Maintain proper balance (density) of dairy animals per unit land and keeping only required number of animals to avoid overgrazing of pasture/grassland.

In new or restructured stables, slatted floors in the walking area must consist of broad slats. Where slatted floors already exist, missing slats in the walking area have to be replaced. Where animals are tied up, their natural habits in lying down and standing up have to be considered. It is prohibited to use electrical aids to condition the cow in its movement. Loose housing stables must provide a feeding and a resting area for every cow. It is permissible to have feeding areas smaller than that which would correspond to the number of animals kept, if the feed is constantly accessible.

Integrate milk farming into the rotational cropping pattern to provide qualitative feeds and fodder as a beneficial effect in terms of achieving a better nutrient balance within the whole farm system. Grow fodder trees on all wastelands and boundaries of land holdings for harvesting good quality fodder. Seasonal available surplus fodders need to be converted into hay or silage for use during dry period. Use herbal preparations for control of parasitic problems and treatment of other disease in animals.

4. ORGANIC POULTRY PRODUCTION

Rear only stress-free and healthy birds. Chickens need high protein diets, such as soybeans and maize for growth. Provide organic feed that is nutritious and of high quality to birds. Avoid drug residues in eggs and meat by raising own birds, but it may increase the cost of production. Eggshells are mainly made up of calcium carbonate, and a lack of calcium will result in thin-shelled eggs/leathery eggs and hence appropriate mineral supplements are necessary. Raising chickens for meat is expensive since they have to be raised on protein rich diet in protected environment.

5. ORGANIC MEAT PRODUCTION

Both goats and sheep should be allowed for naturally grazing on those vacant lands, wastelands and harvested fields which are free from chemical pesticides and fertilizer application. The meat produced from these animals (reared as per national organic standards) can be considered as organic meat.

6. PREVENTION AND TREATMENT OF DISEASES

Health in farm animals is not simply the absence of disease but also the ability to resist infection, parasitic attack and metabolic disorders, as well as the ability to overcome injury by rapid healing. EU Council regulation 2092/91 provides information regarding disease prevention, veterinary treatment and husbandry practices in organic production.

Hygiene/bio-security measures, vaccination against specific diseases, autogenous vaccines can be produced for a specific site from bacterial isolates taken from livestock and good pasture management. Adoption of strict sanitary conditions in and around the shelter, red mite control, application of strategic dewormers like Piperazine in endemic parasitic conditions and herbal products can be used in feed/water for control of endo-parasites.

7. ORGANIC APICULTURE/BEEKEEPING

Apiculture/beekeeping has vast potential for organic farmers. A few hundred bees could make a lucrative commercial activity. Apart from providing rich harvest of honey, bees also help to boost crop production by pollinating the flower. For organic apiculture, colonies of bees have to be reared for at least one year and the production of wax/construction of the honeycombs corresponded to standards before the first nectar is foraged. Beekeeping under organic practices is discussed below:

Location: The location must be such that no significant deterioration of the bee products by contamination from agricultural or non-agricultural sources of pollination can be expected within a radius of 3 km of the hive. Hives should be situated in organically managed fields and/or wild natural areas. Hives should not be placed close to fields or other areas where chemical pesticides and herbicides are used. *Exceptions can be made by certification bodies on a case-by-case basis.*

From the organic point of view, it is preferable to keep the colonies in one location all year round. If it is necessary to move the location because of unavailability of forage, care should be taken that the right moment is chosen so that the colonies are not weakened by, for example, lack of food. At each location only as many colonies should be installed as can be sustained by the supply of pollen and nectar available. If the location is changed, the location of the colonies should be recorded on a migration plan (map), showing also the dates, the place, pasture and number of colonies.

Hives: Each beehive should primarily consist of natural materials. With the exception of connecting sections, small elements, roof covering, mesh flooring, feeding equipment and roof insulation, the hives are to be constructed of natural materials such as wood, straw or clay.

Uses of construction materials with potentially toxic effects are prohibited. Persistent materials may not be used in beehives where there is a possibility of permeation of the honey and where residues may be distributed in the area through dead bees. Wing clipping is not allowed.

Treatment of the hives: External treatment of the hives is permissible if natural, not synthetic, means are used. Pesticides-free paint on the basis of natural matter (e.g. linseed oil or wood oil) as well as glues as free as possible of harmful substances are permissible for external treatment. Internal treatment of the hives is prohibited, except with beeswax, propolis and vegetable oils.

Cleaning and disinfections: Cleaning and disinfections can be performed with heat (flaming out, hot water) or mechanically. In the case of acute infection, NaOH-solutions are permissible to disinfect the hive and to clean it out, if they are then neutralised by organic acids. The use of other chemicals is not permitted.

Wax and honeycombs: In natural beekeeping, as uninterrupted a wax cycle as possible, and therefore the continual renewal of the wax by the bees' own means, is to be aimed at. The colonies should be given the opportunity to construct natural honeycombs. Septums, starter strips and other wax products may only be made from wax originating from organic apiculture/beekeeping. Plastic septums are prohibited. By the end of the conversion period, the wax must have been produced by organic beekeeping methods or be of equivalent quality. This condition can be met by the honeycombs already produced by the colonies, by exchanging the combs completely or by encouraging the bees' own wax-producing cycle. Converted frames and hives are to be marked clearly as such.

If wax from organic beekeeping is not available, uncontaminated beeswax from capping or of a higher quality may be purchased. The wax may only be obtained by heat. It should not come in contact with solvents, bleaches or other additives. When the wax is processed, only instruments and containers made of non-oxidising materials may be used. Combs should be stored at low temperature and well ventilated. It is permissible to use thermal methods, acetic acid or *bacillus thuringiensis* preparations to keep the combs hygienic. The foundation comb should be made from organic wax. When bees are grown in wild areas, consideration should be taken of the indigenous insect population. .

Feeding: It is permissible to feed the bees, provided this is necessary for the healthy development of the colonies. As far as the farm's situation allows, the bees should be fed with honey from its own apiaries. The feeding of colonies should be seen as an exception to overcome temporary feed shortages due to climatic conditions. It is only permissible to feed sugar or sugar syrup of organic origin to tide the colonies over the winter and to stimulate brood rearing. Feeding should only take place after the last harvest before the season when no foraging feed is available. To avoid adulteration of the honey be remains of the winter feed, it must be removed before the bees start foraging again. It is forbidden to feed pollen substitutes.

Breeding and increasing stocks: The aim of breeding is to produce bees which are adapted to the local organic situation. Genetic engineering and the use of genetically manipulated bees are prohibited. Preference should be given to natural breeding and reproduction methods. The swarming instinct is to be taken into account. The use of synthetic chemical means to pacify or expel the bees is prohibited. It is forbidden to mutilate the bees e.g. by clipping their wings.

Stocks should be increased by taking advantage of the bees, swarming habits. It is possible both to anticipate swarming by creating an artificial swarm and dividing the remaining colony to increase stocks further, as well as to re-unite anticipated swarms.

Health care: A bee colony should be reared in such a way that it is capable of correcting any imbalances by itself. For pest and disease control and for hive disinfections the products which are allowed are caustic soda, lactic, oxalic, acetic acid, formic acid, sulphur, etheric oils, *bacillus thuringiensis*, etc.

Veterinary medicine should not be used in beekeeping. When working with the bees (e.g. at harvest) no repellent consisting of prohibited substances should be used. Use of synthetic chemical medication is prohibited.

***Extraction of honey and its storage*:** The equipment and vessels used in the extraction of honey by spinning or pressing have to be such that the honey only comes into contact with materials approved for the use in the processing of foodstuff (e.g. stainless steel, glass, non-contaminating plastics). During spinning, straining, filtering and conservation, or if the honey crystallizes, it should not be heated to more than 38°C. Filtering under pressure is prohibited in all forms. Low pressures such as those which occur in normal operation (e.g. when pumping) are permissible.

Only vessels of non-contaminating materials (e.g. stainless steel, glass) may be used. The honey must be stored in a dark, cool and dry place. Whenever possible, the honey should be filled into jars before it becomes solid. Recycling jars must be used. The admixture of raw materials not produced according to these standards must be excluded.

Table 16.5: Criteria for measurable quality of the organic honey

Parameter	**Minimum requirement** (AOAC analyses)
Water content	18 %
Hydroxymethylfurfural (HMF)	10 mg/kg
Invertase index	10 min. (min)
	7 min (Honey from acacia, lime trees and phacelia
Chemotherapeutic drugs	No traces

8. ORGANIC AQUACULTURE

In fish production it is required that the ecologic balance is not being disturbed, that natural populations are not endangered and that the basic principles of sustainability are observed. The specific needs of fish have to be taken into consideration (pond/installation, structure of the habitat, stock density, water quality etc.). The fish must not be exposed to unnecessary strains or stress during keeping, transport or slaughter.

In principle only local fish species that are adapted to the local conditions are to be raised. Exceptions require authorization and are subject to regulations. The use of genetically modified or triploid fish is forbidden. Parents and young stock must not be or have been fed with antibiotics, growth promoters or hormones.

8.1 General Requirement

- Physical separation between conventional and organic production units must be maintained. For sedentary or sessile organisms not living in enclosures the area should be at an appropriate distance from pollution or harmful influence from conventional aquaculture/agriculture or industry.
- No conversion period is required in the case of open collecting areas for wild, sedentary organisms where the water is free-flowing and not directly or indirectly contaminated by substances prohibited in these standards and where the collecting area can be inspected with respect to water quality, feed, medication, input factors or any other relevant sections of these standards and all requirements are met.
- Construction materials and production equipment should not contain paints or impregnating materials with synthetic chemical agents that detrimentally affect the environment or the health of the organisms in question. Adequate measures should be taken to prevent escapes of cultivated species from enclosures.
- Adequate measures should be taken to prevent predation on species living in enclosures. Conventional, veterinary chemicals should only be used if no other justifiable alternative is available, and/or if the use of such chemicals is required according to national law.
- Prophylactic use of veterinary drugs, except vaccinations in certain cases, is prohibited. Vaccinations are permitted if diseases that cannot be controlled by other management techniques are known to exist in the region. Vaccinations are also permitted if they are mandatory under applicable legislation. Genetically engineered vaccines are prohibited. Synthetic hormones and growth promoters are prohibited.
- Aquatic animals should not be subject to any kind of mutilation. Breeding should allow natural birth. The certification body/standard-setting organization may, however, allow the use of production systems that do not provide for natural birth, for instance hatching of fish eggs. Where available, brought-in aquatic organisms should come from organic sources. Artificially polyploided organisms and genetically engineered species or breeds, are prohibited.
- Aquaculture feeds should contain 100 % certified organic components or wild feed resources. When supplying food collected from the wild, the "Code of Conduct for Responsible Fisheries" (FAO, 1995) should be followed. When certified organic components or wild foods are not available, the standard-setting organization may allow feed of conventional origin up to a maximum 5 % (by dry weight).
- In systems using brought-in feed inputs, at least 50 % of the aquatic animal protein in a diet should come from by-products or other waste and/or other

material that would not be used for human consumption.

- Feed rations should be designed so that plant and/or animal sources supply most of the nutritional needs of the organism. Use of human faeces is restricted.
- The products which must not be included in or added to the feed or in any other way be given to the organisms are - Synthetic growth promoters and stimulants, Synthetic appetizers, Synthetic antioxidants and preservatives, Urea, Feedstuffs subjected to solvent (e.g. hexane) extraction, amino acid isolates, material from the same species/genus/family as the one being fed, synthetic colouring agents, genetically engineered organisms or products thereof.
- Vitamins, trace elements and supplements used should be of natural origin when available. The feed preservatives like bacteria, fungi and enzymes, by-products from the food industry (e.g. molasses) and plant based products may be used. Synthetic chemical feed preservatives are permitted in response to severe weather conditions.
- Chemically synthesized tranquillisers or stimulants should not be given to the animals prior to or during transport or at any time.
- Where applicable, aquatic organisms should be in a state of unconsciousness before bleeding to death. Equipment used for stunning should be in good working order and should quickly remove sensate ability and/or kill the organism.

8.2 Specific Requirement

Design of ponds: On an average, at least 30 per cent of embankment line should represent the natural biotope structure to at least 2 m depth in the form of a helophytic zone, reed and/or overhanging trees/shrubs. The entire fish farm has to produce organic fish. The parallel production of organic and non-organic fish is not allowed.

Pond should be constructed in such manner that water bodies inside the operation retain their ecological functions depending on the geographical conditions (e.g. breeding ground for amphibians and water insects, resting place for migrating birds, migration routes for fish). For this purpose, in particular, adequately large areas showing natural vegetation (e.g. water reeds, higher aquatic plants or helophytes) should be protected or re-planted by the enterprise. While protecting the farm areas from predatory birds and other animal species, measures not harming the animals physically should be preferred.

In pond farms, at least 5 % of production area, the natural vegetation should be allowed to develop undisturbed as a refuge for native animal species. Inlet and outlet of the farm should be protected from invasion by wild fishes as well as from stock escaping. Net cages should be secured by means of firm anchoring, strong net walls and a type of construction taking into account the prevailing conditions against damage and related escaping of stocks.

The culture of fishes in artificial containers (polyester, concrete etc.) is not permitted. Only for the short-term stay of hatching for the initial/starter feeding phase and the post-harvest maintenance of fishes up to a maximum of 8 weeks such containers are permitted.

Species and origin of stock: As stock, where possible, species naturally occurring in the region and local strain should be chosen. The stock (spawn, breed-stock, fingerlings) may be purchased from organic farms only. If not available, the stock may be purchased from conventionally managed farms. Natural reproduction or spawn recovery should be adopted. The use of hormones, even from the same species, is not allowed. If due to extreme climatic and weather conditions, no natural spawn recovery can be expected, conventional measures can be resorted in consultation with certifying body.

Quality of water: Effluent water quality has to be monitored and documented on an at least monthly basis by the farm. The water quality (e.g. temperature, pH, salinity, oxygen, ammonium, nitrate, phosphate, suspended solids, heavy metals etc.) must conform to the natural requirements of the fish species. Adequate measures must be taken to minimize the outflow of nutrients and/or suspended solids, especially during harvesting. Organic sediments should be removed on a regular base from the channels and brought to appropriate utilization.

The water quality of source water bodies (in the case of pond farms) or the surrounding lake or sea regions (in the case of net cages) should not become significantly deteriorated (standard value <10 % of the parameters determined) due to the farming operation. This should be secure by sedimentation ponds and/or filtering plants dimensioned adequately. Settled particulate organic matter (product of metabolism, feed residue) should be removed and brought to adequate re-usages (e.g. as fertilizer in agriculture). The outflow of nutrients from the farm should be kept as low as possible. If water is tapped for a pond farm from a stream, then at least 25 % of the average low water level should remain in the source streambed.

The inflowing water should contain none or slight contamination of anthropogenic origin (e.g. heavy metals) as well as not or only slightly be influenced by sewage water (guideline: BOD< 6 mg). The pH-value should be between 6.0 and 9.0. The artificial illumination is provably necessary, and then the simulated day length should not exceed 16 hours. Release of toxic or otherwise harmful substances in the ponds, the channels or the banks should be prevented from. This refers especially to installation and management of pumping stations (e.g. oil spoilage), harvesting techniques as well as the overall hygienic conditions on the farm.

Protection of ecosystem: In order to stabilise/enhance the ecological system and the natural dynamics on the farm area, at least 50% of total dyke surface should be covered by plants. This state should be reached during a period of maximum three years. Recommended plant species are e.g. leguminous plants, aloe and other for the tops of the dyke, mangrove species, semi-aquatic herbs and floating grasses for the lower parts of the slopes. Farms situated in areas originally free from vegetation (e.g. desert, dunes) are excluded from this requirement.

In order to find an ecological adequate and economically effective management against predatory birds, documentation on foraging predators, estimated harvest losses and type of preventive measures should be kept. It is recommended to raise ducks in the ponds, expelling intruding birds from their breeding territories. Native animals (e.g. ant-eaters, migrating water birds, wild cats) should be protected as indicators for a sane environment.

Stocking density: The stocking density depends on the natural oxygen absorption of the water body. Stocking density in net cages should not exceed 10 kg fish/m^3, based on the anticipated harvest weight. The stocking density should not exceed the state, that at least a 50 per cent of fish yield should be attained via the natural feed availability. Only when feed is administered for augmenting the protein content as well as peas and beans, the upper limits is set (3000 for Carp and 7000 for Trench per hectare) for stocking of the fish species of main commercial interest (*Naturland Standards, 01/2004*).

Feeding: At least a 50 per cent of fish growth should be achieved through natural feed availability in the pond. In order to ensure an optimised utilization of the protein rich pond feed, supplementary feeding is permitted. The type, quantity and composition of feed must take into account the natural feeding methods of the concerned fish specie. All the feedstuffs must be produced in accordance with national standards, or in any case at least in accordance with the IFOAM Basic. Additionally, feed from animal origin in limited amount and defined quality is permitted. For salmonides and other carnivorous fish species, the addition of fish meal (an fish oil) is allowed. It has to be produced either from residues of edible fish processing or come from provably sustainable fishing. Feed from genetically altered organisms or their products are not permitted.

Health and Hygiene: The ponds should be refilled at least once in a year during March/April. Breeding ponds may also be refilled later in the year. If hygienic measures (e.g. for controlling leeches) are necessary, then quick lime is permitted to be applied on to the humid pond bottom (max. 200 kg/ha). Its application into the pond (max. 150 kg/ha) for the purpose of pH stabilization and for precipitating of suspended organic matter is permitted in critical weather situation. Continuous aeration or the use of liquid oxygen for augmenting the stocking densities is not permitted.

Customary smoking techniques are permitted. Only hardwood and spices should be subjected to glowing. The glowing temperature should not exceed, on an average, 500°C (max 650°C). The smoke conduction should be such that a cooling of the smoke takes place, and any entry of substances (fat, protein, drip fluid) from the material to be smoked into the glowing zone is avoided.

Organic fertilization: In order to control plankton growth, organic fertilizer in the form of solid dung, hay or comparable substances may be applied to the pond to the extent of maximum 40 kg N/ha. Numbers of waterfowl cultured on the fishponds should be appropriately taken into this calculation.

Transport, slaughtering and processing: Transport and slaughtering must be done in a way as fast and considerate as possible in order to avoid any unnecessary suffering of the animals. Maintenance of the cold chain from the point of slaughtering up to the sale point must be strictly observed, in order to prevent any deterioration in the product quality. Slaughtering of fishes should be carried out by means of incision of gills or immediate evisceration. Prior to this, fishes should be anaesthetized by means of concussion, electrocution, carbon dioxide and, if need be, by natural plant anaesthetics. Cleaning of factory rooms, devices and machines must ensure a perfect hygiene along with an as high as possible ecofriendliness. Mechano-physical processes should be preferred to chemical processes. Regarding the cleaning and disinfections agents used, a separate book of records should be kept. The wastewater from the slaughtering and processing plants must be subjected to appropriate purification process.

Chapter 17

Organic Food Processing and Handling

Processing of organic food products and handling should be optimised to maintain the quality and integrity of the product and directed towards minimising the development of pests and diseases. Processing and handling of organic products should be done separately in time or place from handling and processing of non-organic products. Pollution sources should be identified and contamination avoided. Flavouring extracts should be obtained from food (preferably organic) by means of physical processes.

1. WHAT DO THE ORGANIC STANDARD SAY?

1.1 General Requirement

Organic products should be protected from co-mingling with non-organic products. All products should be adequately identified through the whole process. Organic and non-organic products should not be stored and transported together except when labelled or physically separated. Certification programme should regulate the means and measures to be allowed or recommended for decontamination, cleaning or disinfections of all facilities where organic products are kept, handled, processed or stored.

Besides storage at ambient temperature, the special storage conditions such as controlled atmosphere, cooling, freezing, drying, humidity regulation, ethylene gas etc. are permitted for agricultural produce.

1.2 Specific Requirement

***Pest and disease control*:** Pests should be avoided by good manufacturing practices. This includes general cleanliness and hygiene. Treatments with pest regulating agents must thus be regarded as the last resort.

Recommended treatments are physical barriers sound, ultra-sound, light, and UV-light, traps, temperature control, controlled atmosphere and diatomaceous earth. A plan for pest prevention and pest control should be developed. For pest management and control the following measures should be used in order of priority:

- Preventive methods such as disruption, elimination of habitat and access to facilities
- Mechanical, physical and biological methods
- pesticidal substances may be used as listed in organic standards
- other substances used in traps etc. Irradiation is prohibited

There should never be direct or indirect contact between organic products and prohibited substances. (e.g. pesticides). Persistent or carcinogenic pesticides and disinfectants are not permitted. The certification programme should set up rules to determine which protection agents and disinfectants may be used.

***Ingredients, additives and processing aids*:** Ingredients of agriculture should be 100 % of organic origin. For the production of enzymes and other microbiological products the medium should be composed of organic ingredients. In cases where an ingredient of organic agriculture origin is not available in sufficient quality or quantity, the certification programme may authorise use of non-organic raw materials subject to periodic revaluation. The same ingredient within one product should not be derived both from an organic and non-organic origin. Water and salt may be used in organic products.

Minerals (including trace elements), vitamins and similar isolated ingredients should not be used. Preparations of microorganisms and enzymes commonly used in food processing may be used, with the exception of genetically engineered microorganisms and their products. The use of additives and processing aids should be restricted.

***Processing methods*:** Processing methods should be based on mechanised, physical and biological processes. The vital quality of an organic ingredient should be maintained throughout each step of its processing. Processing methods should be chosen to limit the number and quantity of additives and processing aids. The approved processes are– mechanical and physical, biological, smoking, extraction, precipitation, filtration etc.

Extraction should only take place with water, ethanol, plant and animal oils, vinegar, carbon dioxide, nitrogen or carboxylic acids. These should be of food grade quality, appropriate for the purpose. Irradiation is not allowed.

Filtration substances should not be made of asbestos nor may they be permitted with substances which may negatively affect the product.

***Packaging*:** The use of material for packaging should be eco-friendly. Unnecessary packaging materials should be avoided. Recycling and reusable systems should be used wherever possible. Biodegradable packaging materials should be used. Material used for packaging must not contaminate food.

2. PROCESSING OF FRUITS AND VEGETABLES

Processing of organic fresh produce requires cleaning, grading followed by peeling, stoning or slicing (horticultural produce). At this stage fruits and some vegetables such as onions and peppers are ready for freezing, but most vegetables need to be blanched with hot water or steam at 80° to 100° C to inactivate enzymes that could otherwise lead to a loss in vitamin C and flavour. Fruit can be coated in sugar or in syrup that contains an antioxidant like ascorbic acid. Coating retards browning avoids the 'cooked' taste after defrosting and increases product quality. The products may be packaged before or after freezing.

2.1 Freezing

Freezing is quite often applied to vegetables but rarely used for fruits, as they do not handle it well. Nutritional quality is generally retained at at-harvest levels when the product is sold frozen. Significant vitamin loss may occur during subsequent thawing and cooking. Colour, odour and taste are retained well by freezing, but superficial dehydration occurs. If properly frozen, stored and defrosted, frozen produce becomes a high quality raw material. The degree of freezing depends on the duration of storage. Some conditions of freezing for horticultural crops are listed below:

Table 17.1: Practical storage life of frozen products

Products	**Practical storage life (month)**		
	-18°C	**-25°C**	**-30°C**
Fruits in sugar	12	18	24
Fruits in sugar with ascorbic acid	18	24	>24
Asparagus	15	24	>24
Carrots	18	24	>24
Cauliflower	15	24	>24
Green beans	15	24	>24
Beans (Lima)	18	>24	>24
Peas	18	>24	>24
Potatoes	24	>24	>24
Spinach	18	>24	>24

Source: IIR, 1990

2.2 Drying

Fruits are the principal imported dried produce and are easy to transport and to store. Dried vegetables are produced in low volumes for the local market but can be useful for e.g. soup mixes. The major risks with dried products are microbiological attack and physiological deterioration. Physiological deterioration is caused by oxidation and enzymatic activity and leads to browning, loss of vitamins and the development of off-flavours. Dried organic products are particularly vulnerable to deterioration since chemical treatment with the normal conservation agent (SO_2) is restricted. No organic substitutes are available which are as effective as SO_2 in conserving the colour of dried fruit. Maximum 0.3 mg/l of SO_2 may be applied as preservative or stabiliser.

2.3 Water content

Dry fruit products have a water content of 8 to 12 % and dry vegetables, around 7 %. Under these conditions, there are no microbiological problems during storage of the product. Semi-dried or soft products, which may be more highly priced in the market, have a water-content of between 20 and 30 %; therefore microorganisms can develop during storage and refrigeration may be required.

2.4 Additives and Processing Aids

Permitted processing aids which help to retain quality of dried produce include:

- conservation agents and anti-oxidants: ascorbic acid, citric acid, tartaric acid, and salt
- texturing agents: calcium chloride.

Ascorbic acid, citric acid, tartaric acid or lemon juice may be used for acidification. The resulting low pH limits the development of microorganisms and limits non-enzymatic browning. The product is treated by dipping in, or spraying with, acids or lemon juice (it is more difficult to control the acidity if lemon juice is used).

Salt can be used as an aid to drying. For example salted dried plums are produced in China for use as a breakfast item in the Japanese market. The salt aids in dehydration and anti-microbial activity.

2.5 Blanching

A brief period at high temperatures destroys most microorganisms present in the product and inactivates enzymes which promote browning and degradation. Details of time, temperature, and solution vary according to the produce being treated; some examples are listed below:

Table 17.2: Recommendations for blanching fruits and vegetables

Fruit	Process
Apricot	Steam for 5 min
Banana	Boiling water for 5 min
Litchi	Steam for 7 seconds, then soak in 5-10 % citric acid and 2 % salt
Mango, Papaya	Hot water (56 °C) for 1 min
Pineapple	Steam for 10 min
Cabbage	Boiling water for 3 min or Steam for 5 min
Carrot	Boiling water for 4-6 min
French and Green beans	Boiling water for 4-6 min
Spinach	Boiling water and salt (30 g/l), 3 min
Sweet potato	Boiling water for 5 min or Steam for 8 min
Turnip	Boiling water for 4-6 min

Source: Desruelles et al. 1997

2.6 Rapid Drying

Sun drying is the processing method most often used for organic fruits such as apricots, figs and bananas. However, the potential risk to quality and the difficulty of maintaining a high degree of sanitation is a problem. A rapid drying process also decreases the contact time between the product and oxygen. The drying conditions and recommended moisture content of the finished product are given below:

Table 17.3: Drying condition and moisture content of food products

Fruit or vegetable	Drying temperature (°C)	Moisture content (%)	Storage life (month)
Banana	55	12	6
Mango	55	14	6
Papaya	55	7-12	5
Onion	50-55	5	3-12
Tomato	55	6	6
Carrot	50-55	6	6-12
French beans	55	6	6

Source: Bureau International du Travail, 1990

Pasteurisation after packaging: Certain processors pasteurise their products at 70°C after packaging to destroy microorganisms that could have contaminated the product after blanching. This treatment applies to soft fruits like apricots, plums.

Supplementary techniques for specific fruits: Dried dates are subject to considerable damage by pyralids (*Ectomyelois ceratoniae*). In order to avoid the use of chemical products (methyl bromide), two methods can be used to destroy this insect. Immediately after the harvest, dates can be placed into a dryer for 2 hours at 60°C in order to destroy larvae and eggs. Alternatively, the dates can be frozen at -40°C. Shock from cold temperatures seems to satisfactorily destroy the pyralid larvae and eggs. This procedure is more widely used.

Biological control has been tested. Among the techniques tried, one of them uses a bio-pesticide (*bactospeine*) and an insect (*Habrobracon hebetor*). When *Bacillus thuringiensis* is ingested, it produces the pesticide bactospeine in the organism. The use of *B. thuringiensis* is authorised by the EEC organic legislation. *H. hebetor* is a parasite to pyralids. When these two organisms are introduced into the storage room the rate of mortality of the pyralids reaches 76 % after 6 days. Although effective, this method is not yet in commercial use.

After storage, dates are placed in a steam bath to stabilize the fruit's moisture level and to pasteurise it. Before packaging, the dates are sometimes coated with hot, pasteurised sugar syrup. This coating, which has a high sugar content (70 to 80° Brix), acts as a barrier against contamination by microorganisms (due to the low absorption of water into the coating) and prevents the products from sticking together. In this manner, despite a high water content (24 %), the development of microorganisms in date packaging seems to be rare, even among organic dates. However, to obtain these results, one must ensure that good hygienic conditions are applied during processing.

Dried organic grapes (raisins) are primarily dried in the sun. It is preferable to wash the grapes with water and oil before drying, and to spray them with organic sunflower oil before packaging. This helps stop the raisins from sticking together and acts as a barrier against the proliferation of microbes.

The *by-products of dried fruits and vegetables* are also used, incorporated into products. The incorporation of organic fruits and vegetables into other types of products is not as yet widely done. To develop the market for dried fruits as a viable product in industry, the product must be available with consistent quality year-round and in sufficient volume to allow for storage.

Water/air tight packaging: The type of packaging to be used for dried organic foods varies with expected storage conditions. The most common types are flexible packages with low permeability to oxygen and water vapour, and vacuum packaging is common.

2.7 Heat Conservation

These techniques use thermal treatments to conserve processed products by destroying or inactivating enzymes, and killing microorganisms. The products concerned are juices (principally from fruit), canned products (fruits and vegetables) and jams and purées.

Juice or oil extraction: Juice is a simple and natural way to process fresh produce. It allows the preservation of the majority of nutritional qualities (vitamins, minerals), and it largely resolves the problem of storage. Juice could be considered as the ideal product for the organic market. The juices that are consumed in largest quantity (traditional and organic, combined) are apple juice, citrus juices, pineapple juice and vegetable juices. Among exotic fruit juices, the most common are from oranges, mangoes and guava. Carrot and tomato juices are important in the European market. There is probably a consumer view that juicing concentrates any residues in the product, which may explain the increasing popularity of pure organic fruit and vegetable juices. The market does not appear to be saturated as yet.

In addition, tropical fruits often offer a much higher source of vitamin C than fruits from temperate zones. This additional nutritional aspect is an advantage for tropical organic juices. It is preferable to opt for transport of juice in suitable bulk packages such as kegs to diminish transportation costs. The juice can be repackaged in the country where it will be consumed. However, to ensure juice stability during handling, it is necessary to use high-temperature pasteurisation. This can alter the organoleptic and nutritional characteristics of the juice. To remedy this problem, it might be preferable to extract the juice at the location where the juice will be consumed. This also allows the sale of 'fresh squeezed' or flash-pasteurised juices. This type of product is even more consistent with organic farming because little or no processing takes place and the final product is the closest thing to fresh. Fresh squeezed juice which has not undergone any type of heat treatment can be stored for only about 10 days under refrigeration. Flash-pasteurised juice (heat treatment of a few seconds at 70°C) has to stay in a cold room and can be stored for about 24 days. These products, the focus of ever-increasing consumer demand, correspond to the heightened interest in food products that are natural and healthy. It is therefore desirable to associate these qualities with the organic market.

The following three factors are responsible for the successful long-term storage of juices:

The acidity of the juice: The pH must be less than 4.2 in order to stop the development of microorganisms; ascorbic acid, citric acid or lemon juice can be added

Thermal treatment: A temperature which destroys microorganisms has to be used. Conditions depend on the characteristics of the final product: viscosity, packaging, type of juice, etc.

The packaging: Juice packaging has to be impermeable to gas and sometimes to light, and has to avoid microorganism contamination.

Certain juices, especially those which are high in pulp content, may need to be clarified. Enzymatic treatments are authorized in Europe (part B of EC legislation), but filtration and decantation methods can be used instead. There is a growing market for organic oils, such as olive oil, which can be produced by simple physical methods from organic produce.

2.8 Canning

Canned produce must be prepared in such a way as to retain as closely as possible the characteristics of fresh produce. The market for canned products is restricted since consumer opinion relates canned products to products that are not wholesome, or are of poor quality.

Authorised additives and processing aids include:

- Anti-oxidants (e.g. ascorbic acid, citric acid, and tartaric acid)
- Texture aids (calcium chloride)

Canning organic fruits and vegetables does not require any specific adjustments that are not already found in conventional canning methods. Fruit are usually canned in sugar syrup. This syrup is defined for 'natural products' (up to 16 % sugar) or 'products with syrup' (over 20 % sugar).

The time required for sterilisation of the cans depends on the characteristics of the product (viscosity, form, etc.), the size of the can and other factors. Oxygen in the produce can react with metals in the can material, so the cans need to have an internal protective coating to avoid corrosion. It is necessary to remove most of the oxygen from the produce by blanching and pre-heating.

Details of blanching recommendations and syrup compositions are given below for canning a variety of common fruit and vegetables:

Table 17.4: Composition of canned fruit and vegetables

Fruit	**Syrup (g sugar per litre water)**
Apricot	Natural: 80
	Syrup: 450
Banana	Syrup: 300 (+ 0.5% citric acid)
Cubic fruits	Syrup: 400
Fig	Natural: 80
	Syrup: 160-480
Grape	Syrup: 400
Grapefruit	Syrup: 400
Guava	Syrup: 650
Litchi	Syrup: 400-500 (+ citric acid)
Mango	Syrup: 400 (+0.25% citric acid)
Orange	Syrup: 500
Papaya	Syrup: 300-500 (+0.5% citric acid)
Pineapple	Syrup: 200-400
Vegetable	Syrup
Green peas	2 % salt, 4 % sugar
Carrot, French beans, Gombo, Lima beans, Spinach	2 % salt
Tomato, sweet pepper	2 % acidified salt
Sweet potato	Water

Source: Desruelles et al. 1997

Table 17.5: Blanching for canned fruit and vegetables

Fruit/Vegetables	**Processing**
Apricot	Boiling water for 1 min
Banana	Boiling water for a few min
Fig	Boiling water for 5 min
Grape	No blanching
Grapefruit	Steam for a few min
Carrot	Boiling water for 2-4 min
Gombo	Boiling water for 2 min
Green pea	Boiling water for 8-10 min
Lima pea	Hot water (87-95 °C) for 3 min
Spinach	Hot water (71-77 °C) for 6 min
Sweet potato	Steam for 9-12 min

Source: Desruelles et al. 1997

Some blanching details for stewed fruit are listed below:

Table 17.6: Blanching conditions for stewed fruit

Fruit	Processing details
Apricot	Hot water (80-85°C) for 2 min
Mango	Steam for 2 min
Orange	Boiling water for 5 min

Source: Desruelles et al. 1997

Conservation with sugar: The goal of this processing technique is to arrive at a sugar content where the product is stable. This method is principally used for fruit. Durability in storage is determined by acidity, sugar content, and the type of packaging. There is a large variety of products: jams, jellies, syrups and fruit pastes. Importing countries may impose standards and definitions for these products, specifying the minimum quantity of fruits or fruit juice that may be used, as well as the optional ingredients.

Conservation with sugar does not require any particular modifications for organic farming although where possible organic sugar sources will be preferred. The process consists of mixing sugar (or sugar syrup) with pulp of fruits (or fruit juice). The mixture is cooked to remove excess water. Pectin solution is added. When the desired end point is reached (65° Brix for jams and jellies, 68° Brix for preserves), the proper amount of acid is added. Once the pH has been adjusted to 3.1 to 3.4, the mix is ready for packaging.

The low pH and water activity of the product limit bacterial development. However the low solids content (65-70 %) is not a guarantee against the growth of certain microorganisms. To ensure total protection, post-filling sterilization could be carried out.

The ingredients and processing acids required for these products are:

Preserving agents and acidity regulator such as lactic acid, citric acid, calcium citrate, tartaric acid, sodium and potassium tartrate;

Thickening and gelling agents such as pectin, flour from carob beans, guar gum, adragante gum, arabic gum, alginic acid, sodium and potassium alginate, agar, carragheenan;

Processing aids such as calcium chloride, coatings (Carnauba and bees wax), and calcium carbonate.

Jams offer a potential market for pulps and purées. As with juice, these are products that have been only slightly processed, are healthy, and that keep the nutritional properties of the original fruit or vegetable and they correspond to the idea of health that is associated with organic products. Jams are primarily sold in fine markets or specialized shops.

Conservation by fermentation: Fermentation is a chemical change brought about by enzymes, bacteria or microorganisms. The chemical changes are acidification, oxidation of nitrogenous organic compounds and decomposition of sugars and starches (alcoholic fermentation). These last fermentation options result in wines and other alcoholic beverages. Organic processing methods for alcoholic beverages are not addressed here.

There are many other important fermentation processes and products. They principally concern vegetables like sauerkraut, pickled vegetables, fermented soya, and fermented roots or tuber (e.g. cassava, yam)

Fermentation is achieved by:

- salt, without water (e.g. sauerkraut)
- brine: salted solution, sometimes with vinegar
- the inoculation of specific microbes
- water (cassava)

Preparations of microorganisms and enzymes commonly used in food processing may be used, with the exception of genetically engineered microorganisms and their products. Microorganisms grown on organic cultures should be used if available.

Sauerkraut: Cabbages are cut and put in fermentation tanks with salt. The salt is mixed with the shredded cabbage and lactic fermentation causes the conversion into sauerkraut. When the acidity is sufficient the fermentation is stopped by pasteurization. The process does not require specific additives or processing aids. Canned sauerkraut is easy to use and is not as subject to deterioration and spoilage as bulk sauerkraut. Sauerkraut used for canning is ordinarily not cured as long as that sold in bulk.

Pickles: There are many vegetables used for pickles. Sometimes vegetables undergo a preliminary fermentation with salt. They are soaked in several changes of cold water until practically free of salt, followed by many hours in hot water (45-65°C). Not used for fragile vegetables like cauliflower. Calcium chloride has a hardening effect and can be added. Afterward, vegetables are placed in vinegar. Pickles are put into heavily lacquered cans with brine or vinegar. The cans undergo thermal treatment and are sealed.

Other fermentation: The fermentation of soya involves the use of microorganisms like *Saccharomyces, Torulopsis, Pediococcus or Rhizopus*. These elements are authorised by EC Regulation. The fermentation of cassava (or other roots and tubers) is a natural fermentation that needs no additives.

Fermentation is a viable process for enhancing the microbiological safety and storage of foods, especially in areas such as the tropics. It conforms to organic legislation. Many types of new fermented products can be created such as vegetable cheeses and yoghurt.

3. HANDLING OF FRUITS AND VEGETABLES

Handlers and processors should handle and process organic products separately in both time and place from non-organic products. Handlers and processors should identify and avoid pollution and potential contamination sources. Codex Alimentarius required the maintenance of organic product integrity and protection against contamination.

***Harvesting*:** There are few specific considerations for harvesting organic produce. Normal attention must be paid to harvesting each crop at its optimum maturity, bearing in mind its intended market; harvesting early in the day and keeping harvested produce in the shade, wherever possible; and removing field heat quickly.

***Curing*:** Some root, tuber and bulb crops require a 'curing' period at ambient or elevated temperature to promote wound healing and ensure optimum storage life. There are no specific requirements for curing organic produce.

***Packinghouse operations*:** Most markets require strict attention to be paid to the size, grade, quality and maturity of produce, whether or not it is organic. Fruit and vegetables must be cleaned and graded to comply with these regulations. Special consideration needs to be paid to the cleaning or sterilisation of grading and processing equipment in an organic operation. Organic produce must be free of substances used to clean, disinfect, and sanitize food processing facilities, which can be achieved by re-washing equipment with hot water after the use of cleaning agents, or passing non-organic produce through the system before the organic produce. Ethylene is permitted for post-harvest ripening.

3.1 Packing and Packaging Materials

The IFOAM Basic Standards state that '*Organic product packaging should have minimal adverse environmental impacts*' and recommend that '*Processors of organic food should avoid unnecessary packaging materials; and organic food should be packaged in reusable, recycled, recyclable and biodegradable packaging whenever possible*'. Thus, although all types of packaging are authorised, there is an expectation that careful thought will have gone into the choice of the packaging with regard to its environmental impact. In the future, restrictions may be put in place concerning the use of packaging materials that are harmful to the environment, especially for those packaging materials that are not recyclable or biodegradable.

Pre-packaging: Pre-packaging is generally defined as packaging the produce in consumer size unit packs. It reduces the transportation cost by eliminating the unwanted parts of fruits, vegetables and flowers. Pre-packaging increases the shelf life, reduces the shopping time of the consumers and makes the produce easy to handle. Among the various types of packaging films, non-contaminating polyethylene film finds the maximum use. It has been reported that seal/shrink wrap packaging of individual citrus fruits, grapefruit, oranges, lemon, mandarins and tomato, brinjal, capsicum etc. with 10 mm high-density polyethylene film considerably increases the storage life compared to control without packaging. The film wrapping reduces weight loss and maintain firmness of the produce. Polymeric films give good results in packaging of fresh fruits and vegetables due to their high gas permeability and impervious nature to bacteria. Wrapping mangoes in HM film delays ripening and increases the self-life during transportation. Similarly, film wrapping coupled with an ethylene absorbent extends the shelf life of banana.

Packaging: Packaging forms an integral part of marketing of fresh horticultural produce. It provides the essential link between producers and consumers. Various types of packaging materials are used in bamboo baskets, gunny bags, cardboard, paper, glass, metal, plastic, wooden boxes, etc. Use of plastic crates in packaging of horticultural crops helps minimize the cost of packaging materials and makes the whole process less dependent on scarce items like wood etc, thereby resulting in conservation of environment. There is a constant search for newer and better protective and cost-effective packaging materials. A ventilated corrugated fibre board (CFB) box has been developed with ventilated partition for packaging and transportation of mangoes. Cotton stick waste has been gainfully utilized for making CFB boxes. In packaging material being used for organic products are discussed below:

Cardboard and paper: Whilst these traditional materials are generally readily available and inexpensive, they have several drawbacks like porous to gas, permeable to water, easily torn or crushed. They protect products only from light impacts. In organic farming, these materials are principally used for fresh fruits and vegetables. In order to limit impacts between products and to limit movement within the packaging, the use of liners between layers of fruits and vegetables, or of individual paper wrapping, can be efficient. Waxing of packaging restricts water permeability but can make the package unsuitable for recycling. Particular consideration needs to be paid to the use for which the packaging is intended. A lightweight cardboard box may be adequate for use in the local market, but a telescopic box with reinforced corners may be necessary for sea freight. It will need to retain its strength during extended periods at low temperature and high humidity to enable stacking over 2 m high on a pallet.

Plastic: Plastic packaging is likely to deliver the best quality produce, minimising wastage. It can be pre-printed for marketing purposes and it is ideally suited to forming flexible, unbreakable packaging matched to the product's needs. It is light in weight, leading to cheaper transport costs and less fuel consumption in transport. Although plastic packaging is often frowned upon because it is commonly derived from fossil fuels and is not always able to be recycled, the alternatives need to be considered carefully. There are many recyclable or biodegradable plastics and some plastics are now produced from starch.

Plastic is selectively permeable to gas and water, depending on the type of polymer. Some polymers are therefore ideally suited to creating a modified atmosphere around fresh fruits and vegetables. During storage, the respiration of the produce emits carbon dioxide and consumes oxygen. The selective permeability of the polymer results in an increase of carbon dioxide inside the packaging. The respiration rate of the vegetable decreases proportionally to this enrichment, and the composition of the atmosphere stabilizes progressively. The resultant atmosphere, if maintained, can lead to significant improvements in storage quality.

This type of modified atmosphere packaging (MAP) is being developed for highly perishable, high-value produce. The problem is that temperature fluctuations dramatically affect the rates of tissue metabolism and the permeability of the plastics. Stable gas compositions can be obtained only in a precisely controlled environment. The risk of anaerobism in MAP is severe. Anaerobism leads to the production of off-flavours and, in extreme cases, favours the growth of toxic organisms and toxin production. MAP is thus too risky to recommend for routine use in the export of fresh fruit and vegetables from developing countries.

Glass: Glass receptacles are principally used for liquid products or solids in liquid. Glass receptacles are well adapted to organic products as they are impermeable to gas, air moisture, microorganisms and resistant to thermal treatments. Against these positive attributes are the bulk and transparency of glass, which can cause problems for products that are sensitive to light. Also, the energy costs of recycling glass are very high (possibly making it less ecologically-friendly than plastic). In addition, glass breakage is extremely serious on packing lines, a single breakage can lead to expensive downtime as equipment is turned off and cleaned to prevent glass shards entering packages.

Metal: Metal offers the same sealing advantages and resistance to thermal treatments as glass. However, the consumer image of this type of packaging is unfavourable; and metal can be subject to corrosion.

Among the four types of packaging, the most appropriate for organic produce storage is often plastic or glass. In practice, combinations of containers are often used, e.g. glass containers in cardboard cartons. For processing in developing countries, packaging materials often have to be imported from industrialised countries. This implies constraints and supplemental costs (management of stock, financial investments) that can hinder the development and marketing of organic products.

Palletization: Loading and unloading are very important steps in the post-harvest handling of fruits and vegetables. All the subsequent handling operations become very easy once the boxes are placed on the pallets. Necessary care must be taken for safe and easy palletization of organic products.

3.2 Cooling System

It is highly advised to place harvested fruits and vegetables as soon as possible in a storage area that is kept at the appropriate temperature and to retain that temperature thereafter. Research has enabled the optimum storage temperatures to be established for many products. These guidelines need to be verified for each product, given the degree of harvest maturity, the variety, and the growing region of the produce. There is a range of cooling techniques available.

Pre-cooling: Pre-cooling is the process of rapidly removing the heat from commodities. It is a separate operation, prior to storage or transportation that requires special cooling facilities. Several cooling methods available are: room cooling, forced-air cooling, hydro-cooling, packaging icing and vacuum cooling for pre-cooling fresh produce. However, forced-air-cooling method is quite popular. Packaging followed by pre-cooling considerably enhances the shelf life of horticulture produce by reducing the physiological and biochemical changes. The pre-cooling technology is being extensively used in the postharvest handling of grapes. In Alphonso mango, it also reduces the incidence of spongy tissue. Significant reduction in respiration rate slow rate of ripening and good surface colour of mangoes can be achieved by hydro-cooling. The environment-friendly, low-cost evaporative cool chamber developed can be gainfully used for pre-cooling of citrus fruits particularly during winter months.

Air-cooling: Air-cooling requires the supply of cold fresh air (RH 85-90%) to bulk or packaged products. This system requires a moderate investment. Controlling parameters is easy and the system is modular; however, the cooling process is very slow and not homogenous. For regions without access to electricity, very low-cost cool stores can be built which rely on evaporative cooling. Water evaporation from e.g. wet sand between porous brick walls can lower the temperature inside a store to 10-15°C below ambient. Storage in pits or caves is a traditional means of taking advantage of the cool and more constant temperatures below ground. Night air can be used to lower the temperature of a well-insulated cool store, which is then sealed for the daytime.

Forced air-cooling: The simplest design is achieved by building parallel stacks of palletised cartons in a refrigerated cold room. The gap between the two parallel rows of pallets is closed off with a cover. A small fan is placed at one end. The exhaust fan removes air from the enclosed space, so that the pressure falls. Cold air then flows through the ventilation slots in each carton. Advantages of forced-air cooling are its capital cost, its flexibility (cartons, bins, before or after packing) and lack of condensation. Forced-air cooling is more rapid and even than air-cooling but not as rapid as hydro cooling or vacuum cooling.

Hydro-cooling: Hydro-cooling involves immersing the produce in cold water.

The advantages of this method are speed, uniform cooling and no weight loss by dehydration. Disadvantages include the necessity of drying the product surface after cooling and avoiding a build-up or transmission of disease in the hydro-cooling water. There are also problems with the requirement for a large quantity of clean water, disposing of waste water, heavy capital cost and it is not applicable to all types of packaging especially cartons. This technique is used for small fruits or vegetables, leafy vegetables and pineapple. Another system which uses less water involves passing the produce through a cold mist at about 5° C.

Vacuum-cooling: Vacuum-cooling is particularly effective for leafy greens. The products are placed in a chilled vacuum chamber and air is exhausted by a vacuum pump. When the pressure is low enough the water in the produce starts to evaporate and cools the tissue. Pre-cooled fruits and vegetables are then placed immediately into a cool room. The decrease in temperature is very rapid but this technique is capital-intensive.

3.3 Pest Control and Decay

There are not a lot of approved organic post-harvest treatments for pests and diseases. The guidelines usually emphasise the need to minimise pest and disease pressure before harvest. Hot-water (45-55°C) immersion, steam and forced hot-air treatments are sometimes used as organic control methods after harvest. Most pathogenic microorganisms are destroyed within a few minutes of hot-water immersion.

The disadvantages are the risk of product damage:

- re-humidification of the surface increasing the risk of degradation of the fruit or vegetable
- acceleration of ripening
- damage to the colour and firmness of the flesh. It is necessary to maintain a very precise temperature (to within a half degree) and treat for a precise time.

For pathogenic microorganisms, there are antagonistic floras. For example, *Bacillus subtilis* slows the development of several moulds. In order to be effective, it is necessary that the antagonistic flora are stable, can develop under storage conditions, and are tolerated by the consumer. These techniques are not yet in commercial use for stored fruits and vegetables.

Vapour heat treatment (VHT): Vapour heat treatment is the only known eco-friendly treatment available at present for controlling the pests of fresh horticultural produce. In this machine, heated air at 95% relative humidity is circulated through crates of freshly harvested fruits. The time and temperature vary from fruit to-fruit.

Two additional physical processing methods exist and can be effective but they are *not authorized* by the legislation on organic agriculture. These methods are microwaves and ionising radiation. Microwaves are a form of electromagnetic energy, transmitted as waves, which penetrates food and once there is converted to heat. The technique is not suitable for fresh produce because the high water content efficiently absorbs microwave energy. Ionisation treatments consist of exposing the fruit or the vegetable to gamma radiation or X-rays. The doses used and permitted on food never render the products radioactive and ionisation does not leave any residues.

3.4 Labelling

Labelling and advertising organic products must be closely regulated in order to avoid fraud and to protect the consumer. Labelling rules of EU (European Union) is given below:

Table 17.7: European Union Labelling rules

Product Category	Labelling	Additional Indication
Raw material or processed product with at least 95% of the ingredients of organic agricultural origin of the product, a conversion period of at least X.. months before the harvest has been complied with.	Organic farming product in principal denomination.	Use of the organic farming logo after X.. years of organic farming. The labelling must be accompanied by a reference to the ingredients of organic agricultural origin.
Processed products with 70% to 95% of the ingredients from organic agricultural origins.	No mention in the principle denomination. The following indication appear in the list of ingredients: "X.. % of the agricultural ingredients were produced in accordance with the rules of organic production"; -Indication referring to organic production methods.	They appear in the same colour and with an identical size and style of lettering as the other indication, in a separate statement in the same visual field as the sales description
Products in conversion to organic production methods, provided that a conversion period of at least X.. months before harvest has been complied with and that the product contains only one ingredient of an agricultural origin.	"Product under conversion to organic farming"	Such indication must appear in a colour, size, and style of letter in which it is not more prominent than the sales description of the product; the words "organic farming" will not be more prominent than the words "product under conversion to"

Labelling and advertising a product may refer to ***organic production methods*** only where:

- The product was produced or imported by an operator who is subject to the inspection measures in accordance with the rules laid down in organic standard.
- The product or its ingredients have not been subjected to treatment involving the use of ionising radiation.

Labelling must list the following indications:

- The name and/or code number of the inspection authority or body which has carried out the most recent inspection
- A clear statement that the words relate to a method of agricultural production
- The content of non-organic agricultural ingredients
- The name and the content of the non-agricultural ingredients.

Any agricultural product that is **sold, labelled, or represented as organic** shall comprise a maximum of 95 percent of the total weight of the finished product, excluding water and salt, except that not more than 5 percent of the total weight of the finished product may consist of one or more of non-organic allowed ingredients.

Any agricultural product that is sold, labelled or represented as **made with certain organic ingredients** on the main display panel should comprise at least 70 percent, but less than 95 percent, of the total weight of the finished product of non-organic allowed ingredients. The other elements listed depend on the percentage of organic products included in the final product.

3.5 Storage of Fruits and Vegetables

Exposure to high temperature is the biggest factor in deterioration of horticultural produce. Preventing heat injury requires careful temperature management and precise monitoring during storage. The ideal condition for storage of fresh fruits, vegetables and flowers is the lowest temperature which does not cause chilling injury to the product.

Organic product needs to be stored and transported in a way that it is properly identified and physically separated from non-organic products. Although individual products have a range of optimal storage temperatures, in practice most produce can be stored at one of three temperatures. These recommendations should always be verified under local conditions for each variety and harvest maturity. For certain fruits and vegetables (e.g. mangoes, bananas) cooling in stages plus intermittent warming allows the produce to resist chilling injury and spoilage.

The refrigeration of fruits and vegetables improves when carried out with the proper relative humidity for each product. Appropriate storage conditions are always a compromise; high humidity levels limit dehydration and loss of weight, but encourage the development of microorganisms leading to rots.

Table 17.8: Recommended conditions for storage of fruits and vegetables using three temperate zones.

Products not or slightly sensitive to cold Store at 0-2 °C		Products somewhat sensitive to cold Store at 5-8°C		Products very sensitive to cold Store at 13°C	
Apple	Kiwifruit	Custard apple	Mandarin	Avocado	Papaya
Apricot	Litchi	Durian	Melon	Banana	Passion fruit
Berry fruit	Orange	Avocado	Olive	Breadfruit	Pineapple
Cherry	Peach	Kumquat	Pomegranate	Guava	Sapote
Coconut	Pear			Grapefruit	Watermelon
Date	Persimmon			Lemon	
Fig	Plum			Lime	
Grape				Mango	
Artichoke	Celery	Green bean		Cucumber	Sweet pepper
Asparagus	Garlic	Potato		Ginger	Sweet potato
Bean sprouts	Lettuce			Kumara	Taro
Beet	Mushroom			Pumpkin	Tomato
Broccoli	Onion			Squash	Okra
Cabbage	Pea				Yam
Carrot	Radish				
Cassava	Spinach				
Cauliflower	Sweet corn				
	Chestnut				

Source: http://postharvest.ucdavis.edu/Produce/ProduceFacts/index.html

Table 17.9: Ethylene synthesis by fresh produce

0.01-0.1 µl/kg/h at 20°C	0.1-1 µl/kg/h at 20°C	1-10µl/kg/h at 20°C	10-100 µl/kg/h at 20°C
Cherry	Berry fruit	Apricot	Apple
Date	Cucumber	Banana	Avocado
Grape	Green bean	Feijoa	Custard apple
Grapefruit	Guava	Fig	Papaya
Mandarin orange	Kiwifruit		Passion fruit
Most vegetables	Okra	Melon	Pear
Mushroom	Olive	Litchi	Sapote
Orange	Pineapple	Mango	
Pomegranate	Pumpkin	Papaya	
Strawberry	Squash	Peach	
Watermelon	Sweet pepper	Plum	
	Tamarillo	Tomato	

Source: http://postharvest.ucdavis.edu/Produce/ProduceFacts/index.html

When different kinds of produce are mixed in storage, consideration must be given to aroma volatiles which may taint other produce (e.g. durian, onion) and ethylene which may ripen or damage other produce. The risk of mutual contamination is most pronounced between fruits which emit large quantities of ethylene and produce which is sensitive to ethylene like avocados, bananas and papayas.

Table 17.10: Ethylene sensitivity of fresh produce

Low		Medium		High	
Beet	Ginger	Apricot	Okra	Apple	Kiwifruit
Berry fruit	Grape	Asparagus	Olive	Avocado	Lettuce
Cassava	Onion	Celery	Papaya	Banana	Pear
Cherry	Pineapple	Grapefruit	Passion fruit	Broccoli	Sapote
Date	Pomegranate	Guava	Peach	Cabbage	Spinach
Fig	Sweet corn	Litchi	Pea	Carrot	Tomato
Garlic	Sweet Pepper	Mandarin	Plum	Cauliflower	Watermelon
	Sweet potato	Mango	Potato	Cucumber	
	Yam	Mushroom	Pumpkin	Custard apple	
			Squash	Honeydew	
				Melon	

Source: http://postharvest.ucdavis.edu/Produce/ProduceFacts/index.html

The prevalent practice is to use mechanical refrigeration to store horticultural produce. This is energy-intensive and expensive. The low-cost, environment-friendly cool chamber developed by IARI, New Delhi, can go a long way in solving storage problems in general and on-farm storage in particular.

Controlled atmosphere and modified atmosphere (CA/MA) technology: Controlled Atmosphere and Modified Atmosphere Technology has been recognized as one of the latest technologies for enhancing shelf life and maintaining the quality of tropical fruits by maintaining temperature, relative humidity and the combination of gases. CA/MA technology is most appropriate as it reduces respirator rate, ethylene production, compositional changes associated with ripening and incidence physiological disorders.

The beneficial effects of CA/MA treatments are retardation of ripening, senescence and physiological changes. In addition, it helps reduce chilling injury of various commodities. The MA can have a direct or indirect effect on post-harvest pathogens and consequently decay incidence and severity. It can be a useful tool for control of certain insects. However, there are certain limitations such as irregular ripening, development of off-flavours and stimulation of sprouting, if not properly carried out. The design and construction of controlled atmosphere store has required precision control of the system. Thus the controlled atmosphere store has to be relatively gas tight, and fitted with reliable refrigeration system with a means of measuring and controlling the concentration of both carbon dioxide and oxygen.

***Cold/Cool chain*:** One of the important reasons for advancement in the trade of fruits, vegetables and flowers in developed countries is the adoption of cold chain during handling of these commodities. The maintenance of low temperature at different stages of handling by means of a cold chain results in reduction of losses and retention of quality of horticultural produce.

3.6 Transportation system

There are no particular legislative requirements for transporting organic fruit and vegetables but as noted above, air travel is extremely wasteful of fossil fuels. This is less of an issue if the produce is travelling with passengers but is a serious argument against the use of cargo-only planes for fresh fruit and vegetables if there are locally-grown equivalents available in the importing country. Sea transport is many times more economical of fossil fuels (per km) than air- or even road-freight. It is worth noting therefore that produce grown under low-input agriculture in a developing country and transported by sea freight can have a lower 'energy overhead' than produce grown in Europe under intensive mechanised agriculture with high fertilizer inputs.

It is important to look at the infrastructure for freight before establishing a perishable crop as every delay in transit to urban centres, ports and airports reduces the potential shelf life of the product. Ideally refrigerated trucks should be used for produce stored at low temperatures to maintain the 'cool chain' from the grower's property to the marketplace. Harvesting should be timed to coincide with air- or sea-freight opportunities if refrigerated storage is limited.

***Containerisation*:** Containers have been introduced recently in developing countries including India in a big way for carrying fresh horticultural produce for internal distribution. In advanced countries, refrigerated containers as well as Controlled/Modified Atmosphere containers are available for shipment of fruits, vegetables and flowers. The advantages of this technique is that it can be placed on truck or rail.

4. COLD STORAGE AND HANDLING FACILITY IN INDIA

India holds a place of pride as the second largest producer of the horticulture products in the world. For the benefit of the exporters Government of India has taken initiative for providing cold storage and cargo handling facility for perishable products at various International Airports. The current status is appended below:

Name of the Airport	Facility Available
IGI Airport New Delhi	- Three cold stores for fruits vegetables flowers and livestock products. - Separate temperature maintenance in three cold rooms between 2°C and 15°C. - Capacity of storing up to 12 built-up aircraft pallets. - A chamber for pre-cooling of the products - Total storage area of 1127 sq mtrs.

Name of the Airport	Facility Available
Bangalore Airport (Air Cargo Complex of M/s Mysore Sales International Ltd.)	- Capacity to store 8 built-up aircraft pallets - Provision for building up the aircraft pallets on a hydraulic operated lowerable workstation. - Total storage area of 500 sq mtrs
Rajiv Gandhi International Airport Hyderabad	- Capacity to store 6 built-up aircraft pallets - Provision for building up the aircraft pallets on a hydraulic-operated lowerable workstation. - Total storage area of 445 sq mtrs
International Airport, Chennai	- Capacity to store 6 built-up aircraft pallets - Provision for building up the aircraft pallets on a hydraulic-operated lowerable workstation. - Total storage area of 445 sq mtrs
Thiruvanthapuram Airport (Kerala State Industries Enterprises Ltd)	- Capacity to store 18 built-up aircraft pallets - Provision for building up the aircraft pallets on a hydraulic-operated lowerable workstation. - Total storage area of 1000 sq mtrs

4.1 Walk in type Cold Storages

The walk in type cold storage has been set up at various location of India viz. Kolkata and Guwahati Airports and a similar facility at Agartala Airport. These facilities are expected to support the transfer of fresh horticulture produce in the cold conditions from areas in the NE India Region to Kolkata and other International airports. APEDA also has the proposal for setting up the modular cold storage and cargo handling facilities at Ahmedabad, Kolkata and Amritsar International Airports.

4.2 Controlled Atmosphere and Modified Atmosphere Refrigeration Technology in India

The refrigeration technology has already been developed for some of the products like Grapes, Kinnow and Pomegranates, Potato etc. and these products are being exported by refrigerated container in bulk from the country. Litchi, which is identified as a potential exportable fruit from the country is also being exported by sea transportation under refer conditions.

APEDA, launched this trial experiment with the products like mango of major commercial varieties like Alphanso, Banganpalli, Kesar, Totapuri, Suvarnarekha and Chausa for initial trials for preparing an appropriate protocol for transportation of these varieties of mangoes by sea.

5. LEGISLATION APPLYING TO ORGANIC PRODUCTS

The development of organic farming increases the risk of fraud and of unfair competition. In order to protect consumers and producers, many countries have put in place requirements that standardize products at the national and the international level. An increasing number of countries have standards and regulations in place. Details for major importing countries can be found at *http://www.organic-research.com/Laws&Regs/legislation.htm*. These regulations are in addition to the requirements of any national certifying or registering body in the country of production and should be read in conjunction with the Codex Alimentarius guidelines and IFOAM basic standards.

One of the first pieces of organic legislation was the European Union's Council Regulation (EEC) No. 2092/91. It defines the production methods for organic farming: rules for labelling, production, quality control systems, and imports from the developing world.

Chapter **18**

Organic Market Opportunities

Organic agriculture offers trade opportunities for farmers in the developing and developed countries. This market of organic products is expected to grow globally in the coming years and high growth rates over the medium term (from 10-15 to 25-30 per cent) are expected by 2010. This organic market expansion makes it possible for farmers to reap the benefits of a trade with relatively high price premiums.

Organic farming is carried on in almost all countries of the world. The share of organic farms in overall farmland and in terms of numbers of holdings is growing continuously. The market for organic products is also growing very rapidly, not only in Europe, North America and Japan, where the largest markets are to be found, but also in many emerging economies and economies in transition. In India, an organic movement is now emerging on different levels i.e. producer groups, trainers and advisors, certification bodies and processors and traders. The general marketing issues are already discussed in Chapter 1: "*Organic Farming: Opportunity and Challenges*". Some specific issues of marketing of organic produce is discussed in this chapter.

1. IMPORTANT ORGANIC FOOD PRODUCTS

Organic food products encompass an enormous diversity of product groups. The main product desired by European, American and Asian countries are- tea, coffee and cocoa; grains, pulses and seeds; vegetable oils and fats; edible nuts; spices and herbs; dried fruits; fruit juices and concentrates; sugars and honey.

Tea: Tea is traded as black tea, green tea, Oolong tea and instant teas.

- *Black tea*: Fully fermented tea.
- *Green tea*: Certain enzymes in the fresh leaves are inactivated through heat treatment. Only then is the product rolled and dried. Fermentation is suppressed by deactivating these enzymes and the leaves retain their green colour.
- *Oolong tea*: The fermentation process is halted at an earlier stage (partly fermented tea).
- *Instant teas*: Instant teas are made either from low-quality teas (fermented and dried), or from non-dried tea in a special process directly after fermentation. Instant teas lose much of their aroma during the extraction and sub-

sequent freeze-drying processes.

Coffee: Economically the most important coffee varieties are *Coffea arabica* (Arabica) and *Coffea canephora* (Robusta). In comparison with Arabica, 30 percent higher yields are gained from Robusta, but the price is around 30 percent lower.

Table 18.1: Economics of arabica and robusta coffee

	Arabica	Robusta
Share of world Production	70 %	30 %
Site requirements	High sites: fluctuations in annual rainfall and temperature	Low sites: steady high temperatures and rainfall
Main growing areas	Latin America, East Africa	Asia, Africa
Caffeine content	0.6-1.5 %	2.0-2.7 %
Diseases/pests	Susceptible to the berry borer and coffee rust	Resistant against the berry borer and coffee rust

Source: FiBL, 2002

Raw coffee is made by processing the ripe red coffee cherries of the bush-like coffee tree. Blending and roasting the raw coffee is mostly carried out in the importing countries. Processing includes dry or wet processing, dehulling, grading, cleaning, sorting and filling.

Cocoa: With its strong, fine flavour the Criollo group produces the highest cocoa quality. However, its yield is low and it is therefore rarely cultivated. The Trinitario is a hybrid of Forastero and Criollo types. Its share of world production is roughly 10 to 15 percent.

Grains: USA and Canada dominate the market for organic commodity grain products (e.g. wheat, maize and barley) and opportunities for exporters in developing countries to export these grains are limited.

Pulses: It includes kidney beans, chickpeas and broad and horse beans. Although the trade is small, they are important for exporters from developing countries.

Seeds: It includes polyunsaturated oils, sunflower seed, oil and meal.

Vegetable oils and fats: Soya oil, palm oil and coconut oil which are important for exporters in developing countries.

Edible nuts: There are two segments for edible nuts, groundnuts (peanuts) and luxury (tree) nuts. The most important for the European trade are almonds, coconuts, cashew nuts, walnuts and Brazil nuts.

Spices and herbs: The main international trade for spices and herbs is dried and in crude form, cleaned but not further processed. It is estimated that about 85 percent of the trade is in this form. The remainder is for crushed or ground spices, essential oils or oleoresins.

Fruit: Apples, apricots, bananas, dates, figs, mangoes, papayas, peaches, pears and prunes are important tree fruits. Next to vine and tree fruit, there are others such as pineapple. Dried fruits are also in demand because fresh fruit consisting of more than 80 percent of water are prone to microorganism attack. Drying of fruit is mainly done in order to stop the multiplication of microorganisms. Dried fruit can be divided into tree fruit and other fruit.

Fruit juices and concentrates: Fruit juice is orange juice. Apple, pineapple and grapefruit are other fruit species which are the basis for popular fruit juices.

Sugars and Molasses: The sugar that results from sugar cane and sugar beet are important as it contains the same sucrose that is found naturally in the original plants and in fruits and vegetables. A by-product of the cane and beet sugar refining process is molasses, which has a multitude of uses. Molasses is important as a raw material for the production of antibiotics, bakers yeast, rum and alcohol, as well as an animal feed supplement.

Honey: Five common types of honey have export market. They are - Acacia (a honey with a light taste and refined scent. It tends not to crystallize); Orange blossom (a honey with a refreshing bittersweet flavour); Buckwheat (a honey with a strong smell and a taste similar to brown sugar); Lotus (a honey with a mellow, sweet flavour and a faint smell of flowers); Clover (the most widely produced and well-known type of honey)

2. THE INDIAN ORGANIC MARKET

2.1 Domestic Market of Organic Products

In India, organic consumers are generally found in the urban upper-middle class or upper class, and to some extent lower-middle class families in smaller towns,

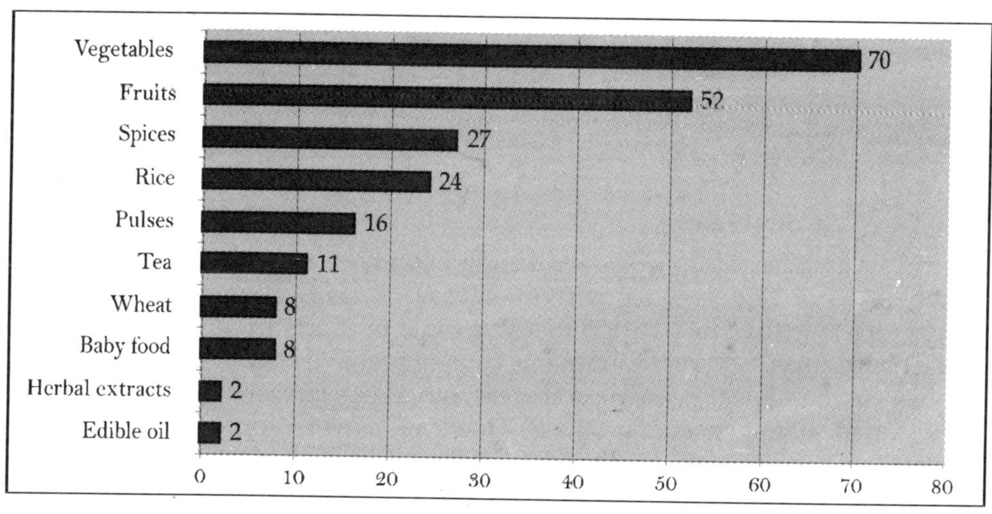

Figure 18.1. Organic Products desired by Organic Food Consumers

especially families with children. Organic vegetables and fruits are the major organic products desired by Indian customers. A list of organic products desired by organic food consumers of India is presented in Figure 18.1:

The major organic food products available in the Indian market are rice, tea, fruits and vegetables, wheat, spices, coffee and pulses. The Table 18.2 shows the contribution of different organic food commodities in India.

Table 18.2: Percent contribution to total production

Crop	Per cent
Rice	24 %
Tea	24 %
Fruits and vegetables	17 %
Wheat	10 %
Cotton	8 %
Spices	5 %
Coffee	4 %
Pulses	3 %
Cashew nuts	3 %
Others	2 %

Source: FIBL (2002)

Major domestic markets are cities like Mumbai, Bangalore, Delhi, Chennai and Hyderabad to name a few.

Distribution of organic products takes place as follows:

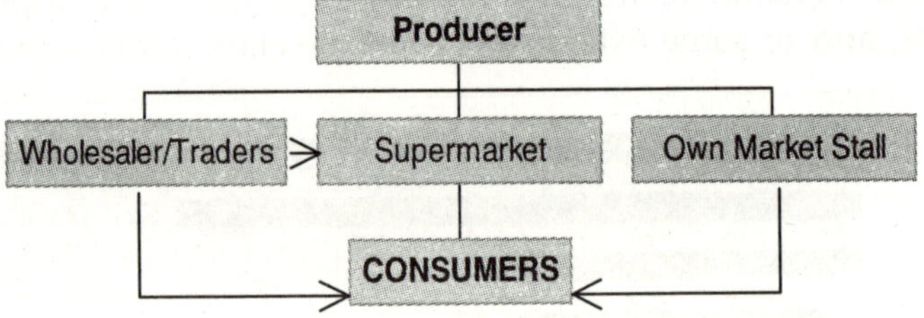

2.1.1 Purchase Strategies

Recently, a number of domestic marketing initiatives have been launched, some of which have registered positive success. A semi-government-operated cooperation initiative has started to sell organic products in a number of outlets in the major cities. Several brand-name companies have begun to include organic products in their lines. Even in rural areas, numerous farmers' groups and NGOs have started selling organic products. However, as yet very little information is available on the size and success of the domestic organic market.

For the organic market, different options are available for the purchase and storage of organic products:

- *Direct purchase*: Purchase from the farmers by the market operators and sale to customers. Purchases could be made at a premium price compared with the prices offered by intermediaries and traders. The market operator is required to maintain the stocks. The farmer representatives are present during the markets and incentive return is additionally provided to the farmers based on sales of their products.
- *Non-purchase strategy*: The market operator facilitates the supply of organic products through the participation of farmers in the market, ensures the organic integrity of products and provides the required extension support and market development activities. The operator receives service charges to cover his costs.
- *Combination of purchase and non-purchase system*: The market operator can combine the direct purchase and non-purchase systems in order to maximise advantages and tide over constraints. In this method, the operator could decide for direct purchase for a limited range of products depending upon the advantages and availability of funds.

2.2 Export Market of Organic Products

Indian organic producers and exporters are well aware of the demand for organic products in developed countries. India's exports of organic food item handled by Agricultural and Processed Food Products Export Development Authority (APEDA) have increased from an estimated Rs.26 million (1999) to an estimated level of Rs.71.23 million (2003).

Products available for the export market are rice, wheat, tea, spices, coffee, pulses, fruits and vegetables, cashew nuts, cotton, oil seeds and medicinal herbs. The channels adopted for the export of organic products are mainly through export companies. The major export markets are Australia, Belgium Canada, France, Germany, Italy, Japan, Netherlands, Sweden, Singapore, South Africa, Saudi Arabia, UAE, the United Kingdom and USA. Export sales of selected organic products are given in Table 18.3:

Table 18.3: Export sales product-wise (2002)

Product	Sales (Tonnees)
Tea	3000
Coffee	550
Spices	700
Rice	2500
Wheat	1150
Pulses	300
Oil Seeds	100
Fruits & Vegetables	1800
Cashew nut	375
Cotton	1200
Herbal Products	250
Total	11,925

Source: Org-Marg, 2002 (Field survey, and secondary sources: APEDA, Tea Board, Coffee Board and Spices Board of India).

3. THE INTERNATIONAL MARKET

One of the factors that promote growth in organic markets worldwide is consumer awareness of health, environment issues and food scandals. Other factors that influence further development of the organic market are the increasing promotions and marketing strategies used by key players, such as retailers. Developing countries are expanding their organic market into developed countries and in parallel are building a domestic market. An overview of world food markets for organic food and beverages (retail sales in 2003) is given below:

Country	Retail sales
Germany	2800-3100 million US$
Europe	10000-11000 million US$
US	11000-13000 million US$
Canada	850-1000 million US$
Japan	350-450 million US$
Worldwide	23000-25000 million US$

3.1 The Japanese Organic Market

The largest Asian market for organic products is found in Japan. The Japanese organic market has been characterized as a market with high demand for organic products and strong purchasing power, and with low domestic supply of organic products. The annual growth rate of the organic market is about 20 per cent.

The major exporters of organic products are Australia, New Zealand, the USA and Canada. The most commonly imported organic products are soybean, organic frozen vegetables, tea and bananas. According to the Japanese Integrated Market Institute, imports of organic products are likely to grow by 40 per cent.

3.2 The US Organic Market

The US is the world's largest market for organic products. Developing countries have sales opportunities in the following product categories:

- Tropical products (coffee, cocoa, tea, tropical fruits and vegetables)
- Off-season products (fruit and vegetables)
- In-season products (fruits and vegetables in temporary or permanent supply)
- Specialty and other products (wines, ethnic foods, herbs, spices, essential oils, sugar, feed grains, and seed grains)

To be certain that organic products will be accepted in the US organic market, the producer or exporter should choose a certification body that provides access to the US market, for example, a certification body accredited by the National Organic Program (NOP) of US; alternatively it is possible to have organic products re-certified by an accredited certification agency.

3.3 The European Organic Market

The European market for organic food is undergoing strong expansion. However, its development varies from country to country. Germany has the largest market volume within the EU, but the market in the United Kingdom is also growing strongly. Switzerland and Denmark have the highest consumption per capita, whereas France, Holland and Italy have low consumption per capita. Some domestic organic products already have a 15 % share of the total market in Europe. In Germany and Switzerland one of the biggest challenges is to increase the diversity of the organic product range, and expand sales channels.

3.4 Demand of Organic Produce by International Market

Of the products mentioned in the chart below, bananas are sold in high quantities followed by wheat and soybean. Tropical organic fruits such as pineapple and mango occupy third place. The rest of the organic products mentioned are sold in quantities ranging from 10 to 260 tonnes per year. The suppliers of these products are from neighbouring countries like Sri Lanka, China, Thailand, or South America, Africa and Europe.

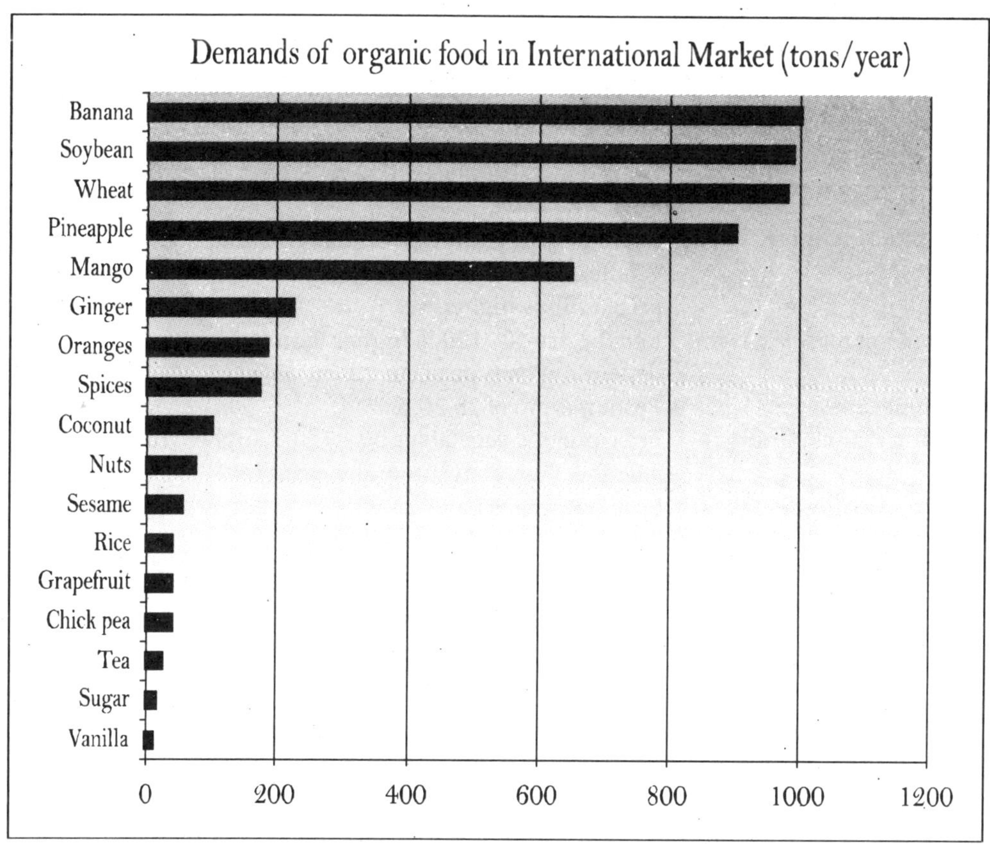

3.5 General Price Policies and Cost Structures

In general, an organic marketing initiative is expected to build in or carry more costs in contrast to conventional marketing. This is largely because of the 'additional' environment management and social responsibility it wants to carry. Organic marketing initiatives therefore have a distinct cost disadvantage against conventional businesses, which do not internalise equivalent environmental and social costs, in the market for similar products categories. Price premiums (2002) in the export markets are given below:

Table 18.4: Price premiums obtained in the export markets

Organic Product	Price premium %
Tea	47
Rice	44
Protein grains	44
Spices	30
Vanilla	53
Mango	25
Pineapple	28
Bananas	31
Nuts	40

Source: FiBL, 2002

Price Premium indicated by some exporters in the Asian Market is given below:

Exporter	Price Premium
Agri-Bio, Singapore	Retail price organic 3.3 times higher than conventional. Wholesale price is 40 % higher than cost price and retail price is 85 % higher.
Hanoi Organics, Vietnam	Retail price 40 - 120 % higher than farm-gate price, depending on type of customer.
Keystone, India	Profit margin of 25-30 %.
Nanjing Planck, China	Price organic vegetables 50 – 150 % higher (100 % on average) than conventional vegetables.
Organic Health, Malaysia	Profit margin is between 20 - 50 %, but most items fall between 30 - 35 %
Thai Organic Food, Thailand	Profit margin is about 30 % to the cost of production, handling, and packing.

Australia, Argentina, Italy, the US and Brazil are respectively the five countries with the most cultivated area that is certified organic. Although, Asia has a very active organic movement (around a quarter of IFOAM's 703 members are from Asia), the area under organic cultivation remains relatively small. Among the more significant countries producing organic products are China, India, Indonesia, Sri Lanka as well as Israel. This is comparable to Canada and France, but smaller than Germany, Italy or the USA.

Table 18.5. Worldwide Land area under organic management

Country	Organic land (hectare)
Australia	11,300,000
Argentina	2,800,000
Italy	1,052,002
USA	930,810
Brazil	803,180
Uruguay	760,000
Germany	734,027
Spain	725,254
UK	695,619
Chile	646,150
India	76,326
Total	26,458,270

Source: SOEL Survey, 2005

Table 18.6: Number of Organic Farms in Asia (*SOEL Survey, 2003*)

Country	Organic farms (no.)	Cultivation area (hectare)	% of total agril. area
China	2910	301,295	0.06
India	5661	41,000	0.03
Indonesia	45,000	40,000	0.09
Israel	N.A.	7,000	1.25
Japan	N.A.	5,083	0.09
Kazakhstan	20	N.A.	N.A.
Rep. of Korea	1,237	902	0.04
Laos	N.A.	150	0.01
Lebanon	17	250	0.07
Malaysia	27	131	0.002
Nepal	26	45	0.001
Pakistan	405	2,009	0.08
Philippines	500	2,000	0.02
Sri Lanka	3,301	15,215	0.65
Syria	1	74	0.001
Thailand	940	3,429	0.02
Vietnam	38	2	N.A.
SUM ASIA	60,083	418,585	N.A.

The total organic area in Asia is now just over 418,585 hectare (of which 75 % is in China). Land area in India under organic management is only 0.05 per cent of total agricultural area (SOEL-Survey, February 2005).

In most Asian countries, the local markets are just emerging. In almost all Asian countries, local distribution is a huge problem and often a price premium cannot be achieved. A number of countries, such as Israel, Japan, Malaysia and the Philippines have specialised organic stores or markets.

The Indian organic products, quantity and their domestic and export market are briefly indicated below in Table 18.7.

Table 18.7: Indian organic products and markets

Product	Quantity (tonnes/yr)	Season	Markets
Tea	3500	Throughout the year	Domestic: Mumbai, Bangalore, Delhi, Hyderabad Export: Australia, Germany, Japan, Netherlands, UK, USA.
Coffee	600	Monsoon (June to September)	Domestic: Bangalore, Chennai, Hyderabad, Mumbai and Delhi Export: Australia, Germany, Japan, Netherlands, Sweden, UK, USA.
Spices	700	Throughout the year	Domestic: Bangalore, Chennai, Hyderabad, Mumbai, Delhi Export: France, Germany, Japan, Netherlands, South Africa, Singapore, UAE, USA
Rice	3500	Kharif: April to September Rabi: November to February	Domestic: Bangalore, Chennai, Hyderabad, Mumbai, Delhi Export: Japan, Singapore, UAE, USA, Canada, Germany.
Wheat	1400	Kharif: April to September Rabi: November to February	Domestic: Mumbai, Bangalore, Delhi, Hyderabad Export: Australia, Germany, Japan, Netherlands, UK, USA.
Pulses	400	March to May	Domestic: Mumbai, Chennai, Bangalore, Delhi, Hyderabad Export: Japan, Singapore, Germany, UAE, Saudi Arabia.
Oilseeds	100	Kharif April to September	Domestic: Negligible Export: European Countries

Product	Quantity (tonnes/yr)	Season	Markets
Fruits & Vegetables	2500	Throughout the year	Domestic: Mumbai, Chennai, Bangalore, Delhi, Hyderabad Export: UK, USA, Australia, France, Germany, Italy, Switzerland, Netherlands
Cashew nut	375	March to June	Domestic: Nil Export: European Countries
Others: Cotton, medicinal herbs & extracts, aloe vera, sapota	Cotton 1200 Herbs 250		Domestic: Mumbai, Bangalore, Delhi, & Hyderabad Export: Australia, Belgium, Germany, Switzerland, Italy, Japan, Netherlands, UK, USA

Source: Org-Marg, 2002 (Source: Fieldwork, various publications from-APEDA, Tea Board, Spice Board, Coffee Board)

4.0 GLOBAL ORGANIC TRADE

Trade with organic products are showing growth rates over 20 % which are rarely found in food markets. Strong opportunities for developing countries in most major markets for organic products that are not produced in Europe or North America, such as coffee, tea, cocoa, spices, tropical fruits, vegetables and citrus fruits. Organic products from developing countries are sold at impressive premiums, often at prices 20 % higher than identical products produced on conventional farms. Although the current organic market is small, this is a big task. It is expected that once national standards are more widely established, bi-lateral equivalency agreements will lead to increased trade opportunities.

Chapter **19**

Organic Farm Record - Keeping System

Documentation of how and where a crop was raised, what products were applied and when, which bin or cooler it was stored in, is of critical importance to establishing the integrity of the product. If the farmer cannot provide reasonable documentation that his or her crop was organically grown, that it has not been contaminated with chemicals, and that it has not commingled with a similar conventional product, then certification may well be denied.

Organic certification is about verifying that you are managing an organic system to grow crops, raise livestock, and/or process food and fibre according to the National Organic Standards. An important part of being prepared is being able to track your product from the field or point of purchase to the consumer, ensuring that the product has been kept separate from non-organic products and has not been contaminated in any way by materials prohibited for use in organic production.

All organic farm operators must develop an organic system plan for certification, which covers all aspects of agricultural production and handling/processing. The certification of the organic products can be considered in conjunction with on-farm records. The National Organic Standards specify that records must *"fully maintain all activities and transactions in sufficient detail as to be readily understood and audited"*.

The organic record-keeping system must accomplish two objectives:

1) trace products as certified organic from the raw ingredients to final sale (for verification of sources and/or sample recall from final destination); and
2) verify the input-output balance of organic ingredients and organic products, including current inventory.

Be prepared to supply samples of paperwork during the inspection to track ingredients to finished products for any item and for any time that may be randomly selected for an input/output audit.

During inspection of the farm or farm produce, the representative of the Certifying body must be given full access to and insight into all relevant information concerning the agricultural operations. Any documents requested concerning the management of the farm have to be shown, and all relevant questions have to be answered.

1. RECORD-KEEPING SYSTEM

1. Field maps
2. Field activity log(s)
3. Field history sheets (previous three years).
4. Documentation of previous land use for rented and/or newly-purchased land.
5. Input records for soil amendments, seeds, manure, foliar sprays and pests control products (keep all labels).
6. Documentation of attempts to source organic seeds and/or planting stock/ organic seedlings.
7. Residue analysis of inputs (i.e. manure sourced off-farm).
8. Compost production records.
9. Monitoring records (soil tests, tissue tests, water tests, quality tests, observational).
10. Equipment cleaning records.
11. Harvest records that show field numbers, date of harvest and harvest amounts.
12. Labour records.
13. Storage records (storage location, storage identification, filed numbers, amount stored and cleaning activities).
14. Clean transport record.
15. Sales records (purchase order, contract, invoice, cash receipt, sales journals, etc.)
16. Shipping records (scale ticket, dump station ticket, bills).

1.1 CROP PRODUCTION DOCUMENTATION

The following records should be maintained for certification of organic food crops:

- **List of crops** being grown, field locations (maps), acreages, and estimated yields.
- **Field history** or land use documentation, if any new land is added this year.
- **Field activity logs** for all practices performed (cultivation, weed control, use of manure or fertilizers, spraying, pruning, beneficial released, etc.).
- **Input purchase/source records** of all inputs used for crop nutrients, pest, disease, or weed control like-
  - Invoices / Receipts
  - Delivery tags

- Receipts or logs recording the pick-up or delivery of free materials
- Labels and/or documentation demonstrating that each material is allowed for use in organic production.
- A generic material (e.g., mined limestone) must be on the National List as allowed.
- A brand name product must either
  - have a label that discloses all ingredients, including inert ingredients, so that they all may be verified as allowed; or
  - be listed as an allowed brand-name material on a list approved by the Certifier.

- Note that manure must either be composted according to National Organic Standards or its date of incorporation documented to comply with the required number of days before harvest of a crop intended for human consumption.

1.1.1 Input application records (material, source/brand name/ manufacturer, regulatory status, field location, date, and rate or quantity used). They are-

- Seeds (crop and cover crop), planting stock, annual seedlings, and transplants
- Seed coatings and inoculants
- Greenhouse materials (e.g., potting soils or soil mix ingredients)
- Crop nutrients and soil amendments
- Pest management materials
- Beneficial insect releases
- Natural, organic, or plastic mulches
- Any other materials applied

*1.1.1.1 **Seed, planting stock, and transplant records***

- Documentation that seeds and annual transplants are certified organic. If any non-organic seed or planting stock used, documentation of:
  - your unsuccessful search for commercially available organic seed or planting stock (most certifiers require documentation of non-availability), and
  - verification that the seed or stock used is not genetically modified or treated with prohibited materials.
- Documentation of compliance of any inoculants or seed coatings (non-GMO status of inoculants organisms and allowed status of all seed coating materials).

1.1.2 Audit Trail Documents that track products from the field of origin to final use or sale. A random audit is part of inspection procedures. It may require the following.

- Field, planting and production records
- Harvest and yield records
- Post-harvest handling records
- Storage records
- Transport records
- Sales records

1.1.3 **Soil Management Activities,** including crop rotation and erosion prevention activities.

1.1.4 **Pest Management Activities** for control of crop pests (insects/mites/invertebrates/vertebrates), diseases, and weeds, including:

- Preventative practices
- Materials used, if any
- Pesticide use reports, as required by law, if applicable

1.1.5 **Certification Documentation** of any organic product purchased for resale.

1.1.6 **Labels and Labelling**

- Printed packaging, bags, boxes, ties, bands, and stickers
- Lot numbering of retail and bulk products, if applicable

1.1.7 **Organic Integrity Documentation** of measures to avoid contamination and commingling, as applicable to your operation. The information to be recorded are-

- Information about neighbouring land use
- Prevention of contamination from borders
- Production, harvest, and sales records for buffer crops, transitional or conventional crops
- Material storage: adequate separation of allowed materials from any non-allowed products
- Irrigation water and system for contamination prevention (i.e., diagram of valves, backflow prevention/documentation of purge or flushing procedures to prevent contamination from shared water systems where fertilizers or other prohibited materials are used)
- Clean-out or purge logs for equipment used for both organic and conventional operations
- Documentation of procedures to verify the absence of sanitizer residues, if sanitizers are used

1.2 LIVESTOCK PRODUCTION DOCUMENTATION

- **Animal lists,** including livestock or poultry descriptions and/or numbers and identification methods

- **Source of poultry and/or livestock**, including breeding, birth, hatching, and/or purchase records
- **Feed harvest and storage records**
- **Feed rations** for each type of animal during each stage of growth and development
- **Feed and feed supplement purchase records** and documentation that they are certified organic or allowed
- **Drinking water**, including source, additives, potential sources of contamination, and results of any water analysis
- **Audit trail** documents that track animals or animal products (harvest or slaughter, processing/post-harvest handling, transport, and sales records)
- **Housing and living conditions**, including grazing management and outdoor access records
- **Animal medications**, including a list of all products used or that may be used (everything in your medicine cabinet or refrigerator, with product names, ingredients, manufacturers, and regulatory status)
- **Health management records**, including vaccinations and all other materials, veterinarian bills, purchase invoices, records of medication used, reason for use, and animal identification
- **Marking and segregating methods** for animals treated with prohibited materials
- **Soil management**, erosion control, crop nutrition, and pasture management
- **Manure management** (must not contribute to contamination of crops, soil, or water)
- **Pest management**, including parasite management
- **Off-site processing records**, including slaughter, cold storage, and meat packing (these activities must take place at facilities that are already certified organic, or they must be inspected as part of your operation)
- **Product or animal sales records**
- **Labels**, if applicable

1.3 HANDLING DOCUMENTATION

- **Product Identification and Composition** for all organic products produced (this must include current formulations, recipes, or batch sheets that support the percentage of organic ingredients in your product label claim—"100 % Organic," "Organic," or "Made with organic....").
- **Facility Map(s)** showing the facility perimeter and buildings, all equipment, and areas used for receiving, raw material storage, processing, packaging, finished product storage, and shipping.

- **Production Flow Chart(s)** that includes equipment used in each step or stage of the process and shows the flow of products through the facility from receiving of raw ingredients to shipping of the final product.

1.3.1 Sources of Ingredients and Processing aids

- *Organic ingredients and processing aids:* You must have on file a copy of the organic certificate from the supplier of any organic ingredient or processing aid, showing that it is certified to National Organic Standards, along with the level of certification that supports the label claim you intend to make For example, if your label makes the claim of 100 % organic, all ingredients and processing aids must be documented to be certified as 100 % organic.
- *Non-organic agricultural ingredients and processing aids:* You must provide documentation affirming that each specific ingredient a) is not commercially available as organic, b) does not contain prohibited inputs and has not been produced using prohibited methods (genetic engineering), c) has not been treated with ionising radiation, and d) is not produced from a crop grown using sewage sludge.
- *Non-agricultural ingredients*: All non-agricultural ingredients must be listed on and consistent with the annotations of the National List of organic standards.

1.3.2 Pest Management

Documentation for preventative practices, procedures, maps, logs, service reports, and incident records must be provided. Whether your pest management is done in-house or by a contracted pest control company, you must document what materials are used, if any, including maintaining product labels on file. If prohibited materials (substances not on the National List) are used inside your facility, be prepared to show documentation of how organic products and materials are protected from contamination during pest control applications.

1.3.3 Sanitation

You will need documentation of standard operating procedures, equipment cleaning, equipment purge logs, and residue testing. Residue test procedures must be appropriate for the sanitation materials used. For example, if chlorine is used as a sanitizer, a chlorine test strip with sensitivity in the low (0-10 ppm) range must be used to show that the level of chlorine remaining is below 4 ppm. Materials that are not listed, but if they are used, they must be completely removed before running organic products. For example, if acid or alkaline sanitizers are used, a pH test with a neutral result (or one that matches the plain water used in the facility) indicates that the sanitizer material has been washed off. Quaternary ammonia is not listed and not allowed, and therefore must be completely removed, such that there are no detectable residues, and residues do not contaminate organic products. Records must be maintained for each area or production line where organic processing occurs, showing how organic products and packaging materials are protected from contamination by conventional product residues and/or sanitation chemicals on food contact surfaces.

1.3.4 Water

You will need documentation of source, use, additives, and any applicable tests results.

1.3.5 Culinary Steam

Provide a list of all boiler additives, all other additives, results from any carryover tests, and explain how the organic product is protected from boiler additive contamination.

1.3.6 Organic Integrity (Organic critical control points)

You will need documentation of systems and procedures to prevent commingling and/or contamination of organic ingredients and products throughout all steps of processing.

1.4 AUDIT TRAIL/AUDIT CONTROL DOCUMENTS

The audit documents for purchase, receiving, storage, production, packaging, handling, transport, and sales may include, but are not limited to, invoices, weight slips, purchase orders for incoming materials, invoices for finished product, descriptions of product tracking or coding, logs for receiving, processing, storage and inventory systems, transport cleaning documentation for incoming and/or outbound materials, and product labels.

The input/output balance audit documents may include, but are not limited to, inventory, purchase, production, and storage records including typical conversion figures for shrinkage, reconditioning, donated products, samples, dumping, shipping, and sales records.

1.4.1 Labels and Labelling

You will need finished product labels (retail and wholesale labels on printed packaging, boxes, etc.), with the proper placement of the phrase identifying the certifier, relative size of certifier logos, lot number, and market destination, as applicable.

1.4.2 Off-site Storage/Contracted Facilities

If your operation uses off-site, contracted warehousing or outside contractors for handling of ingredients or finished products, you will need to provide information about how the off-site facility is used. Depending on what they do, such facilities may need to be certified to operate under the certificate of the entity for whom they provide custom services.

Note: *Sample forms for farm documentation is given in Appendix-13*

2. FARM PLAN INFORMATION

Farm Plan Information is required by the Certifying Agency at the time of submission of application by the farmers/farm operators for Organic certification. The following information need to be provided on the Application form provided by some Certification Body.

2.1 Farm Products and Inventory List: Crop/livestock Product, location. Total areas/head, Projected yield/acre, storage location, Quantity in inventory, year produced, transitional/ conventional, rented/owned.

2.2 Seeds and Seed Treatments

(Seed/Variety/Brand; organic/Untreated/Treated/GMO; Type/Brand of treatment - Fungicide/inoculant; what attempts do you make it use organic/ untreated seeds).

2.3 Source of seedlings and Perennial Stock

Name of suppliers/If certified, by which agents? / On farm seedlings (type and size of green house), potted/ground plants; seedling contains a equipment used, equipment for watering system;

Plant protection methods used to prevent disease/ insects - pests, soil type/ Source of nutrients/ Soil fertility management Programme (Soil testing/ microbiological testing/ tissue testing/ observation of soils/ observation of crops/ health / Comparison of crop yields/ crop quality testing)

Fertility monitoring- weekly/ monthly/ annually/ as needed; rate of effectiveness of your fertility management programme -excellent/satisfactory/needs improvement.

2.4 Crop Fertility Plan

Crop rotation/Green manure ploughing down/Cover crops/ Interplanting/ Incorporation of crop residues/ Sub soiling/ Summer fallow/ Compost/On-farm manure/Off- farm manure/Soil amendment/ Side-dressing /foliar fertilizers/ Bio-dynamic preparations/Soil inoculants other (specify).

Following information are also required to be given:

- Crop residues burning/sewage sludge application
- C:N ratio, temperature maintained
- Forms of nutrient used- liquid/semi-solid/piled/fully decomposed /pelleted
- Types of crops being grown.
- Source of manure- on- farm/off- farm

2.5 Weed Management Plan

Crop rotation/Field preparation/ Prevention of weed seed set/ delayed seedling/

Monitoring soil temperature/ Soil sterilization/ Use of fast emerging varieties/ Mechanical cultivation/ Used of hand tools/ Hand weeding/ Mowing/ Livestock grazing/ Flame weeding/ Electrical/ Smother crops/ Black fallow/ Non- synthetic mulch/ Synthetic mulch/Corn gluter/Soap-based herbicides/ Other (specify plastic or synthetic mulch).

2.6 Pest Management Plan

Problems: insects (list)/rodents/gophers/birds/other animals.

Strategies to control Pest damage: Crop rotation/Selection of plant species/ Development of habitat for natural enemies/ Timing of planting/ Companion planting/ Frog ponds/ Bat hoses/ Bird houses/ Hand-picking / Monitoring/ Trap crops/ Physical barriers/ Physical removal/ Traps/ Lures/ IPM/ Insect repellents/ animal repellents / release of predators/ parasites of pest species/ use of approved products/ use of restricted products/ Limited use of prohibited products/ other.

Monitoring for the effectiveness of pest management programme

- insect monitoring with traps
- observation of crop health
- comparison of crop yields
- crop quality testing
- monitoring records kept
- other, if any

2.7 Disease Management Plan

Problem: List

Disease Prevention strategies: Crop rotation/Field sanitation/ Selection of plant species or varieties/ Timing of planting- cultivation/ Plant spacing/Vector management/ Soil balancing/ Solorization/ Companion planting/ Compost use/ Use of approved materials/ Use of restricted materials/ limited use of prohibited materials/ other (specify).

Monitoring the effectiveness of disease management: Tissue testing/ Observation of soil/ Observation of Crop health/ Comparison of crop yields/ Crop Quality testing/ Monitoring records kept/ Soil testing/ Microbiological testing.

Disease Monitoring: weekly/ monthly/ annually/ as needed.

2.8 Conservation Practices

Terraces/ Contour farming/ Strip cropping/ Under sowing/ Transplanting/ Winter cover crops/ Conservation tillage/ Permanent water-ways/ Windbreaks/ Firebreaks/ Tree lines/ Retention ponds/ Riparian management/ Maintain Wildlife habitat/ Other.

- Soil erosion problem experienced
- Effectiveness of soil conservation Programme
- Soil conservation monitoring - weekly/ monthly/ annually/ as needed.

2.9 Water Management

Water Use: None/ Irrigation/ Livestock foliar sprays/ Washing crops/ Greenhouse/ other

Source of Water: On-site wells/River/Creck/Pond/Spring/ Municipal/ Irrigation channels.

Type of Irrigation: None/Drip/Food/Central pivot/Other (Shared/ independent.

Cleaning process of irrigation lines/nozzles: It is the system flushed and documented between conventional and organic use.

Practice to protect water quality: Fencing livestock from waterways/ Scheduled use of water to conserve its use/ tensiometer monitoring/ laser levelling-land forming/ drip irrigation/ micro-spray/ other.

List of known contaminants in water supply

Effects to minimize water contamination problems

Water Quality monitoring – weekly/monthly/ annually/ as needed.

2.10 Maintenance of Organic Integrity

- Adjoining Land use
- Type of Buffer (Crop land/Tree line/ Hedgerow/ Wildlife Planting/ Grass strip).
- Width of Buffer
- Use of crop harvested from Buffer-sale/ non-organic livestock feed/Seed etc.
- Safeguards to prevent accidental contamination written notification to - highway dept/ Electric dept/ adjoining landowners/ Agriculture dept. Irrigation dept.
- Monitoring of crop contamination
  - Visual observations/Residue analysis/GMO testing/Photographs/ Wind directions/ Speeds data /other.

  Information about the
  - Conventional/ transitional crops being grown.
  - Prohibited soil amendments used on conventional crops.
  - Prohibited herbicides/ pesticides used on conventional crops.
  - List of equipment used for- planting/tillage/cultivation/ spraying/har-

vesting.

- Leakage of fuel/ oil/ hydraulic fluid.
- Harvesting: mechanical/ by hand
- Containers used: gravity wagon/ boxes/ truck boxes/ wooden totes plastic containers cardboard/ waxed boxes.

2.11 Post-Harvest Handling

- Type of packaging material used: bulk/ paper/ cardboard/ wood/ glass/ metal/ foil/ plastic/ waxed paper/ aseptic/ mesh bags/ plastic crates.
- Crop Storage:
- Stored crop inputs to be used: synthetic fumigants/ rodenticides/ sprouting/ inhibitors/ ripeners/ growth regulators/ preservatives/ oils/ colouring agents/ waxes/ other.
- Transportation: Self/buyer/ other.

Steps to be taken to protect the integrity of organic products

- Dedicating organic only
- Inspecting transport units prior to loading
- Cleaning transport units prior to loading
- Use of clean truck Affidavits
- Letter/contracts with transport company starting organic requirements.
- Others (Specify).

3. SOIL TESTING REQUIREMENTS

Organic production system plans must include monitoring practices that measure soil quality. Soil testing provides one method for measuring the quality of the soil and is one way to meet the organic requirements, unless an alternative method can adequately demonstrate that soil quality is being monitored.

Soil test results may be submitted with the organic crop production application, with follow-up testing recommended at least every three years unless your soil test results or conservation practices dictate otherwise. If they are not available, please provide documentation of alternative method to demonstrate your monitoring of soil quality as soon as possible.

The standard soil test, will include an analysis of - Per cent Organic Matter, Cation Exchange Capacity, Soil pH, Neutralizable Acidity (for acid soil), Ammonium Nitrogen test, Phosphorus, Potassium, Calcium, Magnesium, Heavy metals (copper, zinc, iron, manganese, boron, molybdenum, cobalt), Toxic elements, if necessary (arsenic, fluoride, cadmium, nickel, lead, mercury, etc.)

4. MOST COMMON MISTAKES BY CERTIFIED CROP OPERATORS AND/ OR CERTIFICATION APPLICANTS

4.1 Certifier Relations

- Getting a product or practice approved by a certifier, but not getting the approval in writing, and then misunderstanding the "approval".
- Failure to submit requested documentation to the certifier (such as prior land use forms, non-GMO letters, adjoining land use forms, water test results, etc.)
- Not understanding and/or not complying with certification requirements (minor non-compliances) from the previous year.
- Failure to complete required paperwork on time, or at all.
- Not registering with the state organic program, if applicable.
- Failure to pay certification and/or inspection fees.

4.2 Non-approved Inputs

- Use of non-approved substances (including treated seeds), due to negligence and/or not understanding the requirements.
- Use of non-approved substances, due to trusting an input supplier who gave assurances that the material was "approved" for organic farming.
- Failure to inquire about the GMO status of inputs, especially inoculants and Bt products.
- Not having documentation of non-GMO status of inputs, including seeds, inoculants, and Bt products.
- The farmer does not correctly calculate the amount of time from the last date of prohibited inputs used - and the required 24 months (for annuals crops)/36 months (for perennial crops) have not passed. The farmer then wrongly thinks that the present year's crop will be saleable as certified organic, when it is not certifiable.

4.3 Documentation of approved Inputs

- Failure to obtain adequate documentation for purchase of approved inputs.
- Failure to document attempts to source organic seeds/planting material.

4.4 Record Keeping

- Lack of adequate detail or clarity on field maps and/or use of inaccurate maps.
- Field maps which do not show acres, field numbers, and/or adjoining land uses.
- Not keeping field activity records up-to-date.
- Failure to keep seed and input labels and receipts in an organized and accessible manner.

- Failing to keep records for contracted services, such as planting, spraying, harvesting, and/or trucking.
- Failure to keep bin records up-to-date.
- Not recording field numbers on harvest and/or storage records.
- Not using lot numbers or not using a consistent lot numbering system.
- Not providing adequate documentation to buyers when organic products are sold.
- Not keeping records of steps taken to inspect and clean transport units.
- Not maintaining adequate records for operations with both organic and conventional production.

4.5 Commingling and Contamination

- Failure to properly clean harvesting equipment and/or storage units, resulting in commingling or contamination of organic crop.
- Failure to segregate crops harvested from buffer zones.
- Lack of cleaning logs for spray equipment that is also used for prohibited inputs.
- Work area contamination for post harvest handling (e.g. washing vegetables, cutting vegetables, packing vegetables, etc.)
- Mislabelling or mishandling of crop by mistake by workers who are not fully informed of organic certification requirements.
- Misapplication of prohibited materials by workers who are not fully informed of organic certification requirements.
- No GMO drift management plan - not knowing where the nearest GMO fields are located.
- Failure to post no-spray signs when and where these would add protection.

4.6 Organic Plan

- Failure to follow the operation's organic plan.
- Filing "renewal" farm plans with entries marked "No Change", when there have been significant changes, such as new leased or purchased fields, discontinued leases, sub-divided fields, new crops, new inputs, changes to field numbers, changes to lot numbering system, etc.

APPENDICES

(A) PERMITTED/RESTRICTED PRODUCTS (NPOP, INDIA) UNDER CERTIFIED ORGANIC FARMING (source: www.apeda.com)

APPENDIX 1
Products for Use in Fertilizing and Soil Conditioning

Matter Produced on an Organic Farm Unit	
Farmyard & poultry manure, slurry, urine	Permitted
Crop residues and green manure	Permitted
Straw and other mulches	Permitted
Matter Produced Outside the Organic Farm Unit	
Blood meal, meat meal, bone meal and feather meal without Preservatives	Restricted
Compost made from any carbon based residues (animal excrement including poultry)	Restricted
Farmyard manure, slurry, urine *(Preferably after control fermentation and/or appropriate dilution) "Factory" farming sources not permitted.)*	Restricted
Fish and fish products without preservatives	Restricted
Guano	Restricted
Human excrement	Not Allowed
By-products from the food and textile industries of biodegradable material of microbial, plant or animal origin without any synthetic additives	Restricted
Peat without synthetic additives (prohibited for soil conditioning)	
Sawdust, wood shavings, wood provided it comes from untreated wood	
Seaweed and seaweed products obtained by physical processes, extraction with water or aqueous acid and/or alkaline solution	Restricted
Sewage sludge and urban composts from separated sources which are monitored for contamination	Restricted
Straw	Restricted
Vermicasts	Restricted
Animal charcoal	Restricted
Compost and spent mushroom and vermiculate substances	Restricted
Compost from organic household reference	Restricted
Compost from plant residues	

Byproducts from oil palm, coconut and cocoa (including empty fruit bunch, palm oil mill effluent (pome), cocoa, peat and empty cocoa pods)	Restricted
Byproducts of industries processing ingredients from organic agriculture	Restricted
Minerals	
Basic slag	Restricted
Calcareous and magnesium rock	Restricted
Calcified seaweed	
Calcium chloride	
Calcium carbonate of network origin (chalk, limestone, gypsum and phosphate chalk)	
Mineral potassium with low chlorine content (e.g. sulphate of potash, kainite, sylvinite, patenkali)	Restricted
Natural phosphates (e.g. Rock phosphates)	Restricted
Pulverised rock	Restricted
Sodium chloride	
Trace elements (boron, Fe, Mn, molybdenum, Zn)	Restricted
Wood ash from untreated wood	Restricted
Potassium sulphate	Restricted
Magnesium sulphate (Epson salt)	
Gypsum (calcium sulphate)	
Stillage and stillage extract	Ammonium Stillage excluded
Aluminium calcium phosphate	Restricted
Sulphur	Restricted
Stone mill	Restricted
Clay (bentonite, perlite, zeolite)	
Microbiological Preparations	
Bacterial preparations (biofertilizers)	
Biodynamic preparations	
Plant preparations and botanical extracts	
Vermiculate	
Peat	

"Factory" farming refers to industrial management systems that are heavily reliant on veterinary and feed inputs not permitted in organic agriculture.

APPENDIX 2
Products for Plant Pest and Disease Control

Certain products are allowed for use in organic agriculture for the control of pests and diseases in plant production. Such products should only be used when absolutely necessary and should be chosen taking the environmental impact into consideration.

Many of these products are restricted for use in organic production. In this appendix **"restricted"** means that the conditions and the procedure for use shall be set by the certification programme.

I. Substances from plant and animal origin	
Preparation of rotenone from Derris elliptica, Lonchocarpus, Thephrosia spp.	Restricted
Gelatine	
Propolis	Restricted
Plant based extracts (e.g. neem, garlic, pongamia, etc.)	
Preparation on basis of pyrethrins extracted from Chrysanthemum cinerariaefolium, containing possibly a synergist pyrethrum cinerafolium	Restricted
Preparation from Quassia amara	Restricted
Release of parasite predators of insect pests	Restricted
Preparation from Ryania species	Restricted
Lecithin	Restricted
Casein	
Sea weeds, sea weed meal, sea weed extracts,	Restricted
Sea salt and salty water	
Extract from mushroom (Shitake fungus)	
Extract from Chlorella	
Fermented product from Aspergillus	Restricted
Natural acids (vinegar)	Restricted
II. Minerals	
Chloride of lime/soda	Restricted
Clay (e.g. bentonite, perlite, vermiculite, zeolite)	
Copper salts/inorganic salts (Bordeaux mix, copper hydroxide, copper oxychloride) used as a fungicide maximum 8 kg per ha per year depending upon the crop and under the supervision of inspection and certification agency	Restricted
Mineral powders (stone meal, silicates)	Not allowed
Diatomaceous earth	
Light mineral oils	Restricted
Permanganate of potash	Restricted

Lime sulphur (calcium polysulphide	Restricted
Silicates (sodium silicate, quartz)	Restricted
Sodium bicarbonate	
Sulphur (as a fungicide, acaricide, repellent)	Restricted
III. Microorganism /Biocontrol agents	
Viral preparations (e.g., Granulosis viruses, Nuclear polyhydrosis, viruses etc.).	
• Fungal preparations (e.g., Trichoderma spp. etc.)	
• Bacterial preparations (e.g., Bacillus spp etc.)	
• Parasites, predators and sterilized insects."	
IV. Others	
Carbon dioxide and nitrogen gas	Restricted
Soft soap (potassium soap)	
Ethyl alcohol	Not allowed
Homeopathic and Ayurvedic preparations	
Herbal and biodynamic preparations	
V. Traps	
Physical methods (e.g., chromatic traps, mechanical traps, light traps, sticky traps)	
Mulches, nets	

APPENDIX 3
Criteria for the Evaluation of Additional Inputs to Organic Agriculture

Appendices 1 & 2 refer to products for fertilizing of the soil and control of plant pest and diseases in organic agriculture. But there may well be other products which may be useful and appropriate for use in organic agriculture which may not fall under these headings. Appendix 3 outlines the procedure to evaluate other inputs into organic production.

The following checklist should be used for amending the permitted substance list for fertilizing the soil conditioning purposes:

- The material is essential for achieving or maintaining soil fertility or to fulfil specific nutrient requirements, for specific soil-conditioning and rotation purposes which cannot be satisfied by the practices outlined in the standard or of other products included in Appendix 1. The ingredients are of plant, animal, microbial or mineral origin which may undergo the following processes:
- Physical (mechanical, thermal)
- Enzymatic
- Microbial (composting, digestion)

  and,

  Their use does not result in, or contribute to, unacceptable effects on, or contamination of, the environment, including soil organisms.

  Their use has no unacceptable effect on the quality and safety of the final product.

The following checklist should be used for amending the permitted substance list for the purpose of plant disease or pest and weed control:

- The material is essential for the control of a harmful organism or a particular disease for which other biological, physical or plant breeding alternatives and/or effective management techniques are not available

  and,
- The substances (active compound) should be plant, animal, microbial or mineral origin which may undergo the following processes:
  - physical
  - enzymatic
  - microbial

  and, their use does not result in, or contribute to, unacceptable effects on, or contamination of, the environment.
- Nature identical products such as pheromones, which are chemically synthesised, may be considered if the products are not available in sufficient quan-

tities in their natural farm, provided that the conditions for their use do not directly or indirectly contribute to contamination of the environment or the product.

Evaluation

When an input is to be evaluated it must first be investigated by certification programmes to see whether it fulfils the following six criteria. An input must fulfil all 6 requirements before it can be accepted as suitable for use in organic agriculture.

Inputs should be evaluated regularly and weighed against alternatives. This process of regular evaluation should result in organic production becoming ever more friendly to humans, animals, environment and the ecosystem.

1. Necessity

The necessity of each input must be established. This will be investigated in the context in which the product will be used. Arguments to prove the necessity of an input may be drawn from such criteria as yield, product quality, environmental safety, ecological protection, landscape, human and animal welfare.

The use of an input may be restricted to:

- Specific crops (especially perennial crops)
- Specific regions
- Specific conditions under which the input may be used

2. Natures and Method of Production

Nature: The origin of the input should usually be (in order of preference):

- Organic - vegetative, animal, microbial
- Mineral

Non-natural products which are chemically synthesised and identical to natural products may be used.

When there is any choice, renewable inputs are preferred. The next best choice is inputs of mineral origin and the third choice is inputs which are chemically identical to natural products. There may be ecological, technical or economic arguments to take into consideration in the allowance of chemically identical inputs.

Method of Production: The ingredients of the inputs may undergo the following processes:

- Mechanical
- Physical
- Enzymatic

- Action of micro-organisms
- Chemical (as an exception and restricted)

Collection: The collection of the raw materials comprising the input must not affect the stability of the natural habitat nor affect the maintenance of any species within the collection area.

3. Environment

Environmental Safety: The input must not be harmful or have a lasting negative impact on the environment. Nor should the input give rise to unacceptable pollution of surface or ground water, air or soil. All stages during processing, use and breakdown must be evaluated.

The following characteristics of the input must be taken into account:

Degradability: All inputs must be degradable to their mineral form. Inputs with a high acute toxicity to non-target organisms should have a maximum half-life of five days. Natural substances used as inputs which are not considered toxic do not need to be degradable within a limited time.

Acute toxicity to non-target organisms: When inputs have a relatively high acute toxicity for non-target organisms, a restriction for their use is needed. Measures have to be taken to guarantee the survival of these non-target organisms. Maximum amounts allowed for application may be set. When it is not possible to take adequate measures, the use of the input must not be allowed.

Long-term chronic toxicity: Inputs which accumulate in organisms or systems of organisms and inputs which have, or are suspected of having, mutagenic or carcinogenic properties must not be used. If there are any risks, sufficient measures have to be taken to reduce any risk to an acceptable level and to prevent long lasting negative environmental effects.

Chemically synthesised products and heavy metals: Inputs should not contain harmful amounts of man-made chemicals (xenobiotic products). Chemically synthesised products may be accepted only if identical to the natural product.

Mineral inputs should contain as few heavy metals as possible. Due to the lack of any alternative, and long-standing, traditional use in organic agriculture, copper and copper salts are an exception for the time being. The use of copper in any form in organic agriculture must be seen, however, as temporary and use must be restricted with regard to environmental impact.

4. Human Health and Quality

Human Health: Inputs must not be harmful to human health. All stages during processing, use and degradation must be taken into account. Measures must be taken to reduce any risks and standards set for inputs used in organic production.

Product quality: Inputs must not have negative effects on the quality of the product - e.g. taste, appearance and quality.

5. Ethical Aspects - Animal Welfare

Inputs must not have a negative influence on the natural behaviour or physical functioning of animals kept at the farm.

6. Socio-Economic Aspects

Consumers' perception: Inputs should not meet resistance or opposition of consumers of organic products. An input might be considered by consumers to be unsafe to the environment or human health, although this has not been scientifically proven. Inputs should not interfere with a general feeling or opinion about what is natural or organic - e.g. genetic engineering.

APPENDIX 4

List of Approved Ingredients of Non-Agricultural Origin and Processing Aids Used in Food Processing

Food Additives and Carriers [1]

Intl Numbering System	Additive/ Processing aid	Used as		Food category	Functions	Limitation /Note
		Additive	Proc. Aid			
INS 170	Calcium carbonate	*	*	GA	Anticaking, acidity regulator, emulsifier, stabiliser	
INS 220	Sulphur dioxide	*	*	W	Preservative, stabiliser	Max. 0.3mg/l
INS 224	Potassium metabisulphite	*	*	W		
INS 270	Lactic acid	*	*	FV, W, MP	Acidity regulator	
INS 290	Carbon-dioxide	*	*	GA	Carbonating agent, packing gas	
INS 296	Malic acid	*	-	FV, MP	Acidulent	
INS 300	Ascorbic acid	*	-	GA	Antioxidant,	
INS 306	Tocopheroles, mixed natural concentrates	*	-	GA	Antioxidant	
INS 322	Lecithin	*	*	GA	Antioxidant, emulsifier, stabiliser	
INS 330	Citric acid	*	-	FV, MP	Acidity regulator Antioxidant	
				W		Restricted 1 gm/l
INS 335	Sodium citrate	*	-	ME, MP	Acidity regulator Antioxidant	
INS 336	Potassium citrate	*	-	ME, MP	Acidity regulator Antioxidant	
INS 400	Aliginic acid	*	-	FV	Emulsifier, Stabiliser, Thickener	
INS 401	Sodium alginate	*	-			

[1] *Food Additives may contain carriers which shall be evaluated*

INS 402	Potassium alginate	*	-			
INS 333	Calcium citrate	*	-	ME, MP	Acidity regulator Antioxidant	
INS 334	Tartaric acid	*	*	W, MP	Flour treatment, raising agent, emulsifier, antioxidant, preservative	
INS 407 INS 335	Sodium tartrate	*	*	CO/CB		
INS 336	Potassium tartrate	*	*	C/CO/ CB		
INS 341	Mono calcium phosphate	*	-	C	Only for raising flour	
INS 342	Ammonium phosphate	*	-	W	-	0.3 gm/l
INS 406	Agar	*	-	MP, F	Emulsifier, Stabiliser Thickner	Max. 0.5%
INS407	Carrageenan	*	-	MP, F		
INS 410	Locust bean gum	*	-	MP, F		
INS 412	Guar gum	*	-	MP, F		
INS 413	Tragacanth gum	*	-	GA	Emulsifier, Stabiliser Thickner	Max. 0.5%
INS 414	Arabic gum	*	-	MP, F	Emulsifier, Stabiliser Thickner	Max. 0.5%
INS 415	Xanthan gum	*	-	F/FV/ CB	Emulsifier, Stabiliser Thickner	Max. 0.5%
INS 416	Karaya gum	*	-	MP, F		
INS 440	Pectin (Unmodified form)	*	-	FV, F, CB	Emulsifier, Stabiliser, Thickner	
INS 500	Sodium carbonate	*	*	CO/CB	Acidity regulator, Stabiliser, Anticaking agent, Raising agent	
INS 501	Potassium carbonate	*	*	C/CO/ CB	Acidity regulator Stabiliser. For drying grapes	
INS 503	Ammonium carbonate	*	-	C/CO/ CB	Acidity regulator Stabiliser, Raising agent	
INS 504	Magnesium carbonate	*	-	C/CO/CB	Acidity regulator Stabiliser	

INS 508	Potassium chloride	*	-	FV	Stabiliser, Thickner	
INS 509	Calcium chloride	*	-	ME/F/ FV/SO	Coagulation, firming agent	
INS 516	Calcium sulphate	*	*	CB, SO	Acidity regulator, Stabiliser , Flour treatment, coagulation agent.	Restricted Only in baker's' yeast
INS 517	Ammonium sulphate	*		W	Encourage the growth of yeasts	Restricted to 0.3 gm/1
INS 524	Sodium hydroxide	*	*	C, S, Flours & Starches	Acidity regulator, Surface treatment of traditional bakery products. Processing aid for sugar	
INS 526	Calcium hydroxide	*	*	S, C	Additive for maize tortilla flour. Processing aid for sugar	
INS 938	Argon	*	-	GA		
INS 941	Nitrogen	*	*	GA		
INS 948	Oxygen	*	*	GA		
INS153	Wood ash	-	*	MP	Coating agent	
INS 181	Tannin	_	*	W	Clarifying agent	
INS 184	Tannic Acid	-	*	W	Filtration aid	
INS 513	Sulphuric acid	-	*	S	pH adjustment of water in sugar production.	
INS 551	Silicon dioxide	-	*	W/ Dehydra-ted FV Herbs & Spices	Gel or colloidal solution Anticaking agent	
INS 553	Talc	-	*	GA	Lubricant	0.5% -PFA
INS 901	Beeswax	-	*	GA	Releasing agent	
INS 903	Carnauba wax	-	*	GA	Releasing agent	

	Activated carbon	-	*	GA	Decolorizer	
	Bentonite	-	*	FV/W	Filter aid, clarifying agent	
	Casein	-	*	W	Clarifying agent	
	Diatomaceous earth	-	*	S/FV/W	Filter aid	
	Egg white albumen	-	*	W	Clarifying agent	
	Ethanol	-	*	GA	Solvent	
	Gelatine	-	*	MP, FV/W	Emulsifier, Clarifying agent	
	Isinglass	-	*	W	Clarifying agent	
	Kaolin	-	*	GA	Filter aid, Extraction of propolis	
	Perlite	-	*	GA	Clarifying agent	
	Preparations of bark	-	*	S		
	Vegetable oils	-	*	GA	Greasing, releasing agent	
	Glycerol	-	*		Plant extracts Restricted	
	Beet sugar	-	*		Restricted	
	Natural Colours Carotenoids Chlorophyll Annatto Saffron Riboflavin (Lactoflavin) Curcumin Caramel Canthaxanthin	*	-	GA	Colouring agent	

Key - list of abbreviations used in above tables:
* Could be used as:
- Not used as:

GA - Generally Unrestricted
F - Fat products
C - Cereal Products
W - Wine S - Sugar products
CB - Cakes and Biscuits

MP - Milk Products
ME - Meat Products
FV - Fruit/ Vegetable
CO – Confectionery
SO - Soybean products

Flavouring Agents

- Volatile (essential) oils produced by means of solvents such as oil, water, ethanol, carbon dioxide and mechanical and physical processes
- Natural smoke flavour
- Use of natural flavouring preparations should be approved as per national procedure to evaluate additives and processing aids based criteria for evaluation of additional inputs to organic agriculture (Appendix-3)

Preparations of Microorganisms

- Preparations of microorganisms accepted for use in food processing. Genetically modified organisms are excluded.
- Bakers yeast produced without bleaches and organic solvents.

Preparations of Microorganisms and Enzymes

These may be used as processing aids with approval based on the national procedure to Evaluate Additives and Processing Aids for Organic Food Products.

Ingredients

- Drinking water
- Salts (with sodium chloride and potassium chloride as basic component generally used in food processing).
- Minerals (including trace elements) and vitamins, fatty acids, amino acid and other nitrogenous compounds where their use is legally required or where severe dietary or nutritional deficiency can be demonstrated.

APPENDIX-5
Criteria for the Evaluation of Additives and Processing Aids for Organic Food Products

Introduction

List of additives, processing aids, flavouring agents and colours in organic food products are listed in Appendix-4. The following aspects and criteria should be used for evaluation of additives and processing aids in organic food products.

1. Necessity

Additives and processing aids may only be allowed in organic food products if each additive or processing aid is essential to the production wherein the authenticity of the product is respected and the product cannot be produced or preserved without them.

2. Criteria for the approval of additives and processing aids.

The additives and processing aid may be used in the processing of organic foods where:

- There are no other acceptable technologies available to process or preserve the organic product.
- The use of additives or processing aids which minimize physical or mechanical damage to the foodstuff as a substitute for other technologies which if used would result in such damage.
- The hygiene of the product cannot be guaranteed as effectively by other methods (such as a reduction in distribution time or improvement of storage facilities).
- Additives or processing aids do not compromise the authenticity of the product.
- The additives or processing aids do not confuse the customer by giving the impression that the final product is of higher quality than is justified by the quality of the raw material. This refers primarily but not exclusively, to colouring and flavouring agents.
- Additives and processing aids should not detract from the overall quality of the product.
- The additives generally shall have 'GRAS' (Generally Regarded As Safe) status indicating that they are safe when used in accordance with good manufacturing practices.

3. Step-by-step procedure for use of additives and processing aids

a) Instead of using additives or processing aids, the preferred first choice is

   - Food grown under organic conditions which are used as a whole product or are processed in accordance with the IFOAM basic standards - e.g. flour used as a thickening agent or vegetable oil as a releasing agent.

- Foods or raw materials of plant and animal origin which are produced only by mechanical or simple physical procedures - e.g. salt.

b) The second choice is:

- Substance isolated from food and produced physically or by enzymes - e.g. starch, tartrates, pectin.
- Purified products of raw materials of non-agricultural origin and microorganisms - e.g. acerola fruit extract, enzymes and microorganism preparations such as starter cultures.

c) In organic food products the following categories of additives and processing aids are not allowed:

- "Nature identical" substances
- Synthetic substances primarily judged as being unnatural or as a "new construction" of food compounds such as acetylated cross linked starches (modified starches).
- Additives or processing aids produced by means of genetic engineering.
- Synthetic colouring and synthetic preservatives.
- Carriers and preservatives used in the preparation of additives and processing aids shall also be taken into consideration.

APPENDIX 6

Approved Additives for Manufacturing of Packaging Films for Packaging of Organic Foodstuffs

Certain additives are allowed for use in manufacture of packaging films for packaging of foodstuffs. However, many of these are restricted for use in packaging of organic foodstuffs. Restricted means that the conditions and procedures for use shall be set by the accredited certification programme.

The following are approved additives under restriction: -

Use of plastics for packaging of organic foodstuffs

S. No.	Products	Limitation
1.	4, 4'-Bis(2-benzoxazolyl)stilbene	Restricted
2.	9, 9-Bis(methoxymethyl)fluorine	Restricted
3.	Carbonic acid, copper salt	Restricted
4.	Diethyleneglycol	Restricted
5.	2-(4, 6-Diphenyl-1, 3, 5-triazin-2-yl)-5-(hexyloxy)phenol	Restricted
6.	Ethylenediaminetetraacetic acid, copper salt	Restricted
7.	2-(2-Hydroxy-3, 5-di-tert-butyl-phenyl-5-chlorobenzotriazole	Restricted
8.	2-Methyl-4-isothiazolin-3-one	Restricted
9.	Phosphoric acid, trichlorocthylester	Restricted
10.	Polyesters of 1, 2 propanediol and/or 1, 3-and 1, 4 butanediol and/or polypropyleneglycol with adipic acid, also end-capped with acetic acid or fatty acids C10-C18 or n-octanol and/or n-decanol	Restricted
11.	1,1,1-Trimethylolpropane	Restricted
12.	3-hydroxybutanoic acid 3-hydro xypentanoic acid, copolymer	Restricted

APPENDIX 7

List of Approved Feed Materials, Feed Additives and Processing Aids for Animal Nutrition

1. Feed Materials of Plant Origin

- Cereals grains, their products and by-products
- Oilseeds, oil fruits, their products and by-products
- Legume seed, their products and by-products
- Tuber roots, their products and by-products
- Other seeds and fruits
- Forages and roughages
- Molasses as a binding agent

2. Feed Material as an Animal Origin

- Milk and milk products
- Fish, other marine animals, their products and by-products

3. Feed Material from Mineral Origin

- Sea salt, rock salt — Restricted
- Sodium sulphate — Restricted
- Sodium carbonate — Restricted
- Sodium bicarbonate — Restricted
- Sodium chloride — Restricted
- Calcium carbonate — Restricted
- Calcium lactate — Restricted
- Calcium gluconate — Restricted
- Bone dicalcium phosphat precipitate — Restricted
- Defluorinated dicalcium phosphate — Restricted
- Defluorinated monocalcium phosphate — Restricted
- Anhydrous magnesia — Restricted
- Magnesium sulphate
- Magnesium chloride — Restricted
- Magnesium carbonate — Restricted

Restricted

Restricted

4. Trace Elements

- Iron — Feed additives
- Iodine — Feed additives
- Cobalt — Feed additives
- Manganese — Feed additives
- Zinc — Feed additives
- Molybdenum — Feed additives
- Selenium — Feed additives

5. Vitamins — Restricted

6. Enzymes — Restricted

7. **Micro-organisms** Restricted

8. **Preservatives**

- E-336 Formic acid
- E-260 Acetic acid
- E-270 Lactic acid
- E-280 Propionic acid

9. **Binders, anti-caking agent and coagulants**

- E-551b Colloidal silica
- E-551c Kieselgur
- E-553 Sepiolite
- E-558 Bentonite
- E-559 Kaolinitic clays
- E-561 Vermiculite
- E-599 Perlite

10. **Processing aids for silage**

- Sea salt
- Coarse rock salt
- Enzymes
- Yeasts
- Sugar
- Sugar beet pulp
- Cereal flour
- Molasses
- Lactic

APPENDIX 8

Products Authorized for Cleaning and Disinfections of Livestock Buildings and Installations

- Potassium and sodium soap
- Water and steam
- Milk of lime
- Lime
- Quicklime
- Sodium hypochlorite (e.g. as liquid bleach)
- Caustic potash
- Hydrogen peroxide
- Natural essences of plants
- Citric, peracetic acid, formic, lactic, oxalic and acetic acid
- Alcohol
- Nitric acid (dairy equipment)
- Phosporic acid (dairy equipment)
- Formaldehyde
- Sodium carbonate

(B) TRADITIONAL ORGANIC RECIPES FOR SOIL FERTILITY AND PLANT PROTECTION

APPENDIX 9

1. Harnessing cosmic forces and planting calendar

The light and cosmic forces of sun, moon, planets and stars reaches to plants in irregular rhythms. Each contributes to the life, growth and form of the plant. By understanding the gesture and effect of each rhythm, soil preparation, sowing, intercultural operations and harvesting are programmed accordingly to harness their influences. The field preparations, sowing manuring harvesting etc. performed as per constellation are more effective and beneficial. Every constellation has dominant elemental influences and affects 4 specific parts of plants. Manuring, rooting, flowering, growth and fruiting/seed are to be done as per constellation.

Interactions of elements and constellation on plants parts

Element	Plant parts	Constellation
Earth	Roots	Virgo, Capricorn, Taurus
Air	Flowers	Gemini, Libra, Aquarius
Water	Leaf	Cancer, Scorpio, Pisces
Fire	Fruit/seeds	Sagittarius, Aries, Leo

Positions of earth and moon for harnessing cosmic forces

Ascending moon	Descending moon
i) The earth is breathing out: the development occurs in upper parts of the plant	i) The earth is breathing in: development of plant occurs, parts below the ground e.g. root.
ii) Cosmic energy works above the rhizosphere	ii) Cosmic energy works below the rhizosphere
iii) Suitable for - • Foliar applications • Propagation activities • Sowing	iii) Suitable for - • Root development • Transplanting • Manure application • Harvesting of tuber crops

2. Biodynamic preparations

Basically there are two types of biodynamic preparations:

- Biodynamic Compost Preparations (BD –502-507).
- Biodynamic Field Sprays (BD-500-501).

All these preparations are made in descending periods of moon except BD-507, which is best prepared in air/light day. The BD sets are used in Cow Pat Pit

(CPP), BD-compost, Biodynamic liquid manure, and Biodynamic liquid pesticides. These works are needed to regulate the composting process and enable different elements (calcium, nitrogen and phosphorus) for healthy plant growth to be present in a living way.

BD sets used in CPP, compost/liquid manures and pesticides

Preparation	Related planets	Substances from which preparation is prepared	Role
BD-502	Venus	Fermented flower heads of yarrow *(Achillea millefolium)*	Rich in S,K and N
BD-503	Mars	Fermented chamomile blossom *(Matricaria recutita)*	Rich in S,K, and N
BD-504	Mercury	Whole shoot of stinging bettle with flowers, fermented in soil *(Urtica dioica)*	Rich in Fe
BD-505	Moon	Fermented oak bark *(Ouercus robur)*	Rich in Ca
BD-506	Jupiter	Fermented flower heads of Dandelion *(Taraxacum officinale)*	Rich in K, Si
BD-507	Saturn	Valerian flower extract *(Valeriana officinalis)*	Rich in P

Cow Pat Pit (CPP): Also know as "soil shampoo". Cow Pat Pit (CPP) is a strong soil conditioner. It is a biodynamic preparation used during field preparation. It enhances seed germination, promotes rooting in cuttings and grafting, improves soil texture, provides resistance to plants against pests and disease replenishes and rectifies trace element deficiency. It is used to improve soil fertility before sowing, seed treatment and foliar application. It may be prepared throughout the year.

Depending upon the weather and temperature, its preparation is used in 75-90 days. Soaking 0.5–1.0 kg of CPP in 40-45 litres of water overnight and sprinkling on one acre of land before showing improves germination and health of soil.

Biodynamic field sprays (BD 500): These are fundamental biodynamic field spray preparations. Cow horns filled with fresh cow dung from lactating cows are buried in fertile soil. These are buried in descending moon during autumn (October-November) for incubation during whole winter. It is taken out in March–April in descending phase and used or stored in earthen posts at dark and cool place.

For spraying, 25g of BD-500 is dissolved in 13.5 litres of water in plastic bucket by making vortex in clock and anti-clockwise for one hour in evening and the solution is sprayed with the help of natural brush or with a tree twig. Spraying of

BD-500 is done at the time of field preparation in descending period of moon. Microbial activity of BD-500 during stirring showed interesting response.

Microbial analysis of BD 500 during stirring

Stirring interval (minutes)	Bacteria (cfu's/g)	Actinomycetes	Fungi (cfu's/g)
15	26×10^3	22×10^3	10×10^3
30	35×10^3	35×10^3	14×10^3
45	58×10^3	60×10^3	12×10^3
60	66×10^3	88×10^3	35×10^3

BD 501 on horn silica manure: BD 501 is prepared in ascending period of moon by filling cow horn with /mealy/ silica powder and buried in spring (March/ April) after taking out BD-500. Within 6 months, the preparation is ready for use. The solution of prepared by dissolving one gram in 13.5 litres of water. Solution is sprayed on leaves in the form of 'mist' at sunrise and the best constellation is moon opposite to Saturn.

BD 501 works on photosynthesis process. It strengthens the quality of plant products and encourages the development of fruits and seeds. For maximum effect, BD 501 should be applied once at the beginning of a plant's life (at 4-leaf stage) and again at flowering or fruit maturation stage.

Due to enhancement of photosynthesis, starch, sugars and cellulose synthesis improves. It results in improvement of quality and storage life of produce.

Rishi Krishi

This system is used in Maharashtra, Madhya Pradesh, Rajasthanand now in Uttar Pradesh for many years. *Angra Bhomi Sanskar is* done to make the soil fertile, in which 15-20 Kg rhizosphere soil of banyan tree *(Ficus bengalensis- "Vat vriksh")* is broadcasted on one acre of land. It has a lot of earthworms and other beneficial microbes, which improve soil fertility and its biological activity.

Amrit pani: Amrit pani is prepared by mixing 10 kg cowdung, 250g ghee and 500 ml honeys. All ingredients are mixed, stirred and fermented and diluted to 200 litres. This is used after proper stirring to treat seeds *(Beej Sanskar)*, enrich soil *(Bhoomi Sanskar)* and seedlings *(Vanaspati Sanskar)* by spraying on field and plants.

Agnihotra Krishi **(*Homa* farming)**

Agnihotra therapy is practiced to reduce environmental pollution and improve crop production with minimum expenditure. *Homa* farming is a totally revealed science. *Agnihotra* is performed regularly to purify the atmosphere since ancient time. It is based on *Home, yagna* is the technical term describing the process of purification of house and atmosphere through fire, which is tuned to rhythm of nature timed to sunrise and sunset biorhythm. The fire is prepared in small copper pyramid of specific size, and shape, brown rice, dried cowdung patties and cow *ghee* is offered by chanting *mantras*. Radiation effects of astrological combinations and "*mantras*" leads to better capture of cosmic energies from sun and moon. It helps to reset the energy cycle of the planet in natural harmony benefiting all concerns.

Panchgavya Krishi

Panchgavya consists of 5 products of cow/slurry (4kg), cowdung (1kg), urine (10 litres), milk (3 litres), curd (2 litres) and ghee (1 kg). After mixing properly along with sugarcane juice, tender coconut water, ripe banana and toddy. It is incubated for 30-40 days and stirred daily. Three per cent solution is used to treat seed/ seedlings. It encourages vegetative and reproductive growth of plants. A group of farmers in Tamil Nadu are practising in cultivation of a number of cereal and horticultural crops.

(C) ACCREDITED INSPECTION AND CERTIFICATION AGENCIES

APPENDIX 10

Bioinspecta
Director, Ackerstrasse, Postfach CH-5070 Frick, Switzerland,
Branch office in India: Bioinspecta, C/o INDOCERT, Thottumugham P.O. Aluva-683 105, Ernakulam, Kerala State, India.

Ecocert International (Germany)
Country Representative/Managing Director
Ecocert SA Branch Office, 54 A, Kanchan Nagar, Nakshetrawadi Aurangabad - 431 002, Maharashtra State.

Indian Organic Certification Agency (INDOCERT)
Executive Director, Thottumugham P.O. Aluva-683 105, Ernakulam Kerala State, India.

International Resources for Fairer Trade
Director, Sona Udyog (Industrial Estate), Unit No. 7, Parsi Pandhayat Road, Andheri (E), Mumbai - 400 069

IMO Control Private Limited
Director, 26, 17th Main, HAL 2nd 'A', Stage, Bangalore-560 008.

LACON GMBH, Germany
Dr. Heinz Joachim Kopp, M.D.
Weingarten Str. 1.5, 77654, Offenburg, Germany
Branch office in India
Mr. Bobby Issac, LACON, C/o Renewable Energy Centre, Mithradham, Chunangaveli, Alwaye-683 105, Kerala.

SGS India Pvt. Ltd.
Business Manager, M/s SGS India Pvt. Ltd., 250 Udyog Vihar Phase - IV, Gurgaon - 122015

Skal International (Netherlands)
International Inspector, Skal Inspection and Certification Agency
No. 191, 1st Main Road, Mahalaxmi Layout, Bangalore - 560 086

APOG Organic Certification Agency
1st Floor, 5th Main, 9th Cross, Jayamahal Extension, Banglore-46

OneCert Asia Agri Certification Pvt. Ltd
Agrasen Farm, Vatika Road, off. Tonk Road,(P.O.Vatika)
Jaipur - 303 905, Rajasthan, India.

(D) LIST OF INDIAN EXPORTER OF ORGANIC PRODUCTS

APPENDIX 11

Accelerated Freeze Drying Co. Ltd., Amalgam House, Bristow Road, Willingdon Island Cochin-682 003, Kerala.

Amar Singh & Sons, 2, Hari Nagar, Talab Tillo, Jammu Tawi-180 002, Kashmir

Amit Spinning Industries Ltd., 5th Floor, Lotus House, Next to Liberty Cinema, New Marine Lines, Mumbai-400 020, Maharashtra

Mr. D. Jinsi, Senior Manager (Export Marketing), Balmer Lawrie (U.K.) Ltd., P-43, Hyde Road Extension, Kolkata-700 001.

Food & Inns Ltd., Mumbai-400 001
Email: fnl@bom5.vsnl.net.in

Maikaal Fibres Ltd., Madhya Pradesh
Email: mbril@bom4.vsnl.net.in

Mr. Binod Mohan, Director, Tea Promoters (India) Pvt. Ltd., Suite 17, Chowringhee Mansions, 30, Jawaharlal Nehru Road
Kolkata-700 016.

UNTEA, Chamraj Estate, Chamraj Estate P.O, The Nilgiris -643 204,

Tamil Nadu, Email: untetea@md2.vsnl.net.in

Dr. S.R. Maley, Director, Eco Save Systems (P) Ltd.
1, Dwell Inn. St. Anthony's Road, Uakola, Santacruz (E)
Mumbai-400 055.

Mr. A. Kumar, Malwa Vanaspati and Chemicals Co. Ltd.
Mohtanagar, Indore-452 003.

Khadi Gramodyog Bhawan, KVIC, 24 - Regal Building
Connaught Circus, New Delhi

Mr. Balsubramanian, M/s. Swaminathan Research Foundation
3rd Cross Street, Taramani, Institutional Area
Chennai-600 113, Email: bras@mssrf.res.in

IQF Foods Ltd., 78, 11th Cross Road
1st Stage, Indiranagar, Bangalore-560 038, Karnataka.

Kashmir Walnut Trading Company, Krishan Bagh, Talab Tillo
Jammu Tawi-180 002.

Mahesh Agri Exim P. Ltd., 312, Sharda Chamber No.1
3rd Floor, 31, Keshavi Naik road, Bhat Bazar
Mumbai-400 009.

Peermade Development Society, P.O. Box 11, Idukki District
Peermade-685 531, Kerala.

Mr. V.K. Arora, L T Overseas Ltd, A-21, Green Park ,

Aurobindo Marg, New Delhi 110 016.

Giraffe International, 63, Huda Sector 1, Rohtak 124 001.

M/s Yardee & Soree (I) Pvt Ltd, M-13/27, DLF City Phase II
Gurgaon 122 022.

Aryan International, D-184 Freedom-Fighters' Enclave
Neb Sarai, New Delhi 110 068.

Mr. C. Jeyakaran, Managing Director
M/s Kurinji Organic Foods (I) Pvt Ltd, Periyakulam Road
Genguvarpatti 625 203, Madurai.

Grewal's Organic Agriculture Farms, Village: Tehri, Baba Sawan Singh, PO
Moriwala, Dt:Sirsa, Haryana.

Mr. Ashok Lohia, Chairman, Chamong Tee Exports (P) Ltd
2, N.C. Dutta Sarani, Sagar Estate, 5th Floor, Unit 1
Kolkata-700 001

Mr. Omprakash Mor, Executive Director, Eco Farms (I) Pvt. Ltd.
Mor Garden, Dhamangaon Road, Yavatmal, Maharashtra-445 001.

Mr. Sashidaran, Director, Exotic Fruits Private Limited
4009, "Yamuna", 100 ft. Road, HAL II Stage, Bangalore-560 008

Mr. Amit Kumar Sen, General Manager
Godfrey Philips India Ltd, 3, Cooper Street, 1st Floor
Kolkata-700 026.

Mr. C.H. Shah, Director, Premier's Tea India Ltd
6A, Landmark, 228A, AJC Bose Road, Kolkata-700 020

Mr. Mukesh Malhotra, Managing Director
Weikfield Products Co. (I) Pvt. Ltd., Weikfield Estate,
Nagar Road, Pune - 411 014.

Mr. Mohan Chirimar, Managing Director
Ramanugger Tea Estate, Assam, Raghunath Exports Pvt. Ltd.
5F Park Plaza, 71 Park Street, Kolkata-700016

Mr. Victor Keishing, Director, Mata Foundation
Mantripukhri, Imphal-795 002, Manipur.

Achal Industries, 190, Industrial Area,
Baikampady, Mangalore.

Mr. K.G. Nanda, Managing Director, Amaryllis Exports Pvt. Ltd., F-4, Casa Capitol, 17, Wood Street, Ashoknagar, Bangalore-560 025.

Mr. Prem Chona, Director, Atik Private Limited
C - 549, Defence Colony, New Delhi.

Mr. C. Sunil Appaya, Manager-Marketing
BBTC Ltd. Post Box 573, Subramanian Road,
Willingdon Island, Cochin-682 003.

Mr. M. K. Sanyal, Group Manager
BBTC (Elk Hill Estate), Post Box 12, Sidapur P.O.- 571 253
South Coorg, Karnataka.

Kashmir apiaries, Kashmir House
G.T. Road Doraha 141 421 (Ludhiana).

Rajalakshmi Cotton Mills P. Ltd., 234/3A, A.J.C. Bose Road
FMC Fortuna, 4th Floor, Kolkata-700 020.

Mr. Roy Clark, Tradin Organic Agriculture B.V.
No.17, Verem Villas, Verem, Goa-403 114.

Mr. Kaushik, Aventis Biofeeds, 810, Maker Chambers 5
Nariman Point, Mumbai-21.

Mr. R.B. Singh, General Manager, IITC Organic India Pvt. Ltd., Village Kamtar P.O. Chinhat, Lucknow (U.P.).

Mr. J.M. Sahai, K.S. Intertrade Corporation, 719, Indira Prakash Building, 21, Barakhamba Road,New Delhi.

Mr. Ajay Katyal, Sunstars Overseas Ltd., 40, K.M. Stone, G.T. Road
Bahalgarh, Sonipat.

Mr. Sanjay P. Bansal, M/s. Sampad Vikas Ltd.
34 A Metcalfe Street, Kolkata - 700 013.

Mr. Mukesh Varma, Mukesh Varma Green Network
U-3, Green Park Extension, New Delhi-110 016.

KASAM, Contractorpada, Phulbani, Orissa-762 001.
Email: orissakasam@rediffmail.com

M/s. Narayan Ganesh Prabhu Zantye & Co.
Bicholim, Goa-403 504.

Mahima Organic Technology,
202, Kuber House, 162 Kanchan Baug, Indore-1, M.P.

Real Food Bio GmbH, Business Centre , Bredeneyer Str. 2B
45133 Essen, Deutschland.

M/s Picric Limited, Veetee House, 56-57, K.M. G.T. Karnal Road
Village Larsauli, Tehsil Ganaru, Dt :Sonepat 131 001.

Fruit and Vegetable Project, NDDB, Mongolpuri
New Delhi 110 083.

R Thomas and Company, Duke Court, Suite No 5
76 Shakespeare Sarani, Kolkata 700 017.

M/s Ace Naturale Fruits and Food Pvt Ltd
205 Anurag Commercial Complex, R C Dutt Road
Vadodara 390 005.

M/s Ion Exchange India Ltd, Second Floor, Neeta Towers
Opp Sandvik Asia Ltd, Dapodi, Pune 411 012.

Mr. A.K. Sen, Vice President, Assam Company Limited (Kondoli Estate), 52, Chowringhee Road,, Kolkata-700 071.

Ms. Naina Johar, Marketing Executive
Cygnet India Private Limited, 8A, Wood Street,
3rd Floor, Kolkata-700 016.

Mr. N. Balasubramanian, Managing Director
Enfield Agrobase Pvt Ltd, Viswapriya, 3 First Cross Road,
Kasturba Nagar, Adyar, Chennai-600 020.

Mr. Bhimsi Ahir, Director, A/2 Parivaar Apartments, Sundaram Park,Air force II Road, Jamnagar 361 004. Gujarat.

Mr. Sandhir Agarwal, Director, Kamala Tea Co. Ltd., (Selim Hill Tea Estate), 240 B, Acharya Jagadish, Chandra Bose Road, 3rd Floor, Kolkata-700 020.

Mr. S.K. Banerjee, Chairman, T' Classic (Darjeeling) Pvt. Limited
Anandlok Building, (2nd Floor), 227, A. J. C. Bose Road,
Kolkata - 700 020, India.

Mr. Somasundaram. A, Proprietor, SNV Horticultural Farms
Savadipatti, Algapuri (PO), Pin-625 523, Theni Dist., Tamilnadu.

Mr. Jeetender Chhajed, Managing Director
Accolade Overseas Corporation, 73/1336, Samta Nagar
Kandivalli (E), Mumbai-400 010.

Mr. Praveen Jain, Nikunj Chemical Limited

A-1, Shirali Society, Fatehgunj, Baroda - 390 002,Gujarat, India.

Source: www.apeda.com

Address for EM-1 Stock Solution Procurement
Maple Orgtech (I) Pvt. Ltd
New Alipore Block B
Kolkata-700 053.

(E) USEFUL WEBSITES ON ORGANIC AGRICULTURE

APPENDIX 12

1. http://www.ifoam.org (IFOAM-International Federation of Organic Agriculture Movements)
2. http://www.ioia.net (IOIA-Independent Organic Inspectors Association)
3. http://www.organicts.com/organic info/certification/links/index.html
4. http://www.organic-research.com [EU Regulation 2092/91]
5. http://www.ams.usda.gov/nop [State and Federal Organic Programs In The United States]
6. http://www.ota.com [ORGANIC Trade Association-American Organic Standards]
7. http://www.fao.org/organicag [FAO (website on organic agriculture)
8. http://home.prolink.de/~hps/organic/consolid-en.html (EU regulation can be downloaded from this website)
9. http://www.agrecol.de : http://www.agrecolandes.org/
10. http://www.agromisa.org
11. http://www.universalorganics.com.au (general information on organic certification of Australia)
12. http://www.cabi-bioscience.org
13. http://fgf@global.co.za/home
14. http://www.ofa.org.au (organic federation of Australia)
15. http://www.organicherbs.org (organic herb growers of Australia)
16. http://www.sustainablecap.dk (sustainable cap initiative in Denmark)
17. http://www.opam.mb.ca (organic producers association of Manitoba)
18. http://www.gtz.de
19. http://www.zadi.de (German Centre for Documentation and Information in Agriculture)
20. http://www.varanashi.com (varanashi research foundation, India)
21. http://www.dpw.wau.nl/biob (biological farming systems research and teaching group, Wageningen University, The Netherlands)
22. http://www.hdra.org.uk/
23. http://www.geocities.com/rainforest/vines/6274/
24. http://www.cabi.org (abstracts on organic farming in temperate regions)
25. http://www.fibl.ch (Research Institute on Organic Agriculture, Switzerland))
26. http://www.napcl.com (Nadukkara Agro Processing Co. Ltd)
27. http://www.apeda.com (Agricultural and Processed Food Products Export Development Authority, India)
28. http://www.biofach.de (World trade fair for organic foods and natural products)
29. http://www.hortibizindia.nic.in (National Horticulture Board, India)
30. http://www.kinfra.com (KINFRA Food Processing Parks)
31. http://www.iide.org/

32. http://www.kenyaweb.com/agriculture/organic-agri/
33. http://www.hivos.nl (Economic and cultural development in Africa, Asia, Latin America and Southeast Europe)
34. http://www.intercooperation.ch (Intercooperation-IC)
35. http://www.andrewlorand.com (Consulting for Ecological & Biodynamic Agriculture)
36. http://www.indianorganic.info (International conference –Indian organic products)
37. http://www.lifecyclesproject.ca (Life Cycles, Victoria)
38. http://www.ibl.ch-internat/
39. http://www.maela-net.org/
40. http://www.eda.admin.ch/newdelhi (Economic and commercial affairs, embassy of Switzerland, India)
41. http://www.oekoskolen.dk (The Organic Agriculture College, Denmark)
42. http://www.naturland.de (Naturland-organic certifier)
43. http://www.phaladaagro.com (Phalada Agro Research Foundation Ltd, India)
44. http://www.seco-admin.ch (State Secretariat for Economic Affairs, Switzerland)
45. http://www.shl.bfh.ch/
46. http://members.shaw.ca/gardenab (Organic seeds and crops researcher)
47. http://www.indianspices.com (Spices Board, India)
48. http://www.organic-farm.com (Hong Kong organic farming association)
49. http://www.unitec.ac.nz (School of Landscape & Plant Science Sustainable Organic Systems Design Integrated Studies)
50. http://www.worldcom@island.net (World Community Development Education Society)
51. http://forum.yam.org.tw (Homemaker's Union and Foundation, Taiwan)
52. http://www.biosuisse.com (Swiss Inspection and Certification Body)
53. http://www.bio-inspecta.ch/ (The Swiss Organic Certification body
54. http://www.vso.org.uk/
55. http://www.wn.org/
56. http://www.apeda.com/ (Indian Organic Standard NPOP-English and Hindi, can be downloaded from this website)

(F) RECORD-KEEPING FORMS

APPENDIX 13

Form-1

<table>
<tr><td colspan="4" align="center">Field Activity Log
A record of the practices and equipment you use for each field.</td></tr>
<tr><td colspan="4">Farm Name or Unit: Field ID:
Acres: Crop: Year:</td></tr>
<tr><td colspan="4">Field Activities: List date and activity, from pre-plant through post-harvest.</td></tr>
<tr><td>Date</td><td>Activity</td><td>Date</td><td>Activity</td></tr>
<tr><td></td><td></td><td></td><td></td></tr>
<tr><td></td><td></td><td></td><td></td></tr>
</table>

<table>
<tr><td colspan="3">Harvest: Use harvest/storage records to provide more detailed harvest information.</td></tr>
<tr><td>Date</td><td>Yield</td><td>Condition of Harvest</td></tr>
<tr><td></td><td></td><td></td></tr>
<tr><td></td><td></td><td></td></tr>
<tr><td colspan="3">Additional notes and observations:</td></tr>
<tr><td colspan="3"></td></tr>
</table>

Form-2

<table>
<tr><td colspan="3" align="center">Field Inputs Log
A record of the materials you use for each field.</td></tr>
<tr><td colspan="3">Farm Name or Unit: Field ID:
Acres: Crop: Year:</td></tr>
<tr><td colspan="3">Seeds / Transplants</td></tr>
<tr><td>Date</td><td>Crop / Variety
Planted/ Transplanted</td><td>Seeding Rate /
Transplant Spacing</td></tr>
<tr><td></td><td></td><td></td></tr>
<tr><td></td><td></td><td></td></tr>
</table>

<table>
<tr><td colspan="4">Fertilizers / Pest Control</td></tr>
<tr><td>Date</td><td>Material Applied /
Brand or Source</td><td>Rate /
Amount</td><td>Notes</td></tr>
<tr><td></td><td></td><td></td><td></td></tr>
<tr><td></td><td></td><td></td><td></td></tr>
<tr><td colspan="4">Additional notes and observations:</td></tr>
<tr><td colspan="4"></td></tr>
</table>

Form-3

Organic Seed and Planting Stock Search Record Producers may use non-organic seed only when organic seed is not commercially available. Use this form to document companies and individuals you contacted in your search for organic seed and stock.			
Farm Name or Unit: Crop Year:			
Crop/Variety Required:			
Date	Company Name	Contact Information	Outcome of Inquiry
Crop/Variety Required:			
Date	Company Name	Contact Information	Outcome of Inquiry
Crop/Variety Required:			
Date	Company Name	Contact Information	Outcome of Inquiry
Crop/Variety Required:			
Date	Company Name	Contact Information	Outcome of Inquiry
Crop/Variety Required:			
Date	Company Name	Contact Information	Outcome of Inquiry

Form-4

<table>
<tr><td colspan="7">Seed and Planting Stock Record
A record of seed and plants you purchased for use in organic production. Space is provided to record whether seeds/transplants are certified organic (O), untreated non-organic (U), or produced on-farm organic (F); to list seed treatments* used; and to note non-GMO verification analysis, if available. Remember that if you use non-organic seeds** or transplants, you must document your search for the organic equivalent. Non-organic perennial planting stock must be under organic management for a minimum of one year prior to the first organic harvest.</td></tr>
<tr><td colspan="7">Farm Name or Unit:
Crop Year:</td></tr>
<tr><td colspan="4">Seed and Planting Stock Information</td><td rowspan="2">Code (O,U,F)</td><td rowspan="2">Treatment+ Type/ Brand</td><td rowspan="2">Confirmation of non-GMO status? (non-organic only)</td></tr>
<tr><td>Crop</td><td>Variety</td><td>Supplier</td><td>Lot #</td></tr>
<tr><td></td><td></td><td></td><td></td><td></td><td></td><td></td></tr>
<tr><td></td><td></td><td></td><td></td><td></td><td></td><td></td></tr>
<tr><td></td><td></td><td></td><td></td><td></td><td></td><td></td></tr>
<tr><td colspan="7">* "Treatment" refers to natural and synthetic substances included on the National List ONLY.
** under EU Regulation (EEC) No. 2092/91, non-organic seeds/seedlings and planting materials are not allowed w.e.f. January 2006)</td></tr>
</table>

Form-5

<table>
<tr><td colspan="3">Compost Production Record
A record of on-farm compost production practices.</td></tr>
<tr><td colspan="2">Farm Name or Unit:
Production Year:</td><td></td></tr>
<tr><td colspan="2">Compost Pile, Windrow, or Unit I.D.:
Date Started:</td><td></td></tr>
<tr><td colspan="2">Compost Production Method Used:</td><td></td></tr>
<tr><td colspan="2">Feedstocks Used (including inoculants):</td><td>Estimated C/N Ratio:</td></tr>
<tr><td></td><td></td><td></td></tr>
<tr><td></td><td></td><td></td></tr>
<tr><td>Dates</td><td>Temperature</td><td>Turned?</td></tr>
<tr><td></td><td></td><td></td></tr>
<tr><td></td><td></td><td></td></tr>
</table>

Form-6

Fertility / Soil Monitoring Log					
Farm Name or Unit: Acres:	Crop:		Field ID: Year:		
Date of most recent soil test:					
When compared with previous soil tests, are your nutrient levels (circle):					
P (phosphorus)	decreasing	stable	increasing	excessive	not tested
K (potassium)	decreasing	stable	increasing	excessive	not tested
Ca (calcium)	decreasing	stable	increasing	excessive	not tested
Mg (magnesium)	decreasing	stable	increasing	excessive	not tested
S (sulfur)	decreasing	stable	increasing	excessive	not tested
Na (sodium)	decreasing	stable	increasing	excessive	not tested
B (boron)	decreasing	stable	increasing	excessive	not tested
Cu (copper)	decreasing	stable	increasing	excessive	not tested
Mo (molybdenum)	decreasing	stable	increasing	excessive	not tested
Zn (zinc)	decreasing	stable	increasing	excessive	not tested
Mn (manganese)	decreasing	stable	increasing	excessive	not tested
Fe (iron)	decreasing	stable	increasing	excessive	not tested
Organic matter / Humus levels	decreasing	stable	increasing	-----	not tested
pH is:		within or approaching desired range		out of or moving away from desired range	
Crop Monitoring:					
Are there visible signs of nutrient stress? No Yes					
Erosion Monitoring:					
Is there evidence of wind and/or water erosion? No Yes					
Additional Notes on Soil and Crop Monitoring:					

Form-7

Pest / Weed Monitoring Log			
Farm Name or Unit: Acres:	Crop:	Field ID: Year:	
Pest Monitoring: List date, type of insect or pest, and assessment of crop damage you observed.			
Date	Insect / Pest (note monitoring method if desired)	Type of crop damage	Damage Assessment (Low, Medium, High)
Disease Monitoring: List date, type or description of disease, and assessment of damage.			
Date	Disease	Type of crop damage	Damage Assessment (Low, Medium, High)
Weed Monitoring: List date, name/description of problem weed, and assessment of weed pressure.			
Date	Weed	Weed Pressure (Low, Medium, High)	

Form-8

Pest Control Activities and Inputs for Organic Crop Storage A record of the actions and materials you use to prevent/control pests in stored organic crops.		
Farm Name or Unit: Year:		Crop
Storage Unit I.D.:	Location (if off-farm):	
Date	Pest Control Activity / Input	By Whom?

Form-9

Harvest Record for Organic Operations A record of your organic crops harvest for the entire year.				
Farm Name or Unit: Crop Year:				
Harvest Date	Field I.D.	Organic Crop	Quantity / Quality	Where Stored or Sold

Buffer Zone Harvest Record for Organic Operations. A record of your buffer crops harvest for the entire year.				
Farm Name or Unit: Crop Year:				
Harvest Date	Field I.D.	Buffer Crop	Quantity / Quality	Where Stored/Sold/Used

Form-10

Equipment Cleanout Log This sheet should be kept on or near the equipment.			
Machine or Piece of Equipment: Crop Year:			
Cleanout Date	By	Condition of Equipment note any repairs or maintenance needed	Cleanout Performed as per Protocols*? Y / N

* "Protocols" are the routine, step-by-step procedures established to make certain that equipment is properly and completely cleaned each and every time cleaning is required.

Form-11

Equipment Settings and Adjustments for Field Operations			
Farm Name or Unit: Crop Year:			
Machine or Piece of Equipment	Crop	Settings	Notes

Form-12

ON-Farm Bin / Unit Storage Record for Organic Operations A record of your on-farm storage of organic and buffer zone crops.								
Farm Name or Unit:								
Bin or Unit: Capacity: Crop Year:								
Date	Quantity In	Field I.D.	Crop	Organic / Buffer	Lot # Assigned	Quantity Out	Cleanout Date	Cleanout By

Form-13

ON-Farm Cooler / Cold Storage Record for Organic Crops A record of your on-farm storage of organic and buffer zone crops.								
Farm Name or Unit:								
Cold Storage Unit: Capacity: Crop Year:								
Date	Quantity In	Field I.D.	Crop	Organic / Buffer	Lot # Assigned	Quantity Out	Cleanout Date	Cleanout By

Form-14

ON-Farm Cooler / Cold Storage Record for SPLIT Operations
A record of your on-farm cold storage of Organic, Transitional, and Conventional crops in common areas

Farm Name or Unit:

Cold Storage Unit: Capacity:
Crop Year:

Date	Quantity In	Field I.D.	Crop	Organic / Buffer	Lot # Assigned	Quantity Out	Cleanout Date	Cleanout By

Form-15

OFF-Farm Cooler / Cold Storage Record for Organic Crops
A record of your off-farm cold storage of organic crops.

Farm Name or Unit:

Storage Facility Name: Location: Certified Organic?

Cold Storage Unit: Capacity:
Crop Year:

Date	Quantity In	Field I.D.	Crop	Organic / Buffer	Lot # Assigned	Quantity Out	Cleanout Date	Cleanout By

Form-16

Sales Record

Farm Name or Unit:
Crop Year:

Date of Sale	Crop Sold	Storage Unit I.D.	Lot #	Sold as Organic or Conv.	Buyer	Quantity Sold	Price per Unit	Total Price	Balance of Crop Remaining

(G) CONTROL OR INSPECTION

APPENDIX 14

Official Control of foodstuffs
Directives 89/397/EEC of 14-06-1989 & 93/99/EEC of 29-10-1993

"Control" means inspection by Competent Authorities

- of foodstuffs, of food additives, vitamins, minerals, trace elements, and other additives for sale as such of packaging material coming in contact with foods for preventing risks to public health, guaranteeing fair trade practices and protecting consumer interest.
- the export goods shall be inspected in the same manner as domestic goods.
- the inspection shall be carried out regularly or where violation is suspected.
- inspection shall cover all stages of production, manufacture, import, processing, storage, transport, distribution and trade. However, the inspection may be carried out of stages considered necessary. The inspection shall be without prior information.
- the inspection may also be carried out on foods for export.

Inspection includes one or more of following:

- Inspection
- Sampling and analysis
- Inspection of personal hygiene of staff
- Examination of written or other documents
- Examination of verification system and its results.

The following are subject to inspections: -

- The place of manufacture, offices, environment, means of transport, machinery, equipment
- Raw material, ingredients, processing aids
- Semi-finished products
- Packaging material
- Clearing materials, pesticides
- Manufacturing process
- Labelling
- Methods of preservation

Interview with workers, reading of instruments and calibration of instruments shall also be allowed.

Sample in **triplicate** may be taken at any stage of production and the analysis shall be carried out in officially-approved labs or labs, allowed by other members.

The personal hygiene, cleanliness, clothing of workers are subject to examination.

Inspector may take documentary evidence kept with legal or other persons and may take extracts. The necessary measures will be taken if irregularly is suspected.

The legal and other persons shall be obliged to provide all facilities for inspection. The member states shall have a right of appeal against the competent authority and the inspector shall be bound secrecy.

- The Commission shall make arrangements for training of inspectors, quality standards for labs, sampling and methods of analysis wherever necessary.
- The competent authority may draw up an advanced programme for frequency of inspections.

The member states shall inform the Commission regarding:

- Criteria applied for drawing up programmes
- Number and type of inspections carried out
- Number and type of violations

The member shall inform the Commission

- The names of competent authorities and their jurisdictions
- The official labs authorized for analysis

The Directive 93/99/EEC supplement to Directive 89/397/EEC lays down that competent authorities shall have sufficient and suitably qualified and experienced staff in chemistry, food chemistry, veterinary medicine, medicine, food microbiology, food hygiene, food technology, and law etc.

Lab referred to analysis shall meet criteria laid down in European Standard EN 45004, EN 45002 and the criteria for lab accreditations EN 45003.

The method of sampling and analysis, wherever possible, shall comply with Council Directive 85/591/EEC of 23-12-1985

The Commission shall appoint special officials to monitor and evaluate the effectiveness of official food control system operated by a competent authority. The officials so appointed shall be suitably qualified and possess appropriate knowledge and experience for the task. All assistance shall be provided to the officials by the competent authority. These officials shall carry identity and written authorization. The Commission shall present an annual report to member states and to European Parliament.

To facilitate administration, member states may designate a single liaison body and the details of the decision body shall be communicated to Commission.

The Liaison bodies shall provide all assistance and details to another liaison body if so required except to where legal proceedings are going on.

If any violation take place in a country, which comes to the notice during exchange of information, then the competent authority where violation has taken place shall inform all other member states about action taken to deal with violation or to prevent reoccurrence of such violation. A report shall be sent to the Commission as well.

Council Directive 89\45\EEC of 21-12-1988. Damages arising from use of Consumer products and Council Directive 92\59\EEC on general products safety also apply.

Secrecy of information will be covered. The secrecy rules of a member state be duly respected and information is not divulged more widely than provided in international conventions in Criminal affairs. Any refusal to provide information must be justified. The Commission shall be assisted by standing Committees on Foodstuffs set up under 69\414\EEC.

Commission Regulation EC No: - 1000/98 of 13-05-98.

amending EEC NO: - 2377/90 - Residual limits of Vet Drugs.

This is an amendment to EEC 2377/90 amending Annex I & II of EEC 2377/90 reg. Residual limits of vet drugs.

Frequently Asked Questions

There is plenty of information available to assist farmers in converting their land and managing it organically. This chapter answers some of the common questions asked about organic agriculture.

What is organic farming?

Organic refers to agricultural production systems used to produce food and fibre. All kinds of agricultural products are produced organically, including produce, grains, meat, dairy, eggs, fibres such as cotton, flowers, and processed food products. Organic farming management relies on developing biological diversity in the field to disrupt habitat for pest organisms, and the purposeful maintenance and replenishment of soil fertility. Organic farmers are not allowed to use synthetic pesticides or fertilizers. Some of the essential characteristics of organic systems include: design and implementation of an "organic system plan" that describes the practices used in producing crops and livestock products; a detailed record keeping system that tracks all products from the field to point of sale; and maintenance of buffer zones to prevent inadvertent contamination from adjacent conventional fields.

Is organic food more nutritious than conventional food?

The definitive study has not been done, mainly because of the multitude of variables involved in making a fair comparison between organically grown and conventionally grown food. These include crop variety, time after harvest, post-harvest handling, and even soil type and climate, which can have significant effects on nutritional quality. A 2002 report (http://www.omri.org/FAC.html) indicates that organic food is far less likely to contain pesticide residues than conventional food (13 % of organic produce samples vs. 71 % of conventional produce samples contained a pesticide residue, when long-banned persistent pesticides were excluded).

Is organic food safe?

Yes. Organic food is as safe to consume as any other kind of food. Just as with any kind of produce, consumers should wash before consuming to ensure maximum cleanliness. Organic produce contains significantly lower levels of pesticide residues than conventional produce. It is a common misconception that organic food could be at greater risk of *E. coli* contamination because of raw manure application although conventional farmers commonly apply tons of raw manure as well with no regulation whatsoever. Organic standards set strict guidelines on manure use in organic farming: either it must be first composted, or it must be applied at least 90-120 days before harvest, which allows ample time for microbial breakdown of any pathogens.

Is organic food really a significant industry?

Over the past decade, sales of organic products have shown an annual increase of at least 20 %, the fastest growing sector of agriculture. Organic foods can be found at natural food stores and major supermarkets, as well as through grower direct marketing and farmers' markets. Many restaurant chefs across the country are using organic produce because they desire its superior quality and taste. Organic food is also gaining international acceptance, with nations like Japan and Germany becoming important international organic food markets. Approximately 2 % of the U.S. food supply is grown using organic methods.

Why does organic cost more?

The cost of organic food is higher than that of conventional food because the organic price tag more closely reflects the true cost of growing the food: substituting labour and intensive management for chemicals, the health and environmental costs of which are borne by society. These costs include cleanup of polluted water and remediation of pesticide contamination. Prices for organic foods include costs of growing, harvesting, transportation and storage. In the case of processed foods, processing and packaging costs are also included. Organically-produced foods must meet stricter regulations governing all these steps than conventional foods. The intensive management and labour used in organic production are frequently more expensive than the chemicals routinely used on conventional farms. There is mounting evidence that if all the indirect costs of conventional food production were factored into the price of food, organic foods would cost the same, or, more likely, be cheaper than conventional food.

Are organic yields lower?

Organic growers who go through the 2-3 year transition period from conventional to organic management usually experience an initial decrease in yields, until soil microbes are re-established and nutrient cycling is in place, at which point yields return to previous levels.

What are organic standards?

Minimum requirements for a farm or product to be certified as 'organic' are precisely defined by organic standards. There are organic standards on the national as well as international level. For certification, the standards of the target market or importing country are relevant. Certain private labels such as Naturland, Demeter or BIO SUISSE have additional requirements on top of their national standards.

Indian National Standards for Organic Products: In 2000, the Government of India released the National Standards for Organic Products (NSOP) under the National

Programme for Organic Production (NPOP). Products sold or labelled, as 'organic' thereafter need to be inspected and certified by a nationally accredited certification body. A copy of the NSOP is available from info@apeda.com.

European Regulation EC 2092/91: Most relevant for exports to Europe is the European Regulation EC 2092/91. An amended version of this complex regulation is available on http://europa.eu.int/eur-lex/en/consleg/main/1991/en_1991R2092_index.html

IFOAM Basic Standards: Being the 'mother of organic standards', IFOAM Basic standards are not standards for certification but standards for standard setting on the national or international level. They are regularly reviewed and updated in a democratic process by the IFOAM members from all over the world. The latest copy is available from headoffice@ifoam.org

How do organic farmers fertilize crops? How do they control pests, diseases, and weeds?

Organic farmers build healthy soils by nourishing the living component of the soil, the microbial inhabitants that release, transform, and transfer nutrients. Soil organic matter contributes to good soil structure and water-holding capacity. Organic farmers feed soil biota and build soil organic matter with cover crops, compost, and biologically based soil amendments. These produce healthy plants that are better able to resist disease and insect predation. Organic farmers' primary strategy in controlling pests and diseases is prevention through good plant nutrition and management. Organic farmers use cover crops and sophisticated crop rotations to change the field ecology, effectively disrupting habitat for weeds, insects, and disease organisms. Weeds are controlled through crop rotation, mechanical tillage, and hand weeding, as well as through cover crops, mulches, flame weeding, and other management methods. Organic farmers rely on a diverse population of soil organisms, beneficial insects, and birds to keep pests in check. When pest populations get out of balance, growers implement a variety of strategies such as the use of insect predators, mating disruption, traps and barriers. Under the National Organic Rule, growers are required to use sanitation and cultural practices first before they can resort to applying a material to control a weed, pest or disease problem. Use of these materials in organic production is regulated, strictly monitored, and documented. As a last resort, certain botanical or other non-synthetic pesticides may be applied. The details of fertilization and plant protection is explained in this book.

How are organic livestock and poultry raised?

Organic meat, dairy products, and eggs are produced from animals that are fed organic feed and allowed access to the outdoors. They must be kept in living conditions that accommodate the natural behaviour of the animals. Ruminants must have access to pasture. Organic livestock and poultry may not be given antibiotics, hormones, or medications in the absence of illness; however, they may be vaccinated against disease. Parasiticide use is strictly regulated. Livestock diseases and parasites are controlled primarily through preventative measures such as rotational grazing, balanced diet, sanitary housing, and stress reduction.

What does certified organic mean?

Certified organic refers to agricultural products that have been grown and processed according to uniform standards, verified by independent state or private organizations accredited by the IFOAM or National Organic Programme (NOP). All products sold as "organic" must be certified. Certification includes annual submission of an organic system plan and inspection of farm fields and processing facilities. Inspectors verify that organic practices such as long-term soil management, buffering between organic farms and neighbouring conventional farms, and record-keeping are being followed. Processing inspections include review of the facility's cleaning and pest control methods, ingredient transportation and storage, and record-keeping and audit control. Organic foods are minimally processed to maintain the integrity of food without artificial ingredients or preservatives. Certified organic requires the rejection of synthetic agrochemicals, irradiation and genetically engineered foods or ingredients.

Why do I have to become certified?

The labelling and marketing of organic products are controlled by NPOP (India) and an EC Regulation No. 2092/91. Any person or organisation intending to produce or process organic products must be subject to an inspection and certification procedure by an approved inspection body. Anyone contravening this Regulation could be subject to prosecution by the Trading Standards Officers.

What categories of production or processing are involved?

The following operations must be subject to this inspection and certification process:

- Farm production including all arable and horticultural crops and livestock, producing food intended for human consumption.
- Processing involving food preparation and pre-packing out of sight of the final consumer. This includes on-farm processing such as dairy products, butchers shops, etc.
- Organic products imported from countries outside the European Union, known as third countries.
- The re-labelling of products at any stage of the distribution chain.

Who or what has to be certified?

Farm Production: Each farm production unit has to be registered and inspected. The area of land is specified down to the individual fields, OS numbers and areas. Only products from those fields may be marketed as organic. Each production enterprise, such as cereals, dairy cattle and milk or fruits and vegetables is licensed and only products from those enterprises may be marketed as organic. A named individual has to be responsible for the management of the organic unit.

Farm Processing: Where the processing of products on the farm takes place, these will be included in the inspection and certification process. Where the farm's products are stored or processed by contractors or on other farms, the premises used must also be inspected. This can include the drying or storage of organic cereals on another farm or grain store and the cold storage and washing of potatoes or carrots off site.

How do I convert my farm?

Organic Conversion plan for crop is discussed in chapter-4. During the conversion process, only those materials and practices permitted in the National or International Organic Standards may be used. Soluble fertilizers and synthetic pesticides are not permitted. Where fertilizers have been extensively used in the past, their cessation will result in a drop in yield until the organic system starts to take effect. This has to be allowed for and, to spread the loss of yield or income, it is advisable to convert the farm in stages rather than all at once. Crop rotation, long term soil management, buffering between organic farms and conventional farms, etc. are essential to follow. A organic system plan should be prepared for the conversion.

What is the conversion period?

Arable and horticultural land and grassland have to undergo a 24 month monitored conversion period under organic management from the date of application to the date when organic status is achieved for the land. Land with perennial crops such as fruit bushes and fruit trees must undergo a 36 months conversion period. In exceptional cases, such as where land has been under an environmental management agreement for a number of years, and records can confirm the absence of prohibited inputs, a case can be made for reducing the conversion period. Additionally, land intended for non-herbivores (pigs and poultry) may be eligible for a reduced conversion period of up to 12 months.

How are livestock converted?

Beef cattle, sheep and goats can be converted simultaneously with the land used to feed them and, provided that they are finished after the land becomes organic, the animals can be marketed as organic meat animals. Dairy cattle must undergo a conversion period in accordance with the full feeding standards. Sows can be served only after the land has become organic and must be managed in accordance with the Standards from then on for the weaners to be marketed as organic pigs. A flock of laying birds can be introduced after the land has become organic and must be managed in accordance with the Standards. Table birds must be introduced at less than three days old onto organic land and be managed in accordance with the Standards until slaughter. Livestock cannot be sold as 'in-conversion'.

How do I sell my crops during the conversion period?

During the first 12 months of the conversion period, all crops must be sold as non-organic or conventional crops. Crops harvested after the first 12 months of conversion period may be marketed under the label - 'Produced under conversion to organic farming'. There is a strong market for conversion cereals and legumes for stock feed and a developing market for conversion fruit and vegetables. Farm shops can also sell the produce of the farm under this label. The Organic label may only be applied to crops which were sown or planted into land which has achieved organic status.

How can I reach an organic certification agency that serves my area?

Depending on where you live or farm in the country, there may be one or several organic certifications agencies that serve your region. There are more than ten organic certification agencies operating in the India, and these include non-profit organizations, state or county-affiliated agencies, and for-profit corporations. Some agencies work solely within a county or state, while others conduct organic certifications regionally or nationwide.

What is an ICS?

An Internal Control System (ICS) is a documented quality assurance system that allows an external certification body to delegate the annual inspection of individual group members to an identified body/unit within the certified operator. This means in practice that a growers group basically controls all farmers for compliance with organic production rules according to defined procedures. The organic certification body then mainly evaluates whether the Internal Control System is working well and efficiently. The evaluation is done by checking the ICS documentation system and staff qualifications and re-inspecting some farmers.

Who is eligible for group certification?

Smallholder farmer organizations of different types are eligible for group certification. The most common types of smallholder projects are:

1) *Farmers Groups:* an association or cooperative of farmers holds the organic certificate and organizes the ICS (is the ICS operator).
2) *Contract Production:* a trader or processor who is contracting small farmers holds the certificate and organizes the ICS (is the ICS operator)

For smallholder group certification the following conditions must be fulfilled:

- The cost of individual certification is disproportionately high in relation to the sales value of the product sold. Farm units are mainly managed by family labour.

- There is homogeneity of members in terms of geographical locations, production system, size of holding, common marketing system.

In principle only small farmers can be members of the group covered by group certification. Larger farms (i.e., farms bearing an external certification cost that is lower than 2 % of their sales) can also belong to the group but must be inspected annually by the external inspection body. Processors and exporters can be part of the structure of the group but have to be inspected annually by the external inspection body.

How to develop a suitable ICS?

The basic steps are:

- Find qualified personnel and make sure they receive the necessary training in organic production and ICS development.
- Identify farmers. If farmers are not yet familiar with organic principles, awareness creation may be necessary.
- Start developing adapted and suitable ICS forms and (preferably written) procedures. The final ICS manual can initially be a fairly simple document. It is more important that the procedures and forms are actually implemented and understood by all staff than that the manual contains details on every eventuality right from the beginning.
- Either before or during the first inspection, the organic certification body screen and assess the ICS document and most likely offer some comments or conditions for improvement.
- Gradually improve the ICS document (procedures, forms, etc.) and its implementation by the ICS staff.
- There should be **Internal Organic Standards** as per NPOP/EU Regulation. The internal organic standard is written by the ICS operator for the specific local situation of the organic project but under consideration of all applicable certification regulations. The internal organic standard shall address the following topics
    - What units/crops are under organic management and certification plus how to deal with part conversion (i.e. if farmers still grow some non-organic crops as well).
    - Conversion period.
    - Farm production rules for the whole organic production unit (e.g., seeds, fertilization and sustainable soil management, plant-protection, approved inputs, prevention of drift, livestock husbandry).
    - Harvest and post-harvest procedures.

Training of ICS personnel: Each internal inspector needs to receive at least one training per year by a competent person. All internal inspectors and field officers are trained once a year (before the beginning of new control season, usually in April). The training shall include sample field inspections. Content of training is documented and a list of participants kept. It is crucial that all organic staff is aware at all times of the organic procedures. The date of participation and content of the training of all ICS staff needs to be documented in the staff files.

Training of farmers: The most important aim of an organic project is to improve the farmer's knowledge and understanding on how to farm organically and that organic farming is much more than simply not using chemicals. Therefore continuous training of farmers is a very important part of an organic project and is in the responsibility of the ICS operator. The participation and content of the training needs to be documented.

What is healthy soil?

About half of the volume of a healthy soil is composed of mineral particles and organic matter. The other half of the volume is taken up with air and water (about 50/50). Soil maintenance involves a steady source of new organic material and/or fertilizer, pH balance, and aeration. When the soil has sufficient air, water, minerals and organic material, it can support the life of microorganisms which produce valuable plant nutrients and help eliminate thatch through decomposition. By taking care of your soil, you have taken the first step in attaining a healthy farm land.

Why is the structure of my soil important?

Good soil structure is a balanced mixture of space between particles to hold air and water necessary for your crop to thrive. This in turn helps retain moisture and nutrients for good root development, provides drainage of excessive water to avoid disease and the soil is less likely to become compacted or stressed which helps your lawn remain healthy. A healthy soil can better resist insect attack and weed infestation. Also, the necessary amounts of nutrients, such a fertilizer and lime, vary depending on the type of soil they are used on.

What types of soil are there and which is best?

There are four basic types of soil:

- Clay soil is sometimes referred to as 'heavy soil'. It contains little air and water does not drain easily through it.
- Sandy soil is sometimes referred to as 'light soil'. Water flows quickly through it, as well as fertilizer and other nutrients.
- Silt is between that of sand and clay.
- Loam is a balanced mixture of clay, silt and sand with organic matter. It consists of about 50 % open pore space that can be filled with air and water. This is the best soil texture.

How do I know what type of soil I have?

A professional soil test will indicate your type of soil, but to get a rough idea, try using the jar test. Fill a glass jar with a screw-on lid, one-third full of soil that's taken 2 to 4 inches below the surface. Pack it in and mark the soil level on the outside of the jar. Add water until the jar is about three-fourths full. Screw on the cap and shake vigorously for several minutes. Set the jar down to allow the soil to settle overnight or even a few days until the water starts to clear. The sand will go to the bottom first, then silt, then the clay. The top layer will be organic matter. Measure the depth of the silt and sand layers and compare them to the original soil level. This will give you an idea of the percentage of each of these soil types. Subtract the two values from 100 and the remainder is the percentage of clay. General guidelines to determine the soil's composition are:

Sandy soils : 35 % or more sand, less than 15 % silt and clay

Clay soils : 30 % or more clay, less than 50 % silt, less than 50 % sand

Loam soils : less than 20 % clay, 30-50 % silt, 30-50 % sand

How do I improve my soil?

Improving the soil is the first step to having a healthy farm land. By adding amendments to improve soil texture as a top dressing, correcting the pH of the soil to be between 6 and 7 for most crops, and properly aerate when necessary to eliminate soil compaction.

Can soil be aerated naturally?

Yes, by encouraging earthworms. They are natural aerators and are beneficial to the soil. They also help control thatch. As they tunnel through the soil, the worms often leave casts on the surface that make the lawn bumpy and difficult to mow. Let your crop grow a little longer so the castings will be hidden.

Do I need to give my farm land a soil test?

Absolutely! The only way to evaluate your farm land's fertility is by testing the soil. Based on the test results, you may need to add amendments to improve the soil structure, correct soil problems like phosphorus or potassium deficiencies, or change your soil's pH level.

How do I test my soil?

Contact your local county agricultural extension office for sample, forms, bags and instructions or contact a commercial landscape service. Be sure to take several samples around the farm and that each sample comes from a slice of soil that is 4 to 6 inches deep. Avoid areas that have received special treatment or may have been exposed to different environmental circumstances. Soil test should be as often as you get a physical - once a year.

What difference does soil pH make?

The pH level of your soil determines the rate at which nutrients are available to plant roots. The pH scale ranges from 1 to 14, with 7.0 indicating neutral. An extreme pH prevents the plants from getting the nutrients they need. Most grasses do best with a soil pH between 6 and 7.

When pH exceeds 8.0, iron, zinc, copper and manganese are no longer available to the plant. You can correct this condition with soil amendment. To neutralize acidic or "sour" soils, apply lime as per organic standard. However, don't apply lime unless a soil test indicates a real need. Lime can be applied almost any time, but fall or early winter is best. Lime should be applied separately from fertilizers. Apply lime two weeks on either side of organic nitrogen application.

What are the nutrients needed for crop production?

Nitrogen, phosphorus and potassium are primary essential nutrients required by the plant in large quantities. Calcium, magnesium and sulphur are secondary essential nutrients required by the plant in small quantities. Iron, zinc, copper, manganese, boron, molybdenum and cobalt are micronutrients required by the plant in very small quantity.

What is meant by available nitrogen?

Most nitrogen in the soil is present as part of organic matter. Plants can only use this nitrogen after it has been decomposed by soil organisms. The decay of plant residues, plant roots and other organic materials provides some usable nitrogen, but the amount is only about 25 % of what is needed to maintain the vigorous growth desired in most farm land. That's why nitrogen-containing organic fertilizers (FYM, compost, biofertilizers, etc.) are usually required.

Bibliography

Abbott, J.L. (977). Manure in the Home Garden. Publication Q66. University of Arizona, Tucson, AZ: 2

Abrol, I.P. and Palaniappan, S.P. (1987). Green manure crops in irrigated and rainfed lowland rice-based cropping systems in South Asia. *In Sustainable Agriculture-Green manures in Rice Farming Systems.* IRRI, Philippines: 71-87

Acquistucci R., (1999). Technological and Nutritional Quality Indexes of Wheat Produced with Organic Agricultural Techniques. National Institute of Nutrition, Laboratory of Cereal Studies.

Anderson, A. and Bergh, T. (1995). Pesticide residues in some vegetables and berries. Our Food, *Journal of Swedish National Food Administration,* 47(8): 22-24

Anon, (1999). Organic Food and Beverages, World Supply and Major European Markets. International Trade Centre, Geneva.

Anon. (1997). Agriculture issues that affect our health. S.W. *Organic Gardener.* May: 3

Carr, A., M. Smith, L. Gilkeson, J. Smillie, B. Wolf, and F. Marshall Bradley (1991). Document Circular 375, Florida Cooperative Extension Service, Institute of Food and Agricultural Sciences, University of Florida.

Chung, Y.R. and H.A.J. Hoitink (1990). Interactions between *thermophilic fungi* and *Trichoderma hamatum* in suppression of Rhizoctonia damping-off in a bark compost-amended container medium. *Phytopathology* 80: 73-77.

Codex Alimentarius, (2001). Organically produced foods: guidelines for production, processing, labelling and marketing. http://www.codexalimentarius.net/

Couzin, J., (1998). Cattle Diet Linked to Bacterial Growth. *Science* 281:1578-1579.

Dahlsted and Dlouhy (1995). Our Food, *Journal of Swedish National Food Administration,* 1995; 47(8): 39-41.

Desruelles, M.P., Devautour, H., Griffon, D. (1997), CEEMAT-SIARC Montpellier, France.

Eghball, Bahman and Gary W. Lesoing. (2000). Viability of weed seeds following manure windrow composting. Compost Science & Utilization. Winter: 46–53

EU Regulation (EEC) No. 2092/91 and 1804/1999

FAO, (1997). Biological Farming in Europe. REU Technical Series 54. Rome.

FAO, (1999). Organic Agriculture, 15th Session of the Committee on Agriculture, Rome 25-29 January 1999.

FAO/ITC/CTA, (2001). World markets for organic fruit and vegetables. Opportunities for developing countries in the production and export of organic horticultural products. Food and Agriculture Organization of the United Nations/ International Trade Centre/Technical Centre for Agricultural and Rural Co-operation.

FAO/WHO (1999). Codex Alimentarius Commission Guidelines for the Production, Processing, Labelling and Marketing of organically produced foods *cac/gl 32-1999.*

Garibay, Savador, V. and Jyoti, Katke (2003). Market opportunities and challenges for Indian organic products, SECO, Switzerland

Golueke, C.G. (1991). Principles of Composting In: The Biocycle Guide to the Art and Science of Composting. Eds. The Biocycle Staff. JG Press, Inc. Emmaus, Pa: 14-39

Greene C. (2002). Organic Farming and Marketing in the U.S. *Economic Research Service.* U.S.D.A. Washington, DC.

Haapala, JJ. (1997). Risks of chicken manure as fertilizer. *In Good Tilth.* June: 10-12.

Haglund, Å and Johansson, L. (1995). Sensory testing of carrots and tomatoes. Our food, *Journal of Swedish National Food Administration,* 47(8): 52-55.

Hamm U., Gronefeld F. and Halpin D. (2002). Organic Marketing Initiatives and Rural Development: Analysis of the European Market for Organic Food. School of Management and Business. Aberystwyth, United Kingdom.

Hardy, G.E.St.J. and K. Sivasithamparam. (1991) Effects of sterile and non-sterile leachates extracted from composted eucalyptus bark and pine-bark container media on Phytophthora spp. *Soil Biol. Biochem.* 23:25-30.

Hiraga M. (2002). Japanese Organic Market: Market opportunities and characteristics. Bio Market Inc. Japan.

Hoitink, H.A.J., Inbar, Y., and M.J. Boehm (1991). Status of compost-amended potting mixes naturally suppressive to soil-borne diseases of floriculture crops. *Plant Disease* 75:869-873.

Huhnke, Raymond L. (1982). Land Application of Livestock Manure. OSU Extension Facts No. 1710. Oklahoma State University, Stillwater, OK: 4

IFOAM (2000). Statistics Central and Eastern Europe; http://www.ifoam.de/

statistics/statistics_cee.htm

IFOAM Draft Basic Standards (2002). Second revision. Details at: http://www.ifoam.org/standard/ibs_draft2_2002_b.html

International Consultative Group on Food Irradiation. http://www.iaea.org/icgfi/

International Institute of Refrigeration, (1990). Manual of refrigerated storage in the warmer developing countries, Paris: 327.

James M. Stephens, professor, Horticultural Sciences Department, Cooperative Extension Service, Institute of Food and Agricultural Sciences, University of Florida, Gainesville FL 32611.

Jörgensen, K., Rasmussen, G., Thorup I. (1995). Ochratoxin A in Danish cereals 1986-92 and daily intake by the Danish population. *Food Additives and Contaminants.*

Kilcher L., Landau B., Richter T. and Schmid O. (2001). The Organic Market in Switzerland and the European Union: Overview and Market Access Information for Producer and International Trading Companies. Swiss Import Promotion Program and Research Institute of Organic Agriculture. Zürich/Frick, Switzerland.

Kinsey, Neal. (1994). Manure: The good, the bad, the ugly & how it works with your soil. Acres USA. October: 8-13.

Kortbech O.R. (2002). The United States Market for Organic Food and Beverages. International Trade Centre. UNCTAD/WTO.

Kuiper-Goodman, T. (1998). Food safety: mycotoxins and phycotoxins in perspective, in Mycotoxins and phycotoxins - Developments in chemistry, toxicology and food safety, eds M. Miraglia, H. van Egmond, C. Brera and J. Gilbert. IUPAC.

Marx, H., Gedek, B. & Kollarczik, B. (1995). Comparative investigations of mycotoxological status of alternatively and conventionally grown crops. *Z Lebensm Unters Forsch,* 201(1): 83-6.

Matsuda, A. and Shimonagane, K. (1988). Crop disease and soil management skills and outbreak of soil-borne disease. In Soil Health and Material Cycles, *Japanese Society of Soil Science and Plant Nutrition,* Hakuyusha, Tokyo

Mizuno, Shoji (1996). Integrated Soil Building: Concept and Practices. In Proc. Of the national seminar on "Organic Farming and Sustainable Agriculture (Ed. Vareesh, G.K., Shivashankar, K. and Singlachar, M.A.), APOF, Bangalore.

Mori, S. (1986), Effect of organic matter application on food quality. In New Perspective in Organic Matter Research, *Japanese Society of Soil Science and Plant Nutrition,* Hakuyusha, Tokyo

National Programme on Organic Production (NPOP), May 2005, APEDA, Ministry of Commerce, Govt. of India.

Naturland Standards, 01/2004 &1995

Olsen, M. and Möller, T. (1995). Mould and mycotoxines in grain. Our food, *Journal of Swedish National Food Administration*, 47(8):30-33.

Anonymous (2001). Organic and Biodynamic farming Planning Commission, Government of India.

Parrot N. and Mardsen T. (2002). The Real Green Revolution: Organic and Agroecological Farming in the South. Green Peace Environmental Trust. London, United Kingdom.

Pederson, Laura. (1998). Prevent pathogens. *American Agriculturist.* May:26.

Pedigo, L. P. (1999). Entomology & Pest Management. Prentice Hall. New Jersey:742

Anonymous (1995). Regulations concerning the production of animal and vegetable products by ecological methods, Republic of Turkey, Department of Planning and Projects, Ankara,

Sloan (1998). Organics: grown by the book. *Food Technology*: 52 (5):32.

Troop, Don (1989). Is chicken litter really organic? Ozark Cooperative Warehouse Market News (Fayetteville, AR):12.

U.S. Department of Agriculture (2000). Section 205.203(c)(1&2). National Organic Program Standards.

Veeresh, G.K. (2004). Operational Methodologies and Package of Practices in Organic Farming (Edited), National Seminar held on October 7-9, 2004 at Banglore, Association for Promotion of Organic Farming (APOF), Banglore

Vereesh, G.K., Shivashankar, K., and Singlachar, M.A. (1996). Organic Farming and Sustainable Agriculture, Proc. Of the national seminar held at UAS, Bangalore, India, Oct 9-11, 1996, Association for Promotion of Organic Farming (APOF), Bangalore

Voisin, A (1954). Soil, Grass and cancer: the link between human & animal health & the mineral balance of the soil. Louisiana: Acres, USA.

Wallace, H. A. (1998). Beyond the Big Three. A Comprehensive Analysis of the Proposed National Organic Program. Institute for Alternative Agriculture, Docket TMD-94-00-2

Watkins, J.B., McGlasson, W.B., Graham, D., Hall, E.G., (1989). Post harvest. An introduction to the physiology and handling of fruits and vegetables. New South Wales University Press, Kensington, Australia.

Weibel, F.P., Bickel, R., Leuthold, S., Alfoeldi, T. and Niggli, U. (1999). Are organically-grown apples tastier and healthier? A comparative field study using conventional and alternative methods to measure fruit quality. *In publication.*

Williams, Greg, and Pat Williams. (1994). Manure: Is it safe for your garden? *HortIdeas*:23

Woese, K.; Lange, D.; Boess, C. and Bögl, K. W. (1997). A comparison of organically and conventionally grown foods - results of a review of the relevant literature. *J. Sci. Food Agric.*: 74; 281-293.

Young, R.A. and R.F. Holt. (1977). Winter-applied manure: Effects on annual runoff, erosion, and nutrient movement. *Journal of Soil and Water Conservation.* September-October:219–222.

Yussefi M. and Willer H. (2002). Organic Agriculture World Wide 2002, Statistics and Future Prospects. Stiftung Ökologie and Landbau. Bad Dürkheim, Germany.

Glossary

Accreditation Agency: The agency set up by the Steering Committee for National Programme for Organic Production for accrediting Inspection and Certifying Agencies.

Accreditation: Accreditation means Registration by the Accreditation Agency for certifying organic farms, products and processes as per the National Standards for Organic Products and as per the guidelines of the National Accreditation Policy and Programme for organic products. Procedure by which an authoritative body gives a formal recognition that a body or person is competent to carry out specific tasks.

Accredited Programme: Programme of accrediting Inspection and Certification Agencies which have been accredited by the Accreditation Agency and which have agreed to comply with the Accreditation contract.

Active soil: In the long term, only an active soil will bear fruits. Therefore the maintenance and increase of natural soil fertility by appropriate cultivation practices is of central significance. Everything that works against this goal is to be abandoned. Most especially, synthetic chemical fertilizers are forbidden.

Agriculture with a future: Only in harmony with nature has agriculture a future. Organic agriculture must, however, be compatible not only with the environment, but also with human needs. Farms are only capable of survival in the long term if they provide adequate living conditions and appropriate wages.

Appeal: Shall be the process by which an Inspection and Certification Agency can request reconsideration of a decision taken by the Accreditation Agency or an operator can request reconsideration of a decision by the Certification Agency or operator applied for farm certification.

Applicant: The Inspection and Certification Agency that has applied for Accreditation to the Accreditation Agency.

Ayurvedic: Traditional Indian system of medicine.

Biodiversity: The variety of life forms and ecosystem types on Earth. Includes genetic diversity (i.e. diversity within species), species diversity (i.e. the number and variety of species) and ecosystem diversity (total number of ecosystem types).

Breeding: Selection of plants or animals to reproduce and/or to further develop desired characteristics in succeeding generations.

Buffer zone: A clearly defined and identifiable boundary area bordering an organic production site that is established to limit application of, or contact with, prohibited substances from an adjacent area.

Certificate of Registration: The document issued by the Inspection and Certification Agency, declaring that the operator is licensed to use the certificate on specified products.

Certificate: A document issued by an accredited agency declaring that the operator is carrying out the activities or the stated products have been produced in accordance with the specified requirements in accordance with the National Standards for Organic Products.

Certification body: The body that conducts certification, as distinct from standard-setting and inspection.

Certification mark: A certification body's sign, symbol or logo that identifies product(s) as being certified according to the rules of a program operated by that certification body according to the National Standards for Organic Products.

Certification program: System operated by a certification body with its own rules, procedures and management for carrying out certification of conformity.

Certification Transference: The formal recognition by an Inspection and Certification Agency of another Certification programme or Agency or projects or products certified by that programme or Agency, for the purpose of permitting its own certified operators to trade or process under the programme's own certification mark, the products which are certified by the other programme.

Certification: The procedure by which a third party gives written assurance that a clearly identified process has been methodically assessed, such that adequate confidence is provided that specified products conform to specified requirements.

Competent Authority: The official government agency for accreditation.

Conserving diversity: Organic farming has to be integrated into a diverse, self-regulating ecosystem. Hedges, dry swards, edges of fields, mature trees and other bio-topes not only enrich the appearance of the landscape, but also help towards maintaining the diversity of species, and thus also increasing beneficial.

Consideration of value: In turn, it is expected that the consumers appreciate the value to their health and are prepared to pay an acceptable premium for these products.

Consultancy: The advisory service for organic operations, independent from inspection and certification procedures.

Contamination: Pollution of organic product or land; or contact with any material that would render the product unsuitable for organic certification.

Conventional: Farming systems dependent on input of artificial fertilizers and/ or chemicals and pesticides or which are not in conformity with the basic standards of organic production.

Conversion period: The time between the start of the organic management and the certification of crops and animal husbandry as organic.

Conversion: The process of changing an agricultural farm from conventional to organic farm. This is also called transition.

Crop rotation: The practice of alternating the species or families of annual and/or biennial crops grown on a specific field in a planned pattern or sequence so as to break weed, pest and disease cycles and to maintain or improve soil fertility and organic matter content.

Culture: A microorganism, tissue, or organ, growing on or in a medium.

Direct source organism: The specific plant, animal, or microbe that produces a given input or ingredient, or that gives rise to a secondary or indirect organism that produces an input or ingredient.

Disinfect: To reduce, by physical or chemical means, the number of potentially harmful microorganisms in the environment, to a level that does not compromise food safety or suitability.

Evaluation: The process of systematic examination of the performance of an Inspection and Certification Agency to the extent it fulfils specific requirements under the National Accreditation Programme.

Exception: Permission granted to an operator by a certification body to be excluded from the need to comply with normal requirements of the standards. Exceptions are granted on the basis of clear criteria, with clear justification and for a limited time period only.

Farm Unit: An agricultural farm, area or production unit managed organically, by a farmer or a group of farmers, and including all the farming activities or enterprises.

Farming while conserving the land: In the knowledge that healthy soil, pure air and pure water, and a multitude of large and small plants and animals are irreplaceable, organic agriculture strives constantly for a relationship with nature and the environment that conserves both as far as possible.

Food additive: An enrichment, supplement or other substance which can be added to a foodstuff influencing its keeping quality, consistency, colour, taste, smell or other technical property.

Genetic diversity: Genetic diversity means the variability among living organisms from agricultural, forest and aquatic ecosystems; this includes diversity within species and between species.

Genetic engineering: Genetic engineering is a set of techniques from molecular biology (such as recombinant DNA) by which the genetic material of plants, animals, micro-organisms, cells and other biological units are altered in ways or with results that could not be obtained by methods of natural mating and reproduction or natural recombination. Techniques of genetic modification include, but are not limited to: recombinant DNA, cell fusion, micro and macro injection, encapsulation, gene deletion and doubling. Genetically engineered organisms do not include organisms resulting from techniques such as conjugation, transduction and natural hybridisation.

Genetic resources: Genetic material of actual or potential value.

Genetically Modified Organism (GMO): A plant, animal, or microbe that is transformed by genetic engineering.

Green manure: A crop that is incorporated into the soil for the purpose of soil improvement. May include spontaneous crops, plants or weeds.

Guidelines for Organic Production and Processing: Standards for organic production and processing established by the Accreditation Agencies for specific crops in accordance with the National Standards for Organic Products.

Habitat: The area over which a plant or animal species naturally exists; the area where a species occurs. Also used to indicate types of habitat, e.g. seashore, riverbank, woodland, grassland.

HACCP: Hazard Analysis and Critical Control Point. A specific food safety program to identify contamination risks and actions to prevent exposure to such risks.

Homeopathic treatment: Treatment of disease based on administration of remedies prepared through successive dilutions of a substance that in larger amounts produces symptoms in healthy subjects similar to those of the disease itself.

In the interest of the consumers: Organic agriculture offers foodstuffs that contribute greatly to health, along with the greatest possible preservation of the environment, and so is wholly concerned with consumers' interests and their health.

Ingredient: Any substance, including a food additive, used in the manufacture or preparation of a food or present in the final product although possibly in a modified form.

Input Manufacturing: The manufacturing of organic production or processing inputs.

Inputs Banned: Those items, the use of which is prohibited in organic farming.

Inputs Permitted: Those items that can be used in organic farming.

Inputs Restricted: Those items that are allowed in organic farming, in a restricted manner, after a careful assessment of contamination risk, natural imbalance and other factors arising out of their use. Farmers should consult the certifying agency body before use of restricted inputs.

Inspection Agency: The agency that performs inspection services as per the National Accreditation Policy and Programme.

Inspection and Certification Agency: The organization responsible for Inspection and Certification.

Inspection: The site visit to verify that the performance of an operation is in accordance with the production or processing standards.

Inspector: The person appointed by the Inspection and Certifying Agency to undertake the inspection of an operator.

Internal Review: An assessment of the objectives and performance of a programme by the Certification or the Accreditation Agency itself.

Irradiation (ionising radiation): High energy emissions from radio-nucleotides, capable of altering a food's molecular structure for the purpose of controlling microbial contaminants, pathogens, parasites and pests in food, preserving food or inhibiting physiological processes such as sprouting or ripening.

Labelling: Any written, printed or graphic representation that is present on the label of a product, accompanies the product, or is displayed near the product.

Licence: The Accreditation contract that grants a certifier the rights associated with its accredited status in line with the National Program for Organic Production.

Livestock: Any domestic or domesticated animal including bovine (including buffalo and bison), ovine, porcine, caprine, equine, poultry and bees raised for food or in the production of food. The products of hunting or fishing of wild animals shall not be considered part of this definition.

Maintaining quality: Maintenance of quality, and especially of valuable constituents, should also be considered during further processing of the produce from organic farming.

Marketing: Holding for sale or displaying for sale, offering for sale, setting, delivering or placing on the market in any other form.

Media (plural) or medium (singular): The substance in which an organism, tissue, or organ exists.

Multiplication: The growing on of seed stock or plant material to increase supply for future planting.

Natural fibre: A non-synthetic filament of plant or animal origin.

No use of genetic engineering: Organic agriculture (production and processing) renounces on interventions of genetic engineering and the use of genetically modified organisms (GMO) and their follow-up products.

Operator: An individual or business enterprise, responsible for ensuring that products meet the certification requirements.

Organic Agriculture: It is a system of farm design and management to create an eco system, which can achieve sustainable productivity without the use of artificial external inputs such as chemical fertilizers and pesticides.

Organic product: A product which has been produced, processed, and/or handled in compliance with organic standards.

Organic seed and plant material: Seed and planting material that is produced under certified organic management.

Organic: Refers to a particular farming system as described in the IFOAM Basic Standards and not to the term used in "organic chemistry".

Package of Practices: Guidelines for organic production and processing established by the Accreditation Agencies for specific crops, specific to the region.

Parallel production: Any production where the same unit is growing, breeding, handling or processing the same products in both a certified organic system and a non-certified or non-organic system. A situation with "organic" and "in conversion" production of the same product is also parallel production. Parallel production is a special instance of split production.

Part Conversion: When part of a conventional farm or unit has already been converted to organic production or processing and a part is in the process of conversion.

Plant Protection Product: Any substance intended for preventing, destroying, attracting, repelling, or controlling any pest or disease including unwanted species of plants or animals during the production, storage, transport, distribution and processing of food, agricultural commodities, or animal feeds.

Preparation: The operations of slaughtering, processing, preserving and packaging of agricultural products and also alterations made to the labelling concerning the presentation of the organic production method.

Processing aid: Any substance or material, not including apparatus or utensils, and not consumed as a food ingredient by itself, intentionally used in the processing of raw materials, foods or its ingredients, to fulfil a certain technical purpose during treatment or processing and which may result in the non-intentional, but unavoidable presence of residues or derivatives in the final product.

Producing quality: Production of quantity may not be achieved at the expense of intrinsic quality.

Propagation: The reproduction of plants by sexual (i.e. seed) or asexual (i.e. cuttings, root division) means.

Prophylactic crop protection: The health of crop plants is to be governed by the choice of resistant crops and varieties suited to the climate, appropriate fertilization, and suitable methods of cultivation and care. The use of synthetic crop protection products is prohibited.

Raw Materials: All ingredients other than food additives.

Rearing of farm animals by methods appropriate for each: The specific needs of each type of animal must be taken into account. In doing this, ethical and ecological viewpoints must be considered. One should strive for a good life-time performance by the animals, not for maximum output. Embryo transfer is not permitted.

Sanitize: To adequately treat produce or food-contact surfaces by a process that is effective in destroying or substantially reducing the numbers of vegetative cells of microorganisms of public health concern, and other undesirable microorganisms, but without adversely affecting the product or its safety for the consumer.

Split production: Where only part of the farm or processing unit is certified as organic. The remainder of the property can be (a) non-organic, (b) in conversion or (c) organic but not certified. Also see parallel production.

Standards: The standards for National Organic Products established by the Steering Committee for National Programme for Organic Production.

Synthetic: Manufactured by chemical and industrial processes. May include products not found in nature, or simulation of products from natural sources (but not extracted from natural raw materials).

Taking on responsibility: Organic farmers are aware of their responsibility with respect to the natural fundamentals of life, and try to bring their work into harmony with the cycles of nature. Being a human activity, farming is always an intrusion into nature.

Veterinary Drug: Shall mean any substance applied or administered to any food-producing animal, such as meat or milk-producing animals, poultry, fish or bees, whether used for therapeutic, prophylactic or diagnostic purposes or for modification of physiological functions or behaviour.

Index